BIOLOGY
PRINCIPLES & EXPLORATIONS

Test Generator Assessment Item Listing

HOLT, RINEHART AND WINSTON
Harcourt Brace & Company
Austin • New York • Orlando • Atlanta • San Francisco • Boston • Dallas • Toronto • London

For permission to reprint copyrighted material, grateful acknowledgment is made to the following sources:

Holt, Rinehart & Winston, Inc.: Diagrams of coccus, bacillus, & spirillum, DNA segment, ecological pyramid, diagram of the female reproductive system, diagram of a flower, diagram of a grasshopper, diagram of human joints, diagram of organisms in an ocean ecosystem, diagram of the ovarian cycle, diagram of a paramecium, diagram of three protists, diagram of red blood cells, diagram of representatives from each of the five kingdoms, diagrams of a carrot, a pea, and an ear of corn from *Holt Biology: Visualizing Life,* Test Generator Test Item Listing. Copyright © 1994 by Holt, Rinehart & Winston, Inc. Diagram of a fish, diagram of a human heart, diagram of a jellyfish, diagram of the male reproductive system, diagram of a seed, from *Modern Biology,* Graphics Supplement. Copyright © 1993 by Holt, Rinehart & Winston, Inc.

Printed in the United States of America

ISBN 0-03-095275-1

1 2 3 4 5 6 022 99 98 97 96 95

BIOLOGY
PRINCIPLES & EXPLORATIONS

ASSESSMENT ITEM LISTING

CONTAINS:

Over 2,500 test items and answers

Six question types—
analogy, multiple choice, short answer, true/false, essay, and performance-based

PROVIDES YOU WITH:

The ability to **preview, plan,** and **organize** your assessment items *before* you use the *Test Generator* software.

HRW & TEST GENERATOR ASSESSMENT ITEM LISTING

SO CONVENIENT TOGETHER, YOU CAN...

FLIP THE PAGE TO:

- Quickly preview the quantity, quality, and variety of **every** question, answer, and illustration in the *Test Generator* software package
- Assemble a list of questions you intend to use before accessing the *Test Generator* software
- Pick from a variety of questions, including analogy, multiple choice, short answer, true/false, essay, and performance-based
- Select questions by objective, chapter, or specific lesson

&

CLICK THE MOUSE TO:

- Create worksheets, quizzes, mid-terms, and final exams
- Produce make-up exams or alternate tests using our handy revision function
- Scramble question order
- Illustrate your tests with our graphics program
- Select questions by number, preview, random selection—or write them yourself

Chapter 3

TRUE OR FALSE Read each statement. If the statement is true, write *T* in the space provided. If the statement is false, write *F* in the space provided.

1. ______ Proteins are usually polar.
2. ______ Cell surface proteins float on top of phospholipid bilayers.
3. ______ Cell surface markers are composed of proteins with long nucleic acids attached to them.
4. ______ The plasma membrane of the cell determines what moves into and out of the cell.
5. ______ The most common solvent in cells is a form of lipid.
6. ______ Diffusion is an active process that requires a cell to expend a great deal of energy.
7. ______ A cell placed in a strong salt solution would probably burst because of an increase in osmotic pressure.
8. ______ The transport of specific particles through a membrane by channel proteins is known as facilitated diffusion.
9. ______ Exocytosis helps the cell rid itself of wastes.
10. ______ Endocrine glands have gated channels that allow hormones to travel directly to nerve cells.

MULTIPLE CHOICE Circle the letter of the term or phrase that best completes each statement or best answers each question.

11. The interior portion of a cell membrane forms a nonpolar zone that
   a. allows polar molecules to pass through it.
   b. allows food to pass through it.
   c. prevents polar molecules from passing through it.
   d. None of the above

12. Elongated proteins that are found on the surface of cells and that identify the cells are known as
   a. marker proteins. c. receptor proteins.
   b. channel proteins. d. superficial proteins.

The following question(s) refer(s) to the cross section of a plasma membrane.

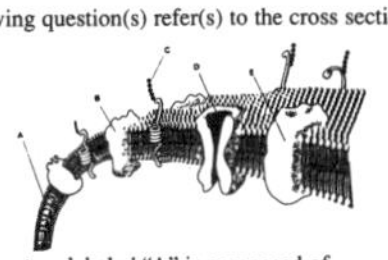

13. The structure labeled "A" is composed of
   a. lipid. c. protein.
   b. carbohydrate. d. None of the above

14. The structure labeled "E" can trigger responses in the cell. This structure is a
   a. marker protein. c. receptor protein.
   b. carbohydrate. d. bipolar lipid.

15. Which type of proteins transmit information into the cell by reacting to other molecules?
   a. channel proteins c. marker proteins
   b. receptor proteins d. end proteins

16. Which type of molecule forms a bilayer within a cell membrane?
   a. protein c. end proteins
   b. phospholipid d. carbohydrate

17. Where are receptor proteins found in a cell membrane?
   a. completely on the outside of the membrane
   b. on the inside of the membrane
   c. protruding through the membrane
   d. only where there are ions

18. Most of the food and waste materials that move into and out of a cell go through
   a. receptor proteins. c. electrical proteins.
   b. marker proteins. d. channel proteins.

The following question refers to the drawing of the lump of sugar and the beaker.

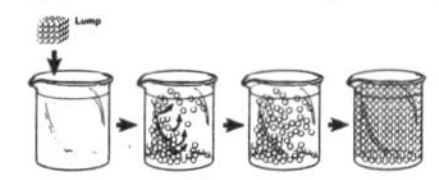

19. The process illustrated in the drawing is called
   a. osmosis. c. active transport.
   b. facilitated diffusion. d. diffusion.

20. Which types of proteins have structures that transmit information into the cell by reacting to the presence of other molecules?
   a. receptor proteins c. electrical proteins
   b. channel proteins d. marker proteins

21. Which of the following does not expend energy?
   a. diffusion c. active transport
   b. chemiosmosis d. a sodium-potassium pump

22. The ability of your body's immune system to distinguish between your cells and bacteria depends upon
   a. selective transport. c. chromosomes.
   b. cell surface markers. d. the type of invading bacteria.

23. The cell membrane
   a. encloses the contents of a cell.
   b. allows material to enter and leave the cell.
   c. is selectively permeable.
   d. All of the above

Hardware requirements:

Macintosh:

Macintosh computer
with a minimum
2MB RAM
System 6.X or later
Hard drive and
diskette drive
Printer

IBM:

IBM PC, PS/2, or
compatible with a
minimum 640K RAM
MS-DOS 3.3 or later
Hard drive and
3.5″ diskette drive
Printer

Table of Contents

About the Biology: Principles and Explorations Test Generator

The *Biology: Principles and Explorations Test Generator* contains over 2,500 test items categorized by chapter of your textbook. It also contains 10 Performance-Based Assessment Activities (a sample activity plus one activity for every unit in the textbook.) For easy reference, this *Assessment Item Listing* contains a complete printout of the questions, answers, and activities that are stored on the Test Items Disks.

The *User's Guide* that is included in the Test Generator package contains directions for installing and using the program.

Chapter Files and Unit Files

The test questions are stored in chapter files. There is a file for each chapter of your textbook. The Performance-Based Assessment Activities are stored in unit files. There is one unit file for each unit of your textbook, plus a sample unit file that contains the sample activity.

Question Attributes

Each test item is classified by its question type, chapter section, and objective code. These are called attributes. The Test Generator lets you use attributes when selecting questions for a test. It also allows you to view and to edit question attributes. Your *User's Guide* describes these features of the program.

Question Types

The *Biology: Principles and Explorations Test Generator* has six types of questions stored on the Test Items Disks. Information about a question type is displayed using a one-letter abbreviation.

Question Types and Codes

Analogies	A	True/False	T
Multiple Choice	M	Essay	E
Short Answer	S	Performance Activity	P

Sections

Each test question is coded with a section number that links the question to one of the sections in the chapter.

Objectives

Each test question is coded with an objective number that ties the question to one of the objectives for a text section in the pupil's edition.

Performance-Based Assessment Activities

The Performance-Based Assessment Activities are found in the ten unit files. There are nine Performance-Based Assessment Activities, one for each unit in your textbook. The sample unit file contains an introductory activity designed to introduce students to the concept of performance-based assessment.

Each unit file contains a set of questions that compose the Performance-Based Assessment Activity for that unit. The first question in each set contains Part I of the activity, which includes guidelines, objectives, background information, and safety precautions for the activity, as well as a brief practice activity. The Answer Key to the first question contains a Scoring Rubric for Part I. The next one to three questions contain Part II of the activity and any alternative versions of Parts I and II. The corresponding Answer Keys contain the Scoring Rubrics for these additional questions.

The last question in each unit file contains a set of Teacher Notes, which includes an introduction to the activity, a Materials List, and Preparation Notes. The Answer Key to this question contains any graphics needed for the activity or Scoring Rubrics. For the activities with graphics, you will need to cut and paste the graphics into the appropriate locations.

The Scoring Rubrics are designed to be used in awarding and tallying points for the activities. As you see fit, you may give students only the questions for Parts I and II of each activity, or you may provide each student with the Scoring Rubrics to show how the student's work was assessed.

Technical Support

The *Test Generator User's Guide* should answer any questions you may have about the program. However, if you need assistance call the Holt, Rinehart and Winston Technical Support Line at 1-800-323-9239.

Chapter 1

TRUE/FALSE Read each statement. If the statement is true, write T in the space provided. If the statement is false, write F in the space provided.

1.______ Since amphibians have been on the Earth for about 100 million years, scientists have concluded that they are not very sensitive to their environments.

2.______ Acid rain has a pH above 8.5.

3.______ Scientific investigations always follow a series of rigidly defined steps.

4.______ A scientist who performs an experiment has no idea what the outcome of the experiment is going to be.

5.______ Amphibians appeared on Earth only after the dinosaurs became extinct.

The following question(s) refer(s) to the drawing of the sand dollar and the paramecium.

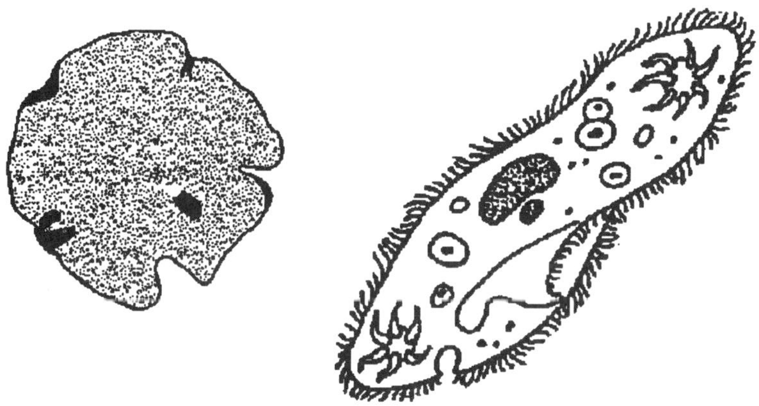

6.______ Reproduction ensures the ongoing success of both species.

7.______ Both species are multicellular.

8.______ Unlike the sand dollar, the paramecium does not have to maintain a stable internal environment.

9.______ The sand dollar and the paramecium both exhibit organization.

10.______ Both species have DNA in their cells.

MULTIPLE CHOICE Circle the letter of the term or phrase that best completes each statement or best answers each question.

11. Acidic solutions have a pH
   a. between 0 and 7.
   b. between 0 and 14.
   c. that is a negative number.
   d. between 7 and 14.

12. Acid rain
   a. has a low pH.
   b. may be produced when smoke mixes with water in the air.
   c. may contain sulfuric acid.
   d. All of the above

13. Scientists usually design experiments
   a. with a good idea of how the experiments are going to come out.
   b. based upon wild guesses.
   c. to learn how to use new laboratory tools.
   d. All of the above

14. A scientific theory is
   a. absolutely certain.
   b. unchangeable.
   c. may be revised as new evidence is presented.
   d. a controlled experiment.

15. A unifying explanation for a broad range of observations is a
   a. hypothesis.
   b. theory.
   c. prediction.
   d. controlled experiment.

16. Generally, the order in which a scientist uses the scientific method is
   a. observations, predictions, hypothesis, verifying prediction, theory, verification.
   b. observations, hypothesis, making predictions, verifying predictions, performing control experiments, theory.
   c. predictions, observations, hypothesis, theory, controlled testing, verification.
   d. observations, hypothesis, predictions, verification, theory.

17. A control
   a. is when a lead scientist controls a group of scientists conducting an experiment.
   b. is always registered with the Food and Drug Administration (FDA).
   c. is the part of an experiment in which a key factor is not allowed to change.
   d. provides the answer to a problem posed by a theory.

18. A hypothesis that is not supported by observations
   a. is known as a theory.
   b. often predicts a different observation.
   c. is rejected.
   d. None of the above

19. A theory in the scientific sense is
   a. the same as a hypothesis.
   b. an uncertain guess.
   c. generally accepted principle.
   d. a prediction.

20. A hypothesis is
   a. a definite answer to a given problem.
   b. a testable possible explanation of an observation.
   c. a proven statement.
   d. a concluding statement.

21. The smallest units that can carry on all the functions of life are called
   a. molecules.
   b. cells.
   c. organelles.
   d. species.

22. Biology is the study of
a. minerals. c. the weather.
b. life. d. energy.

23. All living things maintain a balance within their cells and the environment through the process of
a. growth. c. homeostasis.
b. development. d. evolution.

24. Cell membranes
a. are only found on a small number of cells.
b. contain genes.
c. are made of DNA.
d. are thin coverings that surround cells.

25. Plants and algae use energy that comes from
a. the sun.
b. the movement of the Earth.
c. the moon.
d. heat created deep within the interior of the Earth.

26. Homeostasis means
a. a change over long periods of time.
b. keeping things the same.
c. rapid change.
d. the same thing as evolution.

27. The destruction of the rain forest leads to
a. a lessening of the world's biological diversity.
b. the extinction of many species of plants.
c. the extinction of many species of animals.
d. All of the above

28. The ozone layer in the atmosphere
a. prevents the formation of acid rain.
b. shields the Earth from CFCs.
c. formed because of the activity of people.
d. blocks ultraviolet radiation.

29. Cystic fibrosis is an example of a(n)
a. infectious disease. c. sexually transmitted disease.
b. genetic disorder. d. None of the above

30. Which of the following is false?
Biology will affect the future of life on Earth by
a. exploring ways to fight infectious diseases.
b. applying knowledge to finding ways to cure genetic diseases.
c. finding ways to feed the world's hungry.
d. predicting storms long before they happen.

31. A genetic disorder is
a. passed from parent to child.
b. contracted through kissing.
c. spread by coughing and sneezing.
d. caused by contaminated drinking water.

32. Scientific theory
a. is absolutely certain. c. is called the "truth."
b. can be revised. d. is a reasonable guess.

33. Living things
   a. require energy to carry on life processes.
   b. have the ability to reproduce.
   c. are composed of cells.
   d. All of the above

34. The leading cause of death among American males between the ages of 24 and 44 is
   a. cancer.
   b. heart disease.
   c. AIDS.
   d. accidents.

35. Biologist John Harte wanted to determine why there was a decline in the number of salamanders. Based on his observations, Harte suggested that acid rain might be responsible. This suggestion was a
   a. prediction.
   b. hypothesis.
   c. theory.
   d. scientific "truth."

36. Harte thought that the acid released by snow melt could be killing the salamanders. This idea was a
   a. prediction.
   b. hypothesis.
   c. theory.
   d. scientific "truth."

37. To test his ideas, Harte placed some salamander eggs in containers of pond water and added different amounts of acid to the containers. Harte performed a
   a. hypothesized experiment.
   b. concluding experiment.
   c. control experiment.
   d. theoretical experiment.

38. Harte noticed that in acidic pond water some salamanders developed with curved spines. This was a(n)
   a. hypothesis.
   b. theory.
   c. observation.
   d. control.

SHORT ANSWER Complete each statement by writing the correct term or phrase in the space provided.

39. Frogs, toads, and salamanders are all examples of ________________.

40. Coal miners once took canaries down into the mines to warn of the presence of ____________ gas.

41. The acidity of a solution is described by its ____________________.

42. A unifying explanation for a broad range of observations is a(n) ______________________.

43. An educated guess, or ______________________, may be used in solving a problem.

44. A(n) ______________________ is what you expect to happen if a hypothesis is accurate.

45. A(n) ______________________ is the smallest unit capable of all life functions.

46. All living things require ______________________ to carry out their life activities.

47. All living things maintain a balance within their cells and the environment through the process of ______________________.

48. The sum of all the chemical reactions that your body carries out is known as your ______________________.

49. Molecules of ______________________ inside cells encode information to direct their growth and development.

50. A layer of ______________________ in the atmosphere shields us from some of the sun's harmful rays.

ESSAY Read and answer each question or statement on a separate sheet of paper.

51. The results of an experiment do not support the hypothesis that the experiment was designed to test. Was the experiment a waste of time?

52. Describe some characteristics of a cloud that might lead you to think it is alive.

53. How is genetic engineering being used to try to cure cystic fibrosis?

54. Toads that live in hot, dry regions bury themselves in the soil during the day. How might this be important to the toad?

ANALOGIES An analogy is a relationship between two pairs of words or phrases written as: a:b::c:d. The symbol : is read "is to," and the symbol :: is read "as." Write the letter of the pair of terms that best completes the analogy shown.

55. pH : acidity::
   a. acidity : condition
   b. frog disappearance : environmental condition
   c. high : low
   d. weather : change
   e. salamander disappearance : frogs

____

56. high acidity : salamander eggs::
   a. control : test
   b. acid rain : frogs
   c. pond : experimentation
   d. salamander : frog
   e. snow : rain

____

57. observation : hypothesis::
   a. theory : observation
   b. guess : hypothesis
   c. hypothesis : prediction
   d. investigation : slow
   e. theory : control

____

58. metabolism : chemical reactions::
   a. cell membrane : cells
   b. heredity : homeostasis
   c. reproduction : living
   d. verifying prediction : experimentation
   e. metabolism : cellular organization

____

59. homeostasis : body temperature::
   a. cell : chamber
   b. body : external conditions
   c. heat : cold
   d. external condition : homeostasis
   e. genes : heredity

___

60. cancer : unchecked cell division::
   a. cystic fibrosis : defective gene
   b. extinction : living
   c. CFCs : ozone layer
   d. disposal : dump
   e. acid rain : ozone

___

61. acid rain : living things::
   a. ozone : wind
   b. CFCs : ozone layer
   c. smoking : solid waste
   d. tragedy : cancer
   e. wind currents : tropical forests

___

62. gene transfer : crop resistance::
   a. gene transfer : cystic fibrosis cure
   b. unfit land : tropical soils
   c. poverty : work population
   d. improving : failure
   e. fertilizer : resistance

___

63. biology literacy : you::
   a. science : biology
   b. you : literacy
   c. world : world problems
   d. you : future generations
   e. future : literacy

___

64. frogs : acid rain::
   a. wind : smoke
   b. salamanders : cold weather
   c. theories : predictions
   d. rain forests : fire
   e. pH : water

___

TEST ANSWERS

TRUE/FALSE

1. F
2. F
3. F
4. F
5. F
6. T
7. F
8. F
9. T
10. T

MULTIPLE CHOICE

11. a
12. d
13. a
14. c
15. b
16. b
17. c
18. c
19. c
20. b
21. b
22. b
23. c
24. d

25. a

26. b

27. d

28. d

29. b

30. d

31. a

32. b

33. d

34. c

35. b

36. a

37. c

38. c

SHORT ANSWER

39. amphibians

40. carbon monoxide

41. pH

42. theory

43. hypothesis

44. prediction

45. cell

46. energy

47. homeostasis

48. metabolism

49. DNA

50. ozone

ESSAY

51. No, the experiment was not a waste of time. A scientist works by systematically showing that certain hypotheses are not valid when they are not consistent with the results of experiments. The results of experiments are used to evaluate alternative hypotheses. An experiment can be successful if it shows that one or more of the alternative hypotheses are inconsistent with observations.

52. A cloud can exhibit the properties of movement, sensitivity, development, and complexity. It can grow and change its shape. A cloud has a complex chemical and physical makeup and can move when stimulated by the wind.

53. Cystic fibrosis is a lung disease caused by a defective gene. Scientists are learning to transfer the healthy version of the gene into cystic fibrosis patients. This is called gene transfer therapy.

54. The toad must maintain a constant internal environment (homeostasis) in order to function properly.

ANALOGIES

55. B (indicates)

56. B (is harmful to)

57. C (leads to a(n))

58. D (requires) or (involves)

59. E (regulate(s))

60. A (is caused by)

61. B (is/are harmful to)

62. A (has led to)

63. D (is/are important to)

64. D (are destroyed by)

Chapter 2

TRUE/FALSE Read each statement. If the statement is true, write T in the space provided. If the statement is false, write F in the space provided.

1.______ Most living prokaryotes are multicellular protists.

2.______ Two different structures that all cells have are a cell membrane and a plasma membrane.

3.______ Inside smaller cells, materials and information can be transported more quickly.

4.______ A carbon atom has four outer electrons and can form four ionic bonds with another carbon atom or a different kind of atom.

5.______ The attraction of water to a substance other than water is called adhesion.

6.______ Stable atoms need electrons to fill their outer energy levels.

7.______ Bases have hydrogen ion concentrations that are lower than that of pure water.

8.______ Cellulose is a major source of energy for humans.

9.______ Saturated fats have carbon-carbon double bonds at different points along the fatty acid chain.

10.______ All living things that are not bacteria are eukaryotes.

MULTIPLE CHOICE Circle the letter of the term or phrase that best completes each statement or best answers each question.

11. When the volume of a cell increases, its surface area
 a. increases at the same rate. c. increases at a faster rate.
 b. remains the same. d. increases at a slower rate.

12. Surface area is an important factor in limiting cell growth because
 a. the cell can burst if the membrane becomes too large.
 b. materials cannot enter the cell if it is too large.
 c. the cell may become too large to take in enough food and to remove enough wastes.
 d. waste products cannot leave the cell if it is too large.

13. One difference between prokaryotes and eukaryotes is that
 a. nucleic acids are found only in prokaryotes.
 b. mitochondria are found in larger quantities in eukaryotes.
 c. Golgi vesicles are found only in prokaryotes.
 d. prokaryotes have no nuclear membrane.

14. An example of a prokaryotic cell is a(n)
 a. animal cell. c. bacterium.
 b. tree cell. d. "animalcule."

15. Which of the following is characteristic of prokaryotes?
  a. They have a nucleus.
  b. They were found on Earth before eukaryotes.
  c. The organelles in their cytoplasm are surrounded by membranes.
  d. None of the above

16. The bond formed when two atoms share a pair of electrons is called a
  a. hydrogen bond.
  b. nonpolar bond.
  c. covalent bond.
  d. water bond.

17. Water is a polar molecule because
  a. it contains two hydrogen atoms for each oxygen atom.
  b. it has a charge.
  c. different parts of the molecule have slightly different charges.
  d. it does not have a charge.

18. When placed in the same container, oil and water do not mix because
  a. they are both polar.
  b. water is polar and oil is nonpolar.
  c. they are both nonpolar.
  d. water is nonpolar and oil is polar.

19. A molecule that has a partial positive charge on one side and a partial negative charge on the other side is called a
  a. nonpolar molecule.
  b. polar molecule.
  c. charged molecule.
  d. bipolar molecule.

20. Lipids are
  a. polar molecules.
  b. similar to water molecules.
  c. protein molecules.
  d. nonpolar molecules.

21. Polysaccharides are
  a. carbohydrates.
  b. lipids.
  c. proteins.
  d. unsaturated fats.

22. Which type of molecule makes up a cell membrane?
  a. carbohydrate
  b. nucleic acid
  c. lipid
  d. protein

23. Enzymes
  a. are proteins.
  b. change the rate of chemical reactions.
  c. help organisms maintain internal balance.
  d. All of the above

24. Which of the following pairs is unrelated?
  a. eukaryote-peroxisomes
  b. transport-endoplasmic reticulum
  c. energy-mitochondria
  d. heredity-lysosomes

25. One important organelle that helps maintain homeostasis by moving supplies from one part of the cell to the other is the
  a. endoplasmic reticulum.
  b. mitochondrion.
  c. Golgi body.
  d. cytoplasm.

26. Vesicles that contain the cell's digestive enzymes are called
  a. peroxisomes.
  b. chromosomes.
  c. lysosomes.
  d. vacuoles.

27.A particularly active cell might contain large amounts of
a. chromosomes.
b. vacuoles.
c. mitochondria.
d. walls.

28.The internal transport system of a cell consists of a system of membranes called
a. the nucleus.
b. the mitochondria.
c. the plasmozome.
d. the endoplasmic reticulum.

29.Numerous threadlike organelles that protrude from the surface of a cell and are packed in tight rows are called
a. flagella.
b. microtubules.
c. actin filaments.
d. cilia.

30.Actin filaments are composed of
a. protein.
b. carbohydrate.
c. lipids.
d. glycogen.

31.Proteins are made in cells on the
a. mitochondria.
b. ribosomes.
c. nucleus.
d. cell membrane.

32.The packaging and distribution center of the cell is the
a. nucleus.
b. Golgi apparatus.
c. central vacuole.
d. nuclear envelope.

The following question(s) refer(s) to the drawings of the two molecules.

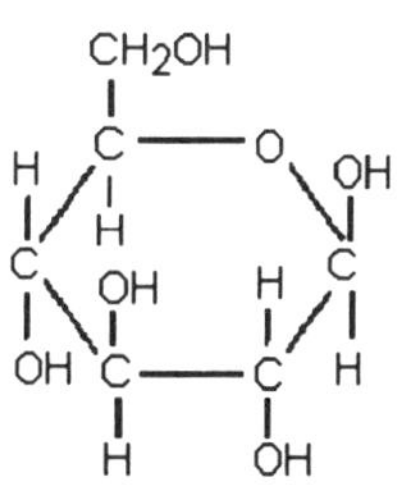

1

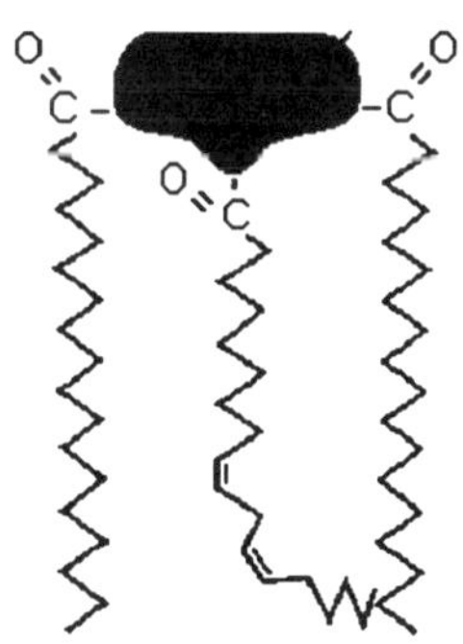

2

33.The molecule illustrated in diagram 1 is a
a. carbohydrate.
b. fat.
c. protein.
d. nucleic acid.

34.The molecule illustrated in diagram 2 is a
a. carbohydrate.
b. fat.
c. protein.
d. nucleic acid.

The following question(s) refer(s) to the drawing of the packaging and distribution of protein inside the cell.

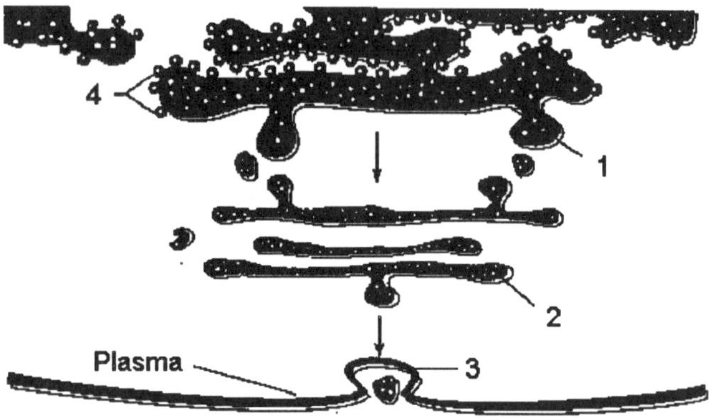

35. The structures labeled 4 are
   a. vesicles. c. ribosomes.
   b. lysosomes. d. chloroplasts.

36. Structure 1 is
   a. part of the endoplasmic reticulum.
   b. a Golgi apparatus.
   c. a mitochondrion.
   d. the nucleus.

37. Which structure packages and distributes proteins and lipids?
   a. structure 1 c. structure 3
   b. structure 2 d. structure 4

38. Structure 3 is a(n)
   a. mitochondrion. c. ribosome.
   b. endoplasmic reticulum. d. vesicle.

39. Structure 3 contains
   a. proteins. c. nucleic acids.
   b. carbohydrates. d. All of the above

SHORT ANSWER Complete each statement by writing the correct term or phrase in the space provided.

40. The fluid portion of the cytoplasm is called the ________________.

41. The statement that "cells are produced only from existing cells" is part of the ________________.

42. The force that holds molecules of water together is called a(n) ________________ bond.

43. Electrons close to the nucleus are at ________________ energy levels.

44. Atoms with unequal numbers of electrons and protons are called ________________.

45. The attraction of water to a substance other than water is ________________.

46. Any substance that ionizes to form hydroxide ions when dissolved in water is called a(n) ________________.

47. Carbohydrates are composed of carbon, hydrogen, and oxygen atoms in the ratio of ________________.

48. Cholesterol is an example of a lipid called a(n) ________________.

49. The meshlike network of protein fibers that supports the shape of the cell is called the ________________.

50. The atoms in a(n) ________________ bond share electrons.

51. A(n) ________________ bond is a weak bond that forms between two polar molecules.

ESSAY Read and answer each question or statement on a separate sheet of paper.

52. Describe one possible hypothesis for the evolution of eukaryotes.

53. We often use soap and water to remove grease and oil from our hands. A molecule of soap has a polar end and a nonpolar end. Why is soap and water more effective than water alone in removing grease and oil from our skin?

54. Why is water a liquid and not a gas at room temperature?

ANALOGIES An analogy is a relationship between two pairs of words or phrases written as: a:b::c:d. The symbol : is read "is to," and the symbol :: is read "as." Write the letter of the pair of terms that best completes the analogy shown.

55. cell : cell membrane::
   a. nucleus : chromosome
   b. nucleus : nuclear envelope
   c. chromosome : DNA
   d. DNA : cell
   e. ribosome : chromosome

____

56. surface area/volume : larger cell::
   a. large : surface
   b. transportation time : smaller cells
   c. surface area : smaller cell
   d. transportation time : door
   e. waste : larger cell

____

57. electrons : energy::
   a. nucleic acids : information
   b. ions : 109
   c. negative charges : positive charges
   d. clouds : protons
   e. electrons : springs

____

58. cohesion : adhesion::
a. hydrogen : polar
b. book : pages
c. hands : people
d. H_2O : water
e. night : day

---

59. hydroxide ions : base::
a. hydrogen ions : ammonia
b. bases : acids
c. hydrogen ions : acid
d. more : less
e. 0–14 : 7

---

60. starch : long::
a. glucose : long
b. glucose : big
c. glycogen : short
d. polypeptides : short
e. glycogen : longer

---

61. enzyme : biochemical reaction::
a. chemical reaction : fast
b. experimentation : prediction
c. observation : hypothesis
d. accelerator : automobile
e. chemical : physical

---

62. endoplasmic reticulum : ribosomes::
a. boat : water
b. liquid : gas
c. happy : tired
d. highway system : trucks
e. baseball : bat

---

63. chromosome : rod::
a. nucleic acid : screen
b. polypeptide : chain
c. carbohydrate : circle
d. sperm : egg
e. egg : 23

---

64. mechanical support : actin filaments::
   a. lysosome : peroxisomes
   b. reproduction : metabolism
   c. flagella : locomotion
   d. stable, internal conditions : homeostasis
   e. protein : chemicals

---

65. peroxisome enzyme : harmful chemicals::
   a. microtubule : cytoskeletons
   b. mouthwash : bacteria
   c. sun : heat
   d. bone : skeleton
   e. chemical : lysosomes

TEST ANSWERS

TRUE/FALSE

1.F

2.F

3.T

4.T

5.T

6.F

7.T

8.F

9.F

10.T

MULTIPLE CHOICE

11.d

12.c

13.d

14.c

15.b

16.c

17.c

18.b

19.b

20.d

21.a

22.c

23.d

24.d

25. a

26. c

27. c

28. d

29. d

30. a

31. b

32. b

33. a

34. b

35. c

36. a

37. b

38. d

39. d

SHORT ANSWER

40. cytosol

41. cell theory

42. hydrogen

43. low

44. ions

45. adhesion

46. base

47. 1: 2: 1

48. steroid

49. cytoskeleton

50. covalent

51. hydrogen

ESSAY

52. Most biologists think that eukaryotes evolved from prokaryotes. Many of the organelles of eukaryotes resemble bacteria that were presumably engulfed by these larger cells. Scientists hypothesize that these bacterial "trespassers" remained inside these cells, gradually losing their ability to live independently. These invading bacteria became organelles, and eukaryotes were the result.

53. Grease and oil are nonpolar molecules. The nonpolar end of the soap molecule is attracted to the nonpolar oils and greases, and the polar end is attracted to water. The soap allows the water molecules to wash away the grease and oil.

54. Water molecules form hydrogen bonds with one another, so water clings to itself in an attraction known as cohesion. Cohesion prevents the molecules from escaping into the atmosphere at room temperature, maintaining water as a liquid.

ANALOGIES

55. B (is surrounded by)

56. B (is less in a)

57. A (carry or have)

58. E (is the opposite of)

59. C (are released by a(n))

60. E (contains glucose chains that are)

61. D (speeds up)

62. D (contains)

63. B (looks like a)

64. D (is provided by)

65. B (removes)

Chapter 3

TRUE/FALSE Read each statement. If the statement is true, write T in the space provided. If the statement is false, write F in the space provided.

1.______ Proteins are usually polar.

2.______ Cell surface proteins float on top of phospholipid bilayers.

3.______ Cell surface markers are composed of proteins with long nucleic acids attached to them.

4.______ The plasma membrane of the cell determines what moves into and out of the cell.

5.______ The most common solvent in cells is a form of lipid.

6.______ Diffusion is an active process that requires a cell to expend a great deal of energy.

7.______ A cell placed in a strong salt solution would probably burst because of an increase in osmotic pressure.

8.______ The transport of specific particles through a membrane by channel proteins is known as facilitated diffusion.

9.______ Exocytosis helps the cell rid itself of wastes.

10.______ Endocrine glands have gated channels that allow hormones to travel directly to nerve cells.

MULTIPLE CHOICE Circle the letter of the term or phrase that best completes each statement or best answers each question.

11.The interior portion of a cell membrane forms a nonpolar zone that
   a. allows polar molecules to pass through it.
   b. allows food to pass through it.
   c. prevents polar molecules from passing through it.
   d. None of the above

12.Elongated proteins that are found on the surface of cells and that identify the cells are known as
   a. marker proteins. c. receptor proteins.
   b. channel proteins. d. superficial proteins.

13.A cell membrane consists of phospholipids and different large molecules of
   a. protein. c. DNA.
   b. fat. d. carbohydrates.

14.Which type of proteins transmit information into the cell by reacting to other molecules?
   a. channel proteins c. marker proteins
   b. receptor proteins d. end proteins

15.Which type of molecule forms a bilayer within a cell membrane?
   a. protein c. end proteins
   b. phospholipid d. carbohydrate

16. Where are receptor proteins found in a cell membrane?
   a. completely on the outside of the membrane
   b. on the inside of the membrane
   c. protruding through the membrane
   d. only where there are ions

17. Most of the food and waste materials that move into and out of a cell go through
   a. receptor proteins.
   b. marker proteins.
   c. electrical proteins.
   d. channel proteins.

18. The ability of your body's immune system to distinguish between your cells and bacteria depends upon
   a. selective transport.
   b. cell surface markers.
   c. chromosomes.
   d. the type of invading bacteria.

19. The cell membrane
   a. encloses the contents of a cell.
   b. allows material to enter and leave the cell.
   c. is selectively permeable.
   d. All of the above

20. Which types of proteins have structures that transmit information into the cell by reacting to the presence of other molecules?
   a. receptor proteins
   b. channel proteins
   c. electrical proteins
   d. marker proteins

21. Which of the following does not expend energy?
   a. diffusion
   b. chemiosmosis
   c. active transport
   d. a sodium-potassium pump

22. The sodium-potassium pump forces
   a. potassium out of the cell.
   b. sodium out of the cell.
   c. potassium into the cell.
   d. a potassium and sugar molecule together.

23. The process by which water passes into or out of a cell is called
   a. solubility.
   b. osmosis.
   c. selective transport.
   d. endocytosis.

24. Diffusion takes place
   a. only through a lipid bilayer membrane.
   b. from an area of low concentration to an area of high concentration.
   c. only in liquids.
   d. from an area of high concentration to an area of low concentration.

25. As a result of diffusion, the concentration of many types of substances
   a. always remains greater inside a membrane.
   b. eventually becomes balanced on both sides of a membrane.
   c. always remains greater on the outside of a membrane.
   d. becomes imbalanced on both sides of a membrane.

26. Molecules that are too large to be moved across a cell membrane can be removed from the cell by
   a. diffusion.
   b. exocytosis.
   c. lipid carriers.
   d. osmosis.

27. An example of a chemical substance that acts as a messenger is a
   a. hormone.
   b. protein.
   c. receptor.
   d. carrier.

28. Protein channels in a cell membrane that are sensitive to electricity are known as
   a. electrical channels.
   b. circuit-sensitive channels.
   c. receptor-sensitive channels.
   d. voltage-gated channels.

29. Molecules that are too large to be moved through the membrane can be transported into the cell by
   a. osmosis.
   b. endocytosis.
   c. lipid carriers.
   d. diffusion.

The following question(s) refer(s) to the cross section of a plasma membrane.

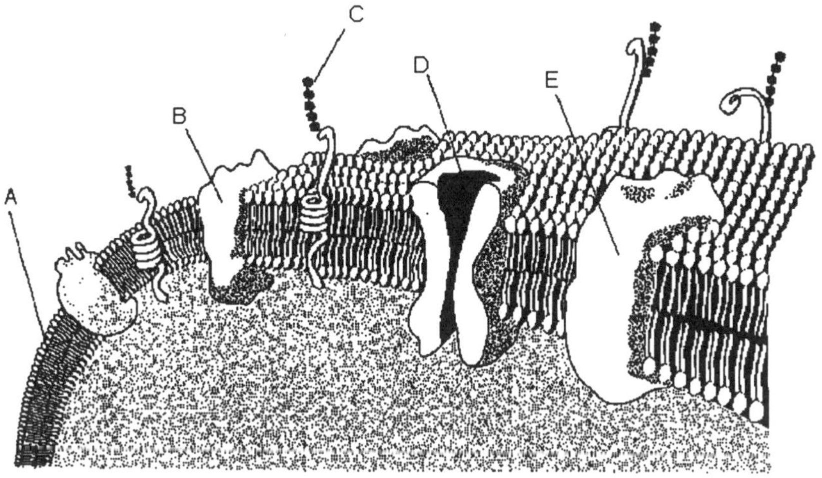

30. The structure labeled "A" is composed of
   a. lipid.
   b. carbohydrate.
   c. protein.
   d. None of the above

31. The structure labeled "E" can trigger responses in the cell. This structure is a
   a. marker protein.
   b. carbohydrate.
   c. receptor protein.
   d. bipolar lipid.

32. The structure that acts as a gate to the cell's interior is labeled
   a. "A."
   b. "B."
   c. "C."
   d. "D."

33. Structure "C" is a
   a. carbohydrate chain.
   b. glycerol molecule.
   c. fatty acid.
   d. nucleic acid.

The following question refers to the drawing of the lump of sugar and the beaker.

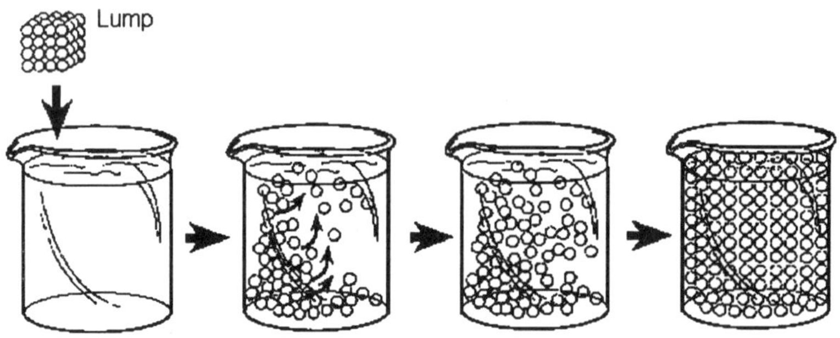

34. The process illustrated in the drawing is called
   a. osmosis.
   b. facilitated diffusion.
   c. active transport.
   d. diffusion.

SHORT ANSWER Complete each statement by writing the correct term or phrase in the space provided.

35. A phospholipid is a molecule with a(n) ______________________ head.

36. Maintaining constant internal conditions in a cell or in the body is called ______________________ .

37. A substance that dissolves in another is called a(n) ______________________.

38. A plasma membrane is said to be ______________________ permeable because it allows the passage of some solutes and not others.

39. ______________________ is the most common solvent in cells.

40. The process whereby protons involved in active transport are used to make ATP is known as ______________________.

41. The transport of food into cells involves the action of the sodium-potassium pump and ______ channels.

42. The process in which an amoeba engulfs its prey and takes it in is known as ______________________.

43. Chemicals known as ______________________ allow a nerve cell to communicate with another nerve cell or a muscle cell.

44. A gated channel that responds to changes in electricity is known as a(n) ______________________-gated channel.

The following question(s) refer(s) to the cell membrane diagrams A and B.

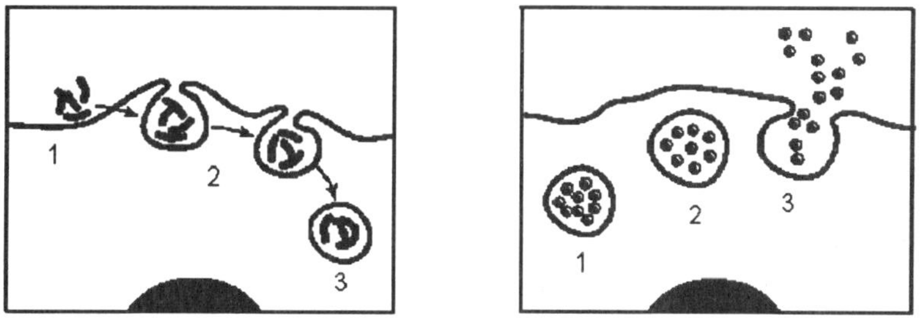

A B

45. Cells often trap extracellular particles and fluid. This is shown in diagram ________________________.

46. The process illustrated in diagram B is called ____________________.

ESSAY Read and answer each question or statement on a separate sheet of paper.

47. Why do you think it's dangerous for humans to drink ocean water?

48. Why does the addition of a solute with polar molecules to one side of a membrane result in the diffusion of water?

49. Describe three ways in which the binding of a signal particle to a receptor can change the cytoplasm in a cell.

ANALOGIES An analogy is a relationship between two pairs of words or phrases written as: a:b::c:d. The symbol : is read "is to," and the symbol :: is read "as." Write the letter of the pair of terms that best completes the analogy shown.

50. receptor proteins : information::
a. markers : hormones
b. channels : particles
c. hormones : triggers
d. cells : tissues
e. channels : cell surface proteins

____

51. voltage-gated channels : nerve cells::
a. saturated fats : unsaturated fats
b. mitochondria : eukaryotic cells
c. algae : prokaryotes
d. chemical gates : solutions
e. Golgi apparatus : bacterium

____

52. diffusion : to a lesser concentration::
   a. osmosis : through a membrane
   b. active transport : passive transport
   c. equilibrium : to a greater concentration
   d. solvent : to a solute
   e. solute : to a solvent

53. solute and solvent : solution::
   a. bound water molecules : hypotonic solution
   b. solutions : solvent
   c. phospholipids and proteins : cell membrane
   d. low concentrations : osmotic pressure
   e. voltage-gated channels : endocrine system

54. endocytosis : exocytosis::
   a. phagocytosis : bacteria
   b. secrete : exocytosis
   c. cold : hot
   d. white blood cell : bacteria
   e. pinocytosis : engulfing

55. heavy rains : flooding::
   a. high : low
   b. osmosis : proton pumping
   c. high solute concentration : isotonic solution
   d. active transport : ATP
   e. concentration difference : osmosis

56. resting : exercising::
   a. diffusion : ATP
   b. active : passive
   c. neurotransmitters : hormones
   d. coupled channels : sodium transport
   e. passive : active

57. proton pump : protons::
   a. ATP : protons
   b. channel : protons
   c. sodium-potassium pump : ATP
   d. sodium-potassium pump : sodium
   e. channel : markers

58. immune system : cell surface markers::

a. respiratory system : glands

b. baby : mother's voice

c. motorcycle : bicycle

d. channels : liver

e. lungs : liver

TEST ANSWERS

TRUE/FALSE

1. T
2. F
3. F
4. T
5. F
6. F
7. F
8. T
9. T
10. F

MULTIPLE CHOICE

11. c
12. a
13. a
14. b
15. b
16. c
17. d
18. b
19. d
20. a
21. a
22. b
23. b
24. d

25. b

26. b

27. a

28. d

29. b

30. a

31. c

32. d

33. a

34. d

SHORT ANSWER

35. polar

36. homeostasis

37. solute

38. selectively

39. Water

40. chemiosmosis

41. coupled

42. phagocytosis

43. neurotransmitters

44. voltage

45. A

46. exocytosis

ESSAY

47. The concentration of salt in ocean water is higher than the concentration of salt in the fluids that surround the cells. Drinking ocean water increases the concentration of salt in the body's fluids. This causes water to leave the cells by osmosis, and without the proper amount of water the cells will be harmed or will die.

48. The addition of solutes to one side of a membrane reduces the number of water molecules that can move freely on that side. The water molecules become bound to the polar molecules in the solute. Water then moves by osmosis from the side where water molecule concentration is greatest to the side where concentration is least.

49. 1. The receptor can act as an enzyme, chemically changing the molecules in the cytoplasm.
2. The receptor may cause the formation of a second messenger that will have an effect in another part of the cytoplasm.
3. The receptor can open a channel through the membrane.

ANALOGIES

50. B (carry)

51. B (are in)

52. A (is the movement of particles)

53. C (make up a)

54. C(is the opposite of)

55. E (results in)

56. E (is the opposite of)

57. D (pumps)

58. B (recognizes/relies on)

Chapter 4

TRUE/FALSE Read each statement. If the statement is true, write T in the space provided. If the statement is false, write F in the space provided.

1.______ Energy of motion is often referred to as potential energy.

2.______ When a person uses food as a source of energy to run a race, energy is converted from chemical to mechanical form.

3.______ The entropy of the universe constantly increases as the energy in the universe becomes more and more organized.

4.______ In organisms, when electrons are transferred from one molecule to another, they are usually carried by hydrogen atoms.

5.______ The products of an endergonic chemical reaction possess more energy than the reactants from which they are produced.

6.______ Without enzymes, chemical reactions necessary for life would not occur at a rate sufficient to sustain life.

7.______ The term *metabolism* refers to the entire collection of exergonic cellular reactions.

8.______ NAD^+ acts as a cell's "energy currency."

9.______ A coupled reaction is one in which an enzyme is simultaneously oxidized and reduced.

10.______ Overproduction of a certain metabolic product by biochemical pathways in a cell can be prevented by feedback inhibition.

MULTIPLE CHOICE Circle the letter of the term or phrase that best completes each statement or best answers each question.

11. Energy is required for a variety of life processes including
    a. growth and reproduction.
    b. movement.
    c. transport of certain materials across cell membranes.
    d. All of the above require energy.

12. Which of the following does not represent a form of energy?
    a. light
    b. heat
    c. water
    d. electricity

13. The statement that "energy can neither be created nor destroyed, but can be changed in form"
    a. is not true for all forms of energy.
    b. is the first law of thermodynamics.
    c. applies only to kinetic energy.
    d. is completely untrue.

14. Which of the following situations is the most stable?
    a. a boulder on top of a hill
    b. a house made out of playing cards stacked on each other
    c. three cheerleaders standing on each other's shoulders
    d. four tennis balls lying on the ground

15. Heterotrophs are organisms that can
   a. produce food from inorganic molecules and sunlight.
   b. survive without energy.
   c. consume other organisms for energy.
   d. carry out either photosynthesis or chemosynthesis.

16. The energy that is stored in food molecules and used by organisms
   a. resides in the nucleus.
   b. is equally distributed among all the atoms' electrons.
   c. is held primarily by electrons shared by carbon and hydrogen atoms.
   d. comes from protons orbiting the nucleus.

17. Oxidation-reduction reactions are important in organisms because they
   a. allow the passage of energy from molecule to molecule.
   b. prevent nuclear reactions from occurring.
   c. allow the creation and destruction of energy.
   d. None of the above; oxidation-reduction reactions do not occur in living organisms.

18. The sun is considered the ultimate source of energy for life on Earth because
   a. all organisms carry out photosynthesis.
   b. all organisms carry out cellular respiration.
   c. either photosynthetic organisms or organisms that have eaten them provide energy for all other organisms on Earth.
   d. the sun heats the Earth's atmosphere.

19. A reaction in which the products have less energy than the reactants is
   a. an endergonic reaction.
   b. an exergonic reaction.
   c. a filamentous reaction.
   d. impossible.

20. Since enzymes are biological catalysts, they
   a. prevent reactions from occurring spontaneously.
   b. lower the activation energy needed for a reaction to occur.
   c. permanently bind to reactants, allowing a reaction to occur.
   d. raise the activation energy until a reaction begins.

21. Factors that may affect the ability of an enzyme to work efficiently include
   a. temperature.
   b. pH.
   c. concentration of substrate.
   d. All of the above affect the enzyme's ability to function.

22. If the products of a chemical reaction contain less energy than the reactants, the reaction must
   a. involve the release of energy.
   b. result in the production of sugar.
   c. be an energy-storing reaction.
   d. occur in the cytoplasm of cells.

23. Changing the course or pathway of a chemical reaction so that it requires less activation energy
   a. is a violation of the laws of nature.
   b. requires higher temperatures than those found within cells.
   c. occurs only when reactants are quickly added to the reaction mixture.
   d. is accomplished by the action of catalysts on reactants.

24. A series of chemical reactions that results in the creation of a new biological molecule
   a. is called heterotrophic consumption.
   b. always involves exergonic reactions.
   c. is called a biochemical pathway.
   d. catalyzes the raising of activation energy.

25. ATP is called a cell's "energy currency" because
   a. it catalyzes all metabolic reactions.
   b. it allows one organelle to be exchanged for another between cells.
   c. all molecules of glucose, a primary source of energy, are made of it.
   d. most of the energy that drives metabolism is supplied by it.

26. When a phosphate group is transferred from an ATP molecule to another molecule,
   a. a substantial amount of energy is transferred as well.
   b. an enzyme is formed.
   c. the ATP molecule undergoes an endergonic change.
   d. activation energy is increased.

27. Energy-bearing electrons may be transferred from one enzyme catalyzing a reaction to a different enzyme catalyzing a different reaction within a cell by a(n)
   a. coenzyme.
   b. autotroph.
   c. ATP molecule.
   d. catalyst.

28. When a molecule gains an electron and an accompanying hydrogen atom, it has been
   a. oxidized.
   b. reduced.
   c. digested.
   d. inactivated.

29. An important example of a coenzyme that functions in metabolism is
   a. ATP.
   b. O_2.
   c. NAD^+.
   d. H_2O.

30. Factors that allow a cell to control when an enzyme is active include
   a. presence or absence of "signal" molecules.
   b. concentrations of specific enzymes.
   c. occurrence of feedback inhibition.
   d. All of the above may control an enzyme's activity.

The following question(s) refer(s) to the graphics showing an exergonic reaction and an endergonic reaction.

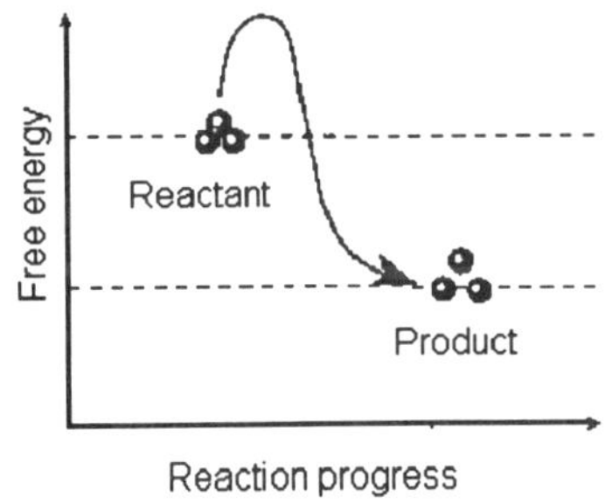

Graph A

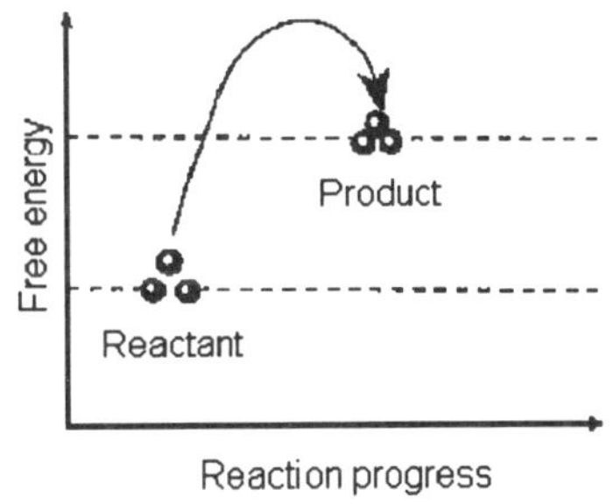

Graph B

31. Which of the graphs illustrates what happens during an exergonic reaction?
   a. Graph A
   b. Graph B
   c. Both graphs; they each show a different stage of an exergonic reaction.
   d. Neither graph shows an exergonic reaction.

32. During which reaction is energy released into the environment?
   a. Graph A
   b. Graph B
   c. Both reactions, since all chemical reactions release energy into the environment.
   d. Neither reaction, since chemical reactions do not involve energy.

The following question(s) refer(s) to the graph showing the energy curves for 4 different reactions.

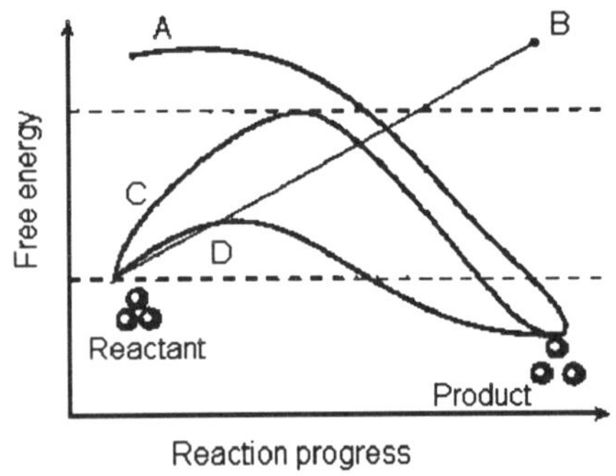

33. Which line in the graph best illustrates a catalyzed reaction?
   a. Line A
   b. Line B
   c. Line C
   d. Line D

SHORT ANSWER Complete each statement by writing the correct term or phrase in the space provided.

34. ______________________ is defined as the ability to cause change, or to do work.

35. Energy that is actively engaged in doing work is called ____________ energy.

36. ______________________ refers to the level of disorder in the universe.

37. Organisms that harvest energy from either sunlight or chemicals in order to make food molecules are called ______________________.

38. Metabolism consists of the capture, transfer, and use of ____________ within the cells of an organism.

39. The loss of electrons from a molecule is called ___________________, while the gain of electrons by a molecule is called _______________.

40. Substances that are changed when they become involved in chemical reactions are called ______________________, while the new substances that are formed are called ______________________.

41. The energy needed to break existing chemical bonds during the initiation of a chemical reaction is called ____________________.

42. Chemical reactions can be sped up by adding a substance called a(n) ______________________, which lowers the amount of activation energy required to start the reaction.

43. Biological catalysts are known as ______________________.

44. The portion of an enzyme molecule into which a specific substrate can fit is called the ______________________.

45. The sum total of all of the chemical reactions that occur within cells is called ______________________.

46. A(n) ______________________ reaction is an endergonic reaction in which an enzyme attaches to both an ATP molecule and a reactant. A single enzyme catalyzes both the splitting of the ATP and the attachment of a phosphate group to the reactant.

47. ______________________ carry energetic electrons from one enzyme to another enzyme within a cell.

48. An enzyme that may change its shape when it combines with a "signal molecule" is called a(n) ______________________ enzyme.

ESSAY Read and answer each question or statement on a separate sheet of paper.

49. Explain the differences between open and closed systems.

50. Briefly explain how plants and certain bacteria may obtain energy from their environments and store it for later use.

51. Describe how an enzyme can function in speeding up a chemical reaction within a cell.

52. Why does an enzyme that catalyzes an endergonic reaction have two active sites?

ANALOGIES An analogy is a relationship between two pairs of words or phrases written as: a:b::c:d. The symbol : is read "is to," and the symbol :: is read "as." Write the letter of the pair of terms that best completes the analogy shown.

53. electrons : chemical bonds::
 a. heterotrophs : autotrophs
 b. sunlight : plants
 c. chemical energy : sun
 d. organization : entropy
 e. nuclei : atoms

____

54. autotroph : heterotroph::
 a. moving object : kinetic energy
 b. series : many
 c. self : other
 d. phosphate : ATP
 e. burning : respiration

____

55. heterotroph : food::
 a. chain : legs
 b. hunter : game
 c. electron : atoms
 d. enzyme : activation energy
 e. catalyst : enzyme

____

56. chemical reaction : activation energy::
a. coenzyme : NAD
b. ignition : spark of fuel
c. NAD : catalyst
d. building : movement
e. pH : alkalinity

---

57. enzyme : catalyst::
a. ATP : enzyme
b. enzyme : ATP
c. NAD : coenzyme
d. ATP : NAD
e. coenzyme : ATP

---

58. activation energy : chemical reaction::
a. second law : thermodynamics
b. endergonic reaction : exergonic reaction
c. product : reactant
d. photosynthesis : heterotrophs
e. splitting of ATP : coupled reaction

---

59. biochemical pathway : energy::
a. barrier : reaction
b. match : burn
c. carrier : protein
d. bicycle : gasoline
e. assembly line : workers

---

60. enzyme : substrate::
a. water : lipid bilayer
b. signal molecule : allosteric site
c. ATP : coupled channel
d. sugar : pore
e. sodium : voltage-gated channel

---

61. temperature : enzyme activity::
a. law : amount
b. pH : relationship
c. bookshelf : reading
d. activation : pathway
e. rainfall : crop growth

---

61. temperature : enzyme activity::
   a. law : amount
   b. pH : relationship
   c. bookshelf : reading
   d. activation : pathway
   e. rainfall : crop growth

TEST ANSWERS

TRUE/FALSE

1. F

2. T

3. F

4. T

5. T

6. T

7. F

8. F

9. F

10. T

MULTIPLE CHOICE

11. d

12. c

13. b

14. d

15. c

16. c

17. a

18. c

19. b

20. b

21. d

22. a

23. d

24. c

25. d

26. a

27. a

28. b

29. c

30. d

31. a

32. a

33. d

SHORT ANSWER

34. Energy

35. kinetic

36. Entropy

37. autotrophs

38. energy

39. oxidation; reduction

40. reactants; products

41. activation energy

42. catalyst or enzyme

43. enzymes

44. active site

45. metabolism

46. coupled

47. Coenzymes

48. allosteric

ESSAY

49. A system is a collection of related objects that can be studied. An open system is one that exchanges energy and matter with its surrounding environment. Closed systems, on the other hand, cannot gain energy or matter from, or lose energy or matter to, their surroundings.

50. Plants and algae use photosynthesis to convert light energy into chemical energy, creating energy-storing macromolecules. Certain bacteria can convert one form of chemical energy, inorganic chemical energy, into another form of chemical energy, organic chemical energy, through a process called chemosynthesis.

51. Enzymes are biological catalysts. They are proteins with specific three-dimensional shapes that allow them to bind to particular substrate molecules. Once attached, the enzymes allow substrates to interact, lowering the activation energy that would otherwise be required for this reaction to occur.

52. An enzyme that catalyzes an endergonic reaction must be able to bind to both the reactant of the reaction and to an ATP molecule. The ATP site detaches a phosphate group from the ATP, releasing energy. The phosphate group is transferred to the reactant along with the energy. This provides the activation energy necessary to start the chemical reaction, of which the reactant is a part, at the second active site.

ANALOGIES

53. B (provide(s) energy for)

54. C (is the opposite of)

55. B (obtains its own)

56. B (requires)

57. C (is a type of)

58. E (is needed for)

59. E (requires)

60. B (binds to a(n))

61. E (determines)

Chapter 5

TRUE/FALSE Read each statement. If the statement is true, write T in the space provided. If the statement is false, write F in the space provided.

1._______ An autotroph is able to make its own organic molecules from inorganic materials and energy.

2._______ Energy moves through food chains from heterotrophs to producers to consumers.

3._______ When light hits a plant, all of the wavelengths are absorbed and used to make sugar.

4._______ Photosystems are clusters of light-absorbing pigments located on the thylakoids of a chloroplast.

5._______ All of the oxygen in the atmosphere of the Earth originally came from the splitting of molecules of carbon dioxide.

6._______ The "light reactions" of photosynthesis can occur only under light conditions, and the "dark reactions" occur only during the dark hours.

7._______ While the rate of photosynthesis is independent of temperature, it has been shown to be markedly affected by changes in light intensity.

8._______ Glycolysis converts two three-carbon pyruvate molecules into a six-carbon glucose molecule.

9._______ Oxidative respiration must follow glycolysis if a cell is to maximize its ATP production.

10._______ Cells that are deprived of oxygen may produce ethyl alcohol and carbon dioxide through fermentation.

11._______ Carbohydrates are important parts of an organism's diet because they are the only molecules that can yield enough energy to produce ATP during cellular respiration.

MULTIPLE CHOICE Circle the letter of the term or phrase that best completes each statement or best answers each question.

12. Based on the cycle of photosynthesis and cellular respiration, one can say that the ultimate original source of energy for all living things on Earth is
    a. carbohydrates. c. the sun.
    b. water. d. carbon dioxide.

13. All organic molecules contain carbon atoms that ultimately can be traced back in the food chain to
    a. the bodies of heterotrophs.
    b. carbon dioxide from the atmosphere.
    c. water absorbed by plants.
    d. the carbon that comes from the sun.

14. Excess sugars are usually produced during photosynthesis and may be stored
    a. as starch in fruit, seeds, and roots.
    b. in modified stems.
    c. by conversion to oils.
    d. All of the above are ways that plants may store excess sugars.

15. Suspended in the fluid stroma of chloroplasts are
   a. organelles called eukaryotes.
   b. numerous mitochondrial membranes.
   c. small coins that provide energy.
   d. stacks of thylakoids called grana.

16. Tiny packets of radiant energy are called
   a. photons.   c. eons.
   b. protons.   d. electrons.

17. When photons of light strike an object, the light may be
   a. reflected.   c. transmitted.
   b. absorbed.   d. All of the above may occur.

18. Chlorophyll is green because
   a. it absorbs green wavelengths of light.
   b. it absorbs blue and yellow wavelengths, which make green.
   c. photons of green wavelengths are reflected.
   d. of an optical illusion caused by transmitted light.

19. What happens when a chlorophyll molecule absorbs a photon of light?
   a. Some of its electrons are raised to a higher energy level.
   b. It disintegrates, giving off huge amounts of heat.
   c. It glows, radiating green light and giving the plant a green appearance.
   d. Red and blue wavelengths are emitted.

20. Flower petals have a variety of colors other than green because they possess
   a. chlorophyll.   c. pigments that reflect green.
   b. carotenoids.   d. chloroplasts.

21. When electrons of a chlorophyll molecule are raised to a higher energy level
   a. they become a photon of light.
   b. they form a glucose bond.
   c. they enter an electron transport chain.
   d. carotenoids are converted to chlorophyll.

22. The electrons of Photosystem I
   a. are eventually replaced by electrons from Photosystem II.
   b. attach to water molecules during the light reaction.
   c. produce molecules of oxygen that enter the atmosphere.
   d. are absorbed by oxygen molecules to form water.

23. Proton pumps found in the thylakoid membranes are directly responsible for
   a. moving hydrogen nuclei out of the grana.
   b. providing the energy to produce ATP molecules.
   c. producing ATP-synthetase.
   d. generating glucose molecules.

24. $NADP^+$ is important in photosynthesis because it
   a. becomes oxidized to form NADP.
   b. is needed to form chlorophyll.
   c. provides additional oxygen atoms.
   d. carries hydrogen atoms and energy for producing organic molecules.

25. The source of oxygen produced during photosynthesis is
   a. carbon dioxide.   c. the air.
   b. water.   d. glucose.

26. The "dark reactions" of photosynthesis
   a. require ATP and NADPH.
   b. can occur in both light and dark conditions.
   c. generate glucose.
   d. All of the above

27. Products of the light reactions of photosynthesis that are required by the dark reactions are
   a. oxygen and ATP.
   b. water and oxygen.
   c. ATP and NADPH.
   d. oxygen and NADPH.

28. Which of the following is *not* part of cellular respiration?
   a. electron transport
   b. glycolysis
   c. Krebs cycle
   d. Calvin cycle

29. Acetyl-coenzyme A
   a. is formed from the breakdown of pyruvate.
   b. enters the Krebs cycle.
   c. can be used in fat synthesis.
   d. All of the above

30. Which of the following is *not* formed during the Krebs cycle?
   a. CO_2
   b. $FADH_2$
   c. NADH
   d. NADPH

31. ATP molecules produced during aerobic cellular respiration
   a. remain in the mitochondria in which they are formed.
   b. are stored in chloroplasts of the same cell in which they are formed.
   c. enter the cell's cytoplasm through the membranes of the mitochondria in which they are formed.
   d. are distributed by the bloodstream to all cell's in the body.

32. Feedback inhibition of cellular respiration
   a. occurs only in plants.
   b. involves allosteric binding of ATP.
   c. is not necessary because a cell can never get enough ATP.
   d. causes the production of CO_2 instead of ATP during glycolysis.

The following question(s) refer(s) to the graph showing an action spectrum.

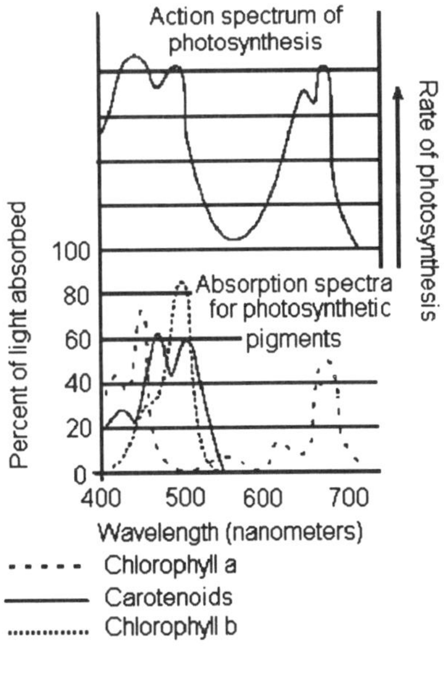

33. Based on the graphs above, which of the following wavelengths of light is least important for photosynthesis?
a. 425 nm
b. 500 nm
c. 550 nm
d. 675 nm

34. Considering that chlorophyll appears green to the human eye, and using the information presented in the graphs above, determine which of the following wavelength ranges includes green light.
a. 400–500 nm
b. 500–600 nm
c. 600–700 nm
d. None of these includes green light.

The following question(s) refer(s) to the graphs showing the effect of light (A) and the effect of temperature (B) on the rate of photosynthesis.

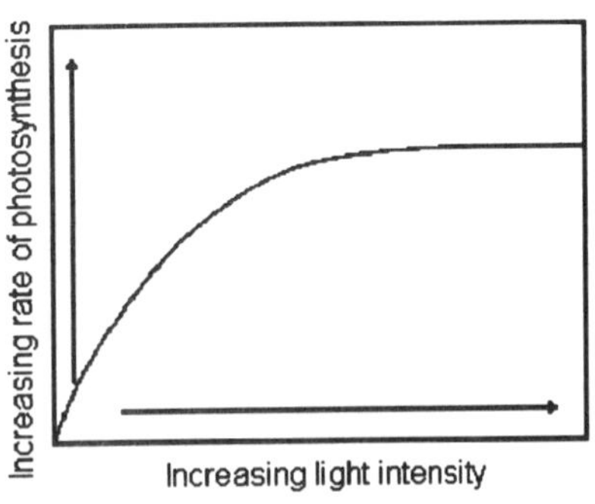

Graph A

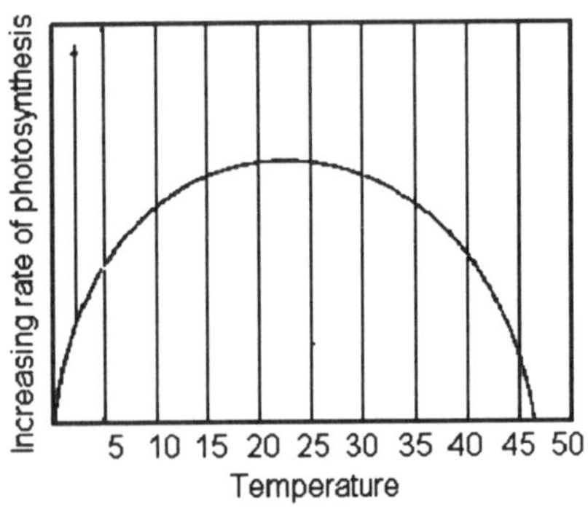

Graph B

35. Graph A demonstrates that the rate of photosynthesis
a. decreases in response to increasing light intensity.
b. increases indefinitely in response to increasing light intensity.
c. increases in response to increasing light intensity, but only to a certain point.
d. is unaffected by changes in light intensity.

36. Taken together, these graphs demonstrate that
a. photosynthesis is independent of environmental influences.
b. increases in light intensity cause increases in temperature.
c. as the rate of photosynthesis increases, the temperature of the plant eventually decreases.
d. the rate of photosynthesis is affected by changes in the plant's environment.

SHORT ANSWER Complete each statement by writing the correct term or phrase in the space provided.

37. The ultimate source of energy for all life on Earth is the ____________________.

38. Heterotrophs are often referred to as ____________________, since they must eat other organisms to obtain the energy and organic molecules that they need for life.

39. Stacks of thylakoids, called ____________________, are found suspended in the stroma of chloroplasts.

40. The complete range of radiant energy forms is called the ____________________.

41. A pigment that absorbs primarily red and blue photons of light for photosynthesis is called ________________________, while pigments that absorb other wavelengths and appear yellow and orange are called ________________________.

42. In the electron transport chain of photosynthesis, electrons travel from molecule to molecule as part of a(n) ________________________ atom.

43. The source of reducing power (attachable hydrogen atoms) that is produced by the action of Photosystem I is a molecule called ________________________.

44. The ________________________ is a cyclic series of biochemical reactions that uses ATP and NADPH to fix carbon from carbon dioxide into glucose molecules.

45. Glucose is split into smaller molecules during a biochemical pathway called ________________________.

46. ________________________ is a biochemical pathway of cellular respiration that is anaerobic.

47. Energetic electrons that provide the energy for the production of most of a cell's ATP are carried to the electron transport chain by ________________________ and ________________________ molecules.

48. ________________________ serves as the final electron acceptor for the electron transport chain of oxidative respiration.

ESSAY Read and answer each question or statement on a separate sheet of paper.

49. The relationship between photosynthesis and cellular respiration is usually described as a cycle. Briefly explain.

50. Explain why the leaves of plants appear green to the human eye.

51. In order for pyruvate to be used for the production of ATP, oxygen must be present. What happens to pyruvate molecules produced during glycolysis if oxygen is not available to a cell?

52. photosynthesis : oxygen::
  a. oxygen : carbon dioxide
  b. cellular respiration : carbon dioxide
  c. cellular respiration : oxygen
  d. cellular respiration : glucose
  e. glucose : oxygen

____

ANALOGIES An analogy is a relationship between two pairs of words or phrases written as: a:b::c:d. The symbol : is read "is to," and the symbol :: is read "as." Write the letter of the pair of terms that best completes the analogy shown.

53. photosynthesis : light::
   a. light bulb : glass
   b. trunk : clothing
   c. automobile : gasoline
   d. country : nation
   e. sound : recording

____

54. electrons : electron transport chain::
   a. carbon compounds : energy
   b. energy : food chain
   c. producer : consumer
   d. chain : flow
   e. storage : molecules

____

55. light reactions : thylakoids
   a. grana : thylakoids
   b. grana : ATP
   c. dark reactions : stroma
   d. stroma : grana of chloroplast
   e. light : electromagnetic spectrum

____

56. electromagnetic spectrum : various wavelengths::
   a. Photosystem I & II : molecule clusters
   b. grana : stroma
   c. carotenoids : chlorophyll
   d. water : oxygen from PS II
   e. Photosystem I : Photosystem II

____

57. electrons released from water : regenerate Photosystem II::
   a. ATP : make NADP
   b. photons : transport NADP
   c. electron transport chain : synthesize electrons
   d. carbon dioxide molecules : synthesize organic molecules
   e. hydrogen atoms : break down ATP

____

58. photon energy : boosting electrons::
   a. entropy : potential energy
   b. proton : electron
   c. ATP & NADPH : fixing carbon
   d. energy : food
   e. oxidation : reduction

____

59. CO_2 concentration : photosynthesis::
a. catalysis : inorganic molecule
b. enzyme : coenzyme
c. organic molecule : chemosynthesis
d. heat : energy
e. enzyme activity : biochemical reactions

____

60. Krebs Cycle : CO_2::
a. glycolysis : glucose
b. acetyl-CoA formation : O_2
c. cellular respiration : O_2
d. electron transport chain : electrons to O_2
e. electron transport chain : ATP

____

61. chloroplast : grana::
a. photosystem : pigment molecules
b. chlorophyll : pigment
c. thylakoids : grana
d. carotenoids : orange
e. chlorophyll : green

____

TEST ANSWERS

TRUE/FALSE

1. T
2. F
3. F
4. T
5. F
6. F
7. F
8. F
9. T
10. T
11. F

MULTIPLE CHOICE

12. c
13. b
14. d
15. d
16. a
17. d
18. c
19. a
20. b
21. c
22. a
23. b
24. d

25. b

26. d

27. c

28. d

29. d

30. d

31. c

32. b

33. c

34. b

35. c

36. b

SHORT ANSWER

37. sun

38. consumers

39. grana

40. electromagnetic spectrum

41. chlorophyll; carotenoids

42. hydrogen

43. NADPH (reduced nicotinamide adenine dinucleotide phosphate)

44. Calvin cycle

45. glycolysis

46. Glycolysis

47. NADH; $FADH_2$

48. Oxygen

ESSAY

49. The relationship between photosynthesis and cellular respiration is often described as cyclic because the products of one process are used as the reactants for the other. Photosynthesis produces carbohydrates from carbon dioxide and water, incorporating light energy into the bonds of glucose. Cellular respiration, on the other hand, releases energy from the bonds of glucose for use by the cell, and in the process produces carbon dioxide and water.

50. When photons of visible light strike the leaves of a plant, red and blue wavelengths are absorbed by chlorophyll. Green wavelengths, however, are reflected. These reflected photons enter the eyes of the observer, stimulating receptor cells in the eye that can respond to these wavelengths. The observer perceives the leaves as being green.

51. Animal cells deprived of sufficient oxygen for oxidative respiration to proceed will convert pyruvate into lactic acid, which builds up in the cells until the oxygen supply is restored. Organisms such as yeast, which generally do not use oxygen in their metabolism of glucose, carry out a process called fermentation. In fermentation, pyruvate made during glycolysis is converted to ethyl alcohol and carbon dioxide.

ANALOGIES

52. B (produces)

53. C (requires)

54. B (travel(s) through)

55. C (take place in the)

56. A (consist of)

57. D (are used to)

58. C (is/are needed for)

59. E (affects the rate of)

60. E (produces)

61. A (contains)

Chapter 6

TRUE/FALSE Read each statement. If the statement is true, write T in the space provided. If the statement is false, write F in the space provided.

1.______ The information needed by a cell to direct its activities and to determine its characteristics is contained in molecules of deoxyribonucleic acid (DNA).

2.______ Each human somatic cell contains two copies of each chromosome for a total of 23 homologous chromosomes.

3.______ After the replication of a cell's chromatids, there are twice as many centromeres as there are chromosomes.

4.______ Gametes are diploid so that when fertilization occurs, the resulting zygote will have the characteristic number of chromosomes for that species.

5.______ Down syndrome occurs as a result of nondisjunction of chromosome 21 during meiosis.

6.______ It is possible that a fragment of DNA may become detached from a chromosome and then reattach in the reverse orientation, resulting in a mutation called inversion.

7.______ One trigger that may cause a cell to divide is an increase in its surface-area-to-volume ratio.

8.______ Chromatids separate from each other during telophase.

9.______ Meiosis results in the formation of haploid cells from diploid cells.

10.______ Nondisjunction, if it occurs, takes place during prophase II of meiosis.

11.______ Cytokinesis only occurs during metaphase II.

MULTIPLE CHOICE Circle the letter of the term or phrase that best completes each statement or best answers each question.

12. When a molecule of DNA is being used to direct a cell's activities,
   a. areas containing specific active genes are extended.
   b. the areas of the molecule containing active genes become shorter.
   c. the entire DNA molecule becomes tightly coiled.
   d. the molecule becomes a rod-shaped structure with two chromatids.

13. Chromatids are
   a. dense patches within the nucleus.
   b. bacterial chromosomes.
   c. joined strands of duplicated genetic material.
   d. prokaryotic nuclei.

14. In order to fit within a cell, DNA becomes more compact by
   a. breaking apart into separate genes.
   b. extending to form very long, thin molecules.
   c. wrapping tightly around associated proteins.
   d. being enzymatically changed into a protein.

15. A protein disk that attaches two chromatids to each other in a chromosome is called a(n)
   a. chloroplast.   c. gamete.
   b. centromere.   d. centriole.

16. A diploid cell is one that
   a. has two homologues of each chromosome.
   b. is designated by the symbol "2*n*."
   c. has chromosomes found in pairs.
   d. All of the above

17. Diploid is to somatic cell, as haploid is to
   a. body cell.   c. gamete.
   b. chromosome.   d. zygote.

18. A student can study a karyotype to learn about the
   a. molecular structure of a chromosome.
   b. genes that are present in a particular strand of DNA.
   c. medical history of an individual.
   d. chromosomes present in a somatic cell.

19. If nondisjunction occurs,
   a. there will be too many gametes produced.
   b. no gametes will be produced.
   c. a gamete will receive too many or too few homologues of a chromosome.
   d. a mutation takes place.

20. A mutation caused by a piece of DNA breaking away from its chromosome and becoming attached to a non-homologous chromosome is called
   a. deletion.   c. inversion.
   b. duplication.   d. translocation.

21. Binary fission
   a. occurs when two cells collide with each other.
   b. produces excess energy.
   c. creates new species.
   d. is the process by which bacteria reproduce.

22. The chromosome of a bacterium
   a. is wrapped around proteins.
   b. has a circular shape.
   c. occurs in multiple pairs within the cell.
   d. is found within the nucleus.

23. Which of the following shows the correct sequence of the cell cycle?
   a. C ---> M ---> G_1 ---> S ---> G_2
   b. S ---> G_1 ---> G_2 ---> M ---> C
   c. G_1 ---> S ---> G_2 ---> M ---> C
   d. None of the above

24. The stage of the cell cycle that occupies most of the cell's life is
   a. G_1.   c. G_2.
   b. M.   d. S.

25. Mitosis is a process by which
   a. DNA is replicated.   c. cells grow in size.
   b. cytokinesis occurs.   d. a cell's nucleus divides.

26. Which of the following statements is *true*?
   a. Prokaryotes divide by mitosis.
   b. Eukaryotes have circular chromosomes.
   c. Animal cells form new cell walls when they divide.
   d. Plant cells and animal cells have different strategies for cytokinesis.

27. During meiosis, two successive divisions
   a. result in the formation of two identical cells.
   b. cause the formation of a zygote.
   c. are responsible for the formation of four haploid cells.
   d. must occur before mitosis can form gametes.

28. Separation of homologues occurs during
   a. mitosis.
   b. meiosis I.
   c. meiosis II.
   d. fertilization.

29. Which stage of meiosis is characterized by the occurrence of crossing-over?
   a. anaphase II
   b. metaphase I
   c. prophase I
   d. prophase II

30. The difference between anaphase of mitosis and anaphase I of meiosis is that
   a. the chromosomes line up at the equator in anaphase I.
   b. centromeres do not exist in anaphase I.
   c. chromatids do not separate at the centromere in anaphase I.
   d. crossing-over occurs only in anaphase of mitosis.

31. Meiosis I is often called "reduction division" because
   a. chromosomes gain electrons and hydrogen atoms.
   b. gametes are much smaller than the cells from which they are produced.
   c. the number of cells is reduced from four to two.
   d. diploid cells divide to become haploid cells.

32. The diploid number of chromosomes in a human skin cell is 46. The number of chromosomes found in a human ovum is
   a. 46.
   b. 92.
   c. 23.
   d. 12.5.

The following question(s) refer(s) to the steps of mitosis in diagrams A, B, C, and D.

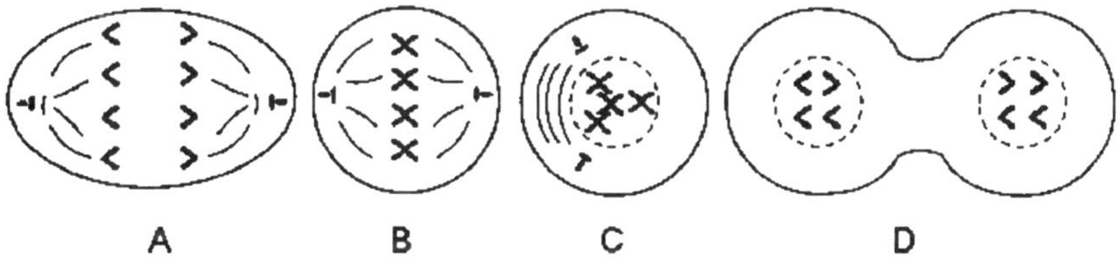

33. Which of the following correctly indicates the order in which these events occur?
   a. A, B, C, D
   b. B, A, C, D
   c. C, B, A, D
   d. A, C, B, D

34. During which stage do the centromeres divide?
   a. A
   b. B
   c. C
   d. D

The following question(s) refer(s) to diagrams A and B showing cells from an organism with a diploid number of 4.

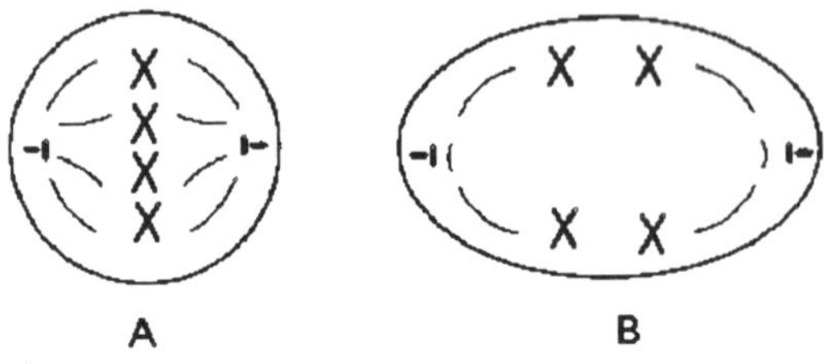

35. Which of these cells is in the process of dividing to form gametes?
   a. A   c. both
   b. B   d. neither

36. Which of the cells will be a diploid cell at the completion of division?
   a. A   c. both
   b. B   d. neither

SHORT ANSWER Complete each statement by writing the correct term or phrase in the space provided.

37. A(n) ______________________ is a segment of DNA that transmits information from parent to offspring.

38. Following replication of its DNA, each chromosome contains two ______________________, which are attached to each other by a centromere.

39. A fertilized egg cell is called a(n) ______________________.

40. A mutation called ______________________ involves a fragment of a chromosome breaking off and becoming lost when a cell divides.

41. ______________________ is the process by which bacteria split asexually into two identical organisms.

42. Collectively, the time spent in $G_1 + S + G_2$ is called ______________________.

43. Chromosomes coil up to become shorter and thicker during ______________________.

44. "Cables" made of microtubules that extend from the poles of a cell to the centromeres during cell division are called ____________________.

45. The process called ______________________ guarantees that the number of chromosomes in gametes is half the number of chromosomes in body cells.

46. When the arms of two homologous chromosomes exchange reciprocal segments of DNA, __________ is said to have occurred.

47. The stage of meiosis during which homologues line up along the equator of the cell is called ______________________.

48. The cells resulting from meiosis in either males or females are called ______________________.

49. After a new nuclear membrane forms during telophase of mitosis or meiosis, the ____________ divides, resulting in two cells.

50. Another name for meiosis I, in which the daughter cells become haploid instead of diploid, is ______________________.

51. Fertilization of the haploid sperm and egg results in the restoration of the ____________ number of chromosomes in the zygote.

ESSAY Read and answer each question or statement on a separate sheet of paper.

52. Discuss how a karyotype can be used to diagnose Down syndrome.

53. Briefly describe the five stages of the cell cycle.

54. Compare the features of mitotic metaphase, meiotic metaphase I, and meiotic metaphase II.

55. Explain why crossing-over is an important source of genetic variation.

ANALOGIES An analogy is a relationship between two pairs of words or phrases written as: a:b::c:d. The symbol : is read "is to," and the symbol :: is read "as." Write the letter of the pair of terms that best completes the analogy shown.

56. Monosomy : nondisjunction::
   a. chromatids : centromere
   b. male : XY chromosomes
   c. haploid : mitosis
   d. meiosis : diploid
   e. female : XY chromosomes

____

57. human chromosome number : chimpanzee chromosome number::
   a. karyotype : trisomy
   b. haploid : diploid
   c. diploid : haploid
   d. zygote : gametes
   e. trisomy : monosomy

____

58. female : XX::
   a. female : gametes
   b. female : eggs
   c. male : YY
   d. male : XY
   e. male : sperms

____

59. growth : G_1 ::
   a. mitosis : C
   b. mitosis : meiosis
   c. mitochondria replication : S
   d. cytokinesis : M
   e. DNA copying : S

____

60. centrioles : formation of spindle fibers::
   a. spindles : centriole attachment
   b. centromeres : centrioles separating
   c. spindles : microtubules
   d. centromeres : chromosome spindle attachment
   e. microtubules : formation of centromeres

____

61. metaphase : prophase::
   a. photon : light particle
   b. G_2 : S
   c. thylakoid : grana
   d. autotroph : producer
   e. carbon fixation process : Calvin cycle

____

62. meiosis I : mitosis::
   a. berry : apple
   b. resist : treacherous
   c. tall : high
   d. battle : copper
   e. front : back

____

63. crossing over : genetic variation::
   a. dividing squares : circles
   b. card shuffling : new combinations
   c. stacking pennies : disorder
   d. measuring lines : lack of exactness
   e. folding paper : increasing

____

64. sperm : gamete::
   a. gamete : sex
   b. gamete : sperm
   c. spore : haploid cell
   d. spore : diploid cell
   e. chromosome : chromatid

____

65.5 : cell cycle::
a. 6 : prophase
b. 9 : cytokinesis
c. 3 : meiosis
d. 4 : mitosis
e. 6 : telophase

66.centromeres : splitting::
a. centrioles : spinning
b. spindle fibers : rotating
c. chromatids : migrating
d. microtubules : replicating
e. cells : condensing

TEST ANSWERS

TRUE/FALSE

1.T

2.F

3.F

4.F

5.T

6.T

7.F

8.F

9.T

10.F

11.F

MULTIPLE CHOICE

12.a

13.c

14.c

15.b

16.d

17.c

18.d

19.c

20.d

21.d

22.b

23.c

24.a

25. d

26. d

27. c

28. b

29. c

30. c

31. d

32. c

33. c

34. a

35. b

36. a

SHORT ANSWER

37. gene

38. chromatids

39. zygote

40. deletion

41. Binary fission

42. interphase

43. prophase

44. spindle fibers

45. meiosis

46. crossing over

47. metaphase I

48. gametes

49. cytoplasm

50. reduction division

51. diploid

ESSAY

52. A karyotype is a photograph that shows the collection of chromosomes found in an individual's cells. Analysis of this collection of chromosomes can reveal abnormalities in chromosome number. Down syndrome is associated with trisomy 21—an extra chromosome # 21 in the cells of a person. Such an abnormality can be detected by observing a karyotype of an affected person.

53. The G_1 stage of the cell cycle is the phase of cell growth. This is followed by the S stage in which DNA is copied. G_2 involves the cell preparing for cell division. The M phase is when mitosis occurs. The cell cycle concludes with the C stage during which cytokinesis takes place. The newly formed cells then enter into a new cell cycle, repeating these stages again.

54. During metaphase of mitosis, the diploid number of chromosomes of the cell line up single-file across the equator of the cell. Meiotic metaphase I is characterized by the homologous chromosomes lining up as pairs (double file) along the equator. Metaphase II of meiosis appears similar to mitotic metaphase, except that the number of chromosomes is the haploid number rather than the diploid number. These chromosomes line up single-file across the equator.

55. Crossing-over occurs when two homologous chromosomes exchange reciprocal segments of DNA during prophase I of meiosis. This results in chromosomes in which the two chromatids no longer have identical genetic material. When meiosis is completed, the resulting gametes carry new combinations of genes.

ANALOGIES

56. B (is the result of)

57. B (is less than)

58. D (has the sex chromosomes in chromosome pair)

59. E (occurs during)

60. D (are the site of)

61. B (follows)

62. C (is similar to)

63. B (results in)

64. C (is a type of)

65. D (steps are in)

66. C (separate by)

Chapter 7

TRUE/FALSE Read each statement. If the statement is true, write T in the space provided. If the statement is false, write F in the space provided.

1._______ Genetics is the branch of biology that involves the study of how different traits are transmitted from one generation to the next.

2._______ Heterozygous individuals have two of the same alleles for a particular gene.

3._______ The physical appearance of an individual organism, as determined by the genes it has inherited from its parents, is called its genotype.

4._______ The law of segregation states that two or more pairs of alleles separate independently of one another during gamete formation.

5._______ Crosses involving a study of one gene are called monohybrid crosses.

6._______ In codominance, two dominant alleles are expressed at the same time.

7._______ All genes have only two alleles.

8._______ The expression of all genes is dependent only upon the inheritance of alleles from an individual's parents, and is independent of environmental influences.

9._______ If a mutated piece of DNA is carried in a recessive gene, the only way for this mutated gene to be expressed in a child born to two normal parents would be for both parents to be heterozygous carriers of this mutated gene.

10._______ An autosomal trait will occur with equal frequency in both males and females.

MULTIPLE CHOICE Circle the letter of the term or phrase that best completes each statement or best answers each question.

11. The scientific study of heredity is called
   a. meiosis. c. genetics.
   b. crossing over. d. pollination.

12. The "godfather" of genetics was
   a. T.A. Knight. c. Gregor Mendel.
   b. Hans Krebs. d. None of the above

13. Mendel's work was considered more advanced than Knight's because
   a. Knight did not record any of his results.
   b. Mendel quantified his findings.
   c. Mendel used pea plants that were evolutionarily advanced.
   d. Knight's pea plants and Mendel's pea plants had different colored flowers.

14. A genetic trait that appears in every generation of offspring is called
   a. dominant. c. recessive.
   b. phenotypic. d. superior.

15. An organism that has inherited two of the same alleles of a gene from its parents is called
   a. hereditary. c. homozygous.
   b. heterozygous. d. a mutation.

16. The phenotype of an organism
   a. represents its genetic composition.
   b. reflects all the traits that are actually expressed.
   c. occurs only in dominant pure organisms.
   d. cannot be seen.

17. The law of segregation states that
   a. alleles of a gene separate from each other during meiosis.
   b. different alleles of a gene can never be found in the same organism.
   c. each gene of an organism ends up in a different gamete.
   d. each gene is found on a different molecule of DNA.

18. Mendel's finding that the inheritance of one trait had no effect on the inheritance of another became known as the
   a. law of dominance.
   b. law of universal inheritance.
   c. law of separate convenience.
   d. law of independent assortment.

19. Tallness (T) is dominant to shortness (t) in pea plants. Which of the following represents a genotype of a pea plant that is heterozygous for tallness?
   a. T
   b. TT
   c. Tt
   d. tt

20. An individual heterozygous for a trait and an individual homozygous recessive for the trait are crossed and produce many offspring that are
   a. all the same genotype.
   b. two different phenotypes.
   c. three different phenotypes.
   d. all the same phenotype.

21. The difference between a monohybrid cross and a dihybrid cross is that
   a. monohybrid crosses involve traits for which only one allele exists, while dihybrid traits involve two alleles.
   b. monohybrid crosses involve self-pollination, while dihybrid crosses involve cross-pollination.
   c. monohybrid crosses involve one gene; dihybrid crosses involve two genes.
   d. dihybrid crosses require two Punnett squares; monohybrid crosses only need one.

22. In pea plants, yellow seeds are dominant over green seeds. What would be the expected genotype ratio in a cross between a plant with green seeds and a plant that is heterozygous for seed color?
   a. 1:3
   b. 1:2:1
   c. 4:1
   d. 1:1

23. The fact that a man and woman, both of whom have wavy hair, could have children with curly hair, wavy hair, or straight hair is best explained by the phenomenon called
   a. codominance.
   b. dominance.
   c. incomplete dominance.
   d. None of the above; this would be impossible.

24. What would be the blood type of a person who inherited an A allele from one parent and an O allele from the other?
   a. type A
   b. type B
   c. type AB
   d. type O.

25. The code contained in a particular gene results in the production of
   a. a variety of type of chemicals.
   b. a specific protein.
   c. certain types of DNA molecules.
   d. particular carbohydrate molecules.

26. Both sickle cell anemia and hemophilia
   a. are caused by genes coding for defective proteins.
   b. are seen in homozygous dominant individuals.
   c. provide resistance to malaria infections.
   d. are extremely common throughout the world.

27. A diagram in which several generations of a family and the occurrence of certain genetic characteristics are shown is called a
   a. Punnett square. c. pedigree.
   b. monohybrid cross. d. family karyotype.

28. If a characteristic is sex-linked, it
   a. occurs most commonly in males. c. can never occur in females.
   b. occurs only in females. d. is always fatal.

29. While studying several generations of a particular family, a geneticist observed that a certain disease was found equally in males and females and that all children that had the disease had parents who also had the disease. The gene coding for this disease is probably
   a. sex-linked recessive. c. autosomal recessive.
   b. sex-linked dominant. d. autosomal dominant.

30. Since the allele for colorblindness is located on the X chromosome, colorblindness
   a. cannot be inherited. c. is sex-linked.
   b. occurs only in adults. d. None of the above is true.

31. In humans, the risks of passing on a genetic disorder to one's children can be assessed by
   a. analysis of a pedigree. c. prenatal testing.
   b. genetic counseling. d. All of the above

32. Which of the following is an example of gene technology?
   a. A genetic counselor studies a pedigree.
   b. A student studies the colors of flowers in pea plants.
   c. A geneticist explains the inheritance of albinism using a Punnett square.
   d. A physician transfers a normal gene into the DNA of a person with a genetic disease.

The following question(s) refer(s) to the diagram showing the cross between two rabbits. In rabbits, black fur (B) is dominant to brown fur (b). Consider the following cross between two rabbits.

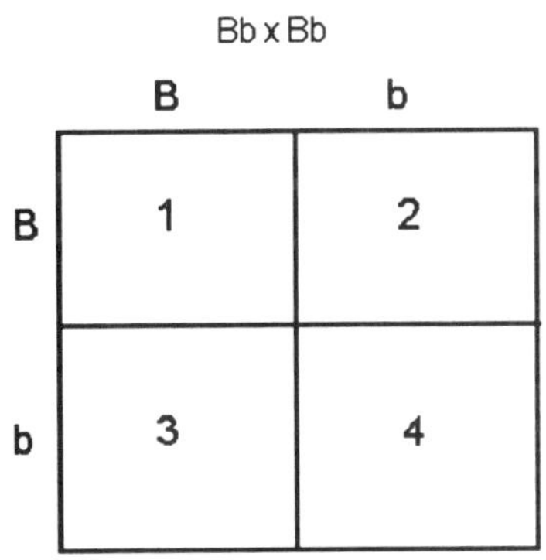

33. The device illustrated above, which is used to organize genetic analysis, is called a
   a. Mendelian box. c. Genetic graph.
   b. Punnett square. d. Phenotypic paradox.

34. Both of the parents in the cross are

a. black.
b. brown.
c. homozygous dominant.
d. homozygous recessive.

35. The phenotype of the offspring indicated by Box 3 would be

a. brown.
b. a mixture of brown and black.
c. black.
d. The phenotype cannot be determined.

36. The genotype ratio of the F_1 generation would be

a. 1:1.
b. 3:1.
c. 1:3.
d. 1:2:1.

SHORT ANSWER Complete each statement by writing the correct term or phrase in the space provided.

37. ______________________ refers to the transmission of traits from parent to offspring in sexually reproducing organisms.

38. The portion of a DNA molecule containing the coded instructions that result in an individual characteristic of an organism is called a(n) ______________________.

39. When two members of the F_1 generation are allowed to breed with each other, the offspring are referred to as the ______________________ generation.

40. A trait that is not expressed in the F_1 generation resulting from the crossbreeding of two genetically different, true-breeding organisms is called ______________________.

41. Different forms of a particular gene are called ____________________.

42. An organism's ______________________ refers to the set of alleles it has inherited.

43. The statement that the members of each pair of alleles separate when gametes are formed is known as the ______________________.

44. The likelihood that a specific event will occur is called __________.

45. A cross involving two pairs of contrasting traits is a(n) ______________________ cross.

46. A situation in which a heterozygous individual has a phenotype that is intermediate between the phenotypes of its two homozygous parents is caused by a phenomenon called ______________________.

47. Spontaneous changes in genetic material are called ________________.

48. A trait that is determined by a gene that is only found on the X chromosome is said to be a(n) ______________________ trait.

49. Fragile blood cells with an irregular shape that may block blood vessels is a symptom of a genetic disease known as ________________.

50. A genetic disorder resulting in defective blood clotting is ______________________.

51. Identifying patterns of inheritance within a family over several generations is possible by studying a diagram called a(n) ______________________.

52. A genetic disorder in which an individual lacks an enzyme responsible for converting the amino acid phenylalanine into the amino acid tyrosine is called ______________________.

ESSAY Read and answer each question or statement on a separate sheet of paper.

53. Briefly discuss the reasons that Mendel chose the pea plant, *Pisum sativum*, as the organism to study in his experiments.

54. Explain how continuous variation of a trait can occur.

55. Discuss the inheritance pattern that would be seen in a pedigree designed to study a recessive sex-linked characteristic.

ANALOGIES An analogy is a relationship between two pairs of words or phrases written as: a:b::c:d. The symbol : is read "is to," and the symbol :: is read "as." Write the letter of the pair of terms that best completes the analogy shown.

56. F_2 : F_1 ::
a. P : F_1
b. F_1 : F_2
c. F_1 : P
d. dominant trait : recessive trait
e. recessive trait : dominant trait

____

57. Mendel : ratios of traits::
a. monosomy : trisomy
b. nondisjunction : prenatal testing
c. XY : XX
d. karyotype : numbers of chromosomes
e. autosomes : sex chromosomes

____

58. 2000 yellow seeds : 8000 total seeds::
a. 1 : 6
b. 16 : 1
c. 1 : 8
d. 1 : 3
e. 1 : 4

____

59. codominance : both traits are displayed::
a. probability : crosses
b. heterozygous : alleles are the same
c. homozygous : alleles are same
d. Punnett square : chromosomes combine
e. albinism : pedigrees are used

____

60. a typical genetic disorder : recessive allele::
   a. hemophilia : sickle-cell anemia
   b. albinism : dominant allele
   c. protection from malaria : sickle cell hemoglobin
   d. genetic disorder : genetic counseling
   e. albinism : autosome

____

61. hemophilia : blood-clotting genes::
   a. shadows : darkness
   b. relaxation : rest
   c. loneliness : friends
   d. loudness : sound
   e. organization : planning

____

62. chorionic villi sampling : genetic counseling::
   a. F_2 generation : self-pollination
   b. continuous variation : codominance
   c. ratio : F_2 generation
   d. microscope : observing cells
   e. multiple alleles : environmental influences

____

63. homozygous : heterozygous::
   a. heterozygous : Bb
   b. probability : predicting chances
   c. dominant : recessive
   d. factor : gene
   e. PKU : inability to break down phenylalanine

____

64. several genes : trait::
   a. phenotype : multiple alleles
   b. environmental conditions : phenotype
   c. short hair : brown hair
   d. X chromosome : Y chromosome
   e. sex-linked trait : non-sex-linked trait

____

TEST ANSWERS

TRUE/FALSE

1. T
2. F
3. F
4. F
5. T
6. T
7. F
8. F
9. T
10. T

MULTIPLE CHOICE

11. c
12. c
13. b
14. a
15. c
16. b
17. a
18. d
19. c
20. b
21. c
22. d
23. c
24. a

25. b

26. a

27. c

28. a

29. d

30. c

31. d

32. d

33. b

34. a

35. c

36. d

SHORT ANSWER

37. Heredity

38. gene

39. F_2

40. recessive

41. alleles

42. genotype

43. law of segregation

44. probability

45. dihybrid

46. incomplete dominance

47. mutations

48. sex-linked

49. sickle-cell anemia

50. hemophilia

51. pedigree

52. phenylketonuria; PKU

ESSAY

53. The pea plant, *Pisum sativum*, is an ideal organism for genetic studies for several reasons. There are a number of traits that are easily identified and tracked from generation to generation, each of which has two forms, one of which regularly disappears and reappears in alternate generations. Also, this species is easy to grow and matures quickly. Finally, gametes of both sexes are found in the same flower, so cross-pollination is easy to accomplish by removing the anthers from some and transferring pollen from others to the remaining pistils.

54. Some traits are controlled by several genes, rather than just by one. Since these genes may segregate independently of each other, various combinations of the alleles may occur in individuals. The result may be a wide range of phenotype expressions that occur on a continuum from one extreme to another.

55. Sex-linked characteristics are carried as alleles on the X chromosome. As a result, sex-linked recessive traits are rarely seen in females, unless she is the offspring of an affected male and a female who is a carrier or is affected. Males born to a female who is either a carrier or affected may inherit the sex-linked allele. If the female is affected, both of her X chromosomes will possess the gene under study and the male is sure to inherit it. If she is a carrier, only one of her X chromosomes will possess the sex-linked allele and the male will have a 50-50 chance of inheriting the gene. Since males only have one X chromosome, he will not have a dominant allele in his genotype to counteract the effect of the sex-linked allele.

ANALOGIES

56. C (is produced by the)

57. D (showed/shows)

58. E (is to)

59. C (occurs when)

60. C (results from)

61. C (is the absence of)

62. D (is used in)

63. C (is different from)

64. B (may influence)

Chapter 8

TRUE/FALSE Read each statement. If the statement is true, write T in the space provided. If the statement is false, write F in the space provided.

1.______ It has been discovered that proteins are the genetic material, rather than DNA, because proteins are more complex molecules.

2.______ Even though they contain inactivated or killed infectious organisms, vaccines still will cause an immune response when injected into an organism.

3.______ Thymine and cytosine are pyrimidines.

4.______ The strands of a DNA molecule are held together by hydrogen bonding of adenine with guanine molecules and cytosine with thymine molecules.

5.______ Before a DNA molecule can replicate itself, it must make itself more compact. This is accomplished by the double helix coiling up on itself.

6.______ Multiple DNA polymerases may become attached to more than one portion of a DNA molecule at once, speeding up the rate of replication.

7.______ A gene can affect the phenotype of an individual because the gene contains a code that specifies the nature of a certain protein.

8.______ Introns can be shuffled between genes to create new genes over time.

9.______ That more than one version of a gene may exist on an organism's chromosomes can best be explained by the existence of multigene families.

10.______ While transposons can cause mutations when they enter a gene, other genes have been found that can have a protective effect against mutations.

MULTIPLE CHOICE Circle the letter of the term or phrase that best completes each statement or best answers each question.

11. Griffith's transformation experiments
   a. changed proteins into DNA.
   b. caused non-harmful bacteria to become deadly.
   c. resulted in DNA molecules becoming proteins.
   d. were designed to show the effect of heat on bacteria.

12. Griffith's experiments showed that
   a. dead bacteria could be brought back to life.
   b. harmful bacteria were hardier than harmless bacteria.
   c. heat caused the harmful and harmless varieties of bacteria to fuse.
   d. genetic material could be transferred between dead bacteria and living bacteria.

13. Avery and his coworkers demonstrated that treating bacteria with DNA-destroying enzymes
   a. also inactivated proteins in the cells.
   b. caused them to undergo transformation.
   c. prevented harmless bacteria from transforming into deadly bacteria.
   d. prevented DNA from transforming into protein molecules.

14. Oswald Avery and his research team concluded that
   a. RNA was the genetic material.
   b. protein bases were the genetic material.
   c. DNA and RNA were found in the human nucleus.
   d. DNA was the genetic material.

15. Hershey and Chase investigated the nature of the genetic material by conducting experiments that involved
   a. bacteria and mice.
   b. viruses that infect bacteria.
   c. bacteria that infect viruses.
   d. human DNA injected into mice.

16. Using radioactive tracers to determine the interactions of bacteriophages and their host bacteria, Hershey and Chase demonstrated without question that
   a. genes are composed of protein molecules.
   b. DNA and proteins are actually the same molecules located in different parts of cells.
   c. bacteria inject their DNA into the cytoplasm of bacteriophages.
   d. genes are made up of DNA.

17. Molecules of DNA are composed of long chains of
   a. amino acids.
   b. fatty acids.
   c. monosaccharides.
   d. nucleotides.

18. Which of the following is *not* part of a molecule of DNA?
   a. deoxyribose
   b. nitrogenous base
   c. phosphate
   d. ribose

19. A nucleotide consists of
   a. a sugar, a protein, and adenine.
   b. a sugar, a protein, and starch.
   c. a sugar, a phosphate group, and a nitrogenous base.
   d. a starch, a phosphate group, and a nitrogenous base.

20. Chargaff's rules, or the base-pairing rules, state that in DNA
   a. the amount of adenine equals the amount of thymine.
   b. the amount of guanine equals the amount of cytosine.
   c. the amount of guanine equals the amount of thymine.
   d. Both a and b are correct.

21. The scientists credited with establishing the structure of DNA are
   a. Avery and Chargaff.
   b. Hershey and Chase.
   c. Mendel and Griffith.
   d. Watson and Crick.

22. X-ray diffraction photographs by Wilkins and Franklin suggested that
   a. DNA and RNA are the same molecules.
   b. DNA is composed of either purines or pyrimidines, but not both.
   c. DNA molecules are arranged as a tightly coiled helix.
   d. DNA and proteins have the same basic structure.

23. It is said that a molecule of DNA may resemble a twisted ladder. If this analogy is used, then the sides of the ladder are formed by
   a. alternating sugar and phosphate units.
   b. pairs of purines.
   c. pairs of pyrimidines.
   d. combined alternating pairs of purines and pyrimidines.

24. Continuing the ladder analogy in the previous question, the rungs of the ladder are made of
   a. paired purines and pyrimidines.
   b. adenine paired with thymine.
   c. guanine paired with cytosine.
   d. All of the above are correct.

25. During DNA replication, a complementary strand of DNA is made for each original DNA strand. Thus, if a portion of the original strand is CCTAGCT, then the new strand will be
   a. TTGCATG.
   b. AAGTATC.
   c. CCTAGCT.
   d. GGATCGA.

26. The enzymes responsible for adding nucleotides to the exposed DNA template bases are
   a. replicases.
   b. DNA polymerases.
   c. helicases.
   d. nucleotidases.

27. A gene may be described as
   a. a sequence of amino acids.
   b. special proteins found in chromosomes.
   c. a sequence of nucleotides that controls the production of a certain protein.
   d. a sequence of nucleotides coding for the production of starches and sugars.

28. Non-coding sequences of nucleotides that are found between the actual genes of a DNA molecule are called
   a. introns.
   b. exons.
   c. transposons.
   d. codons.

29. The portions of DNA molecules that actually code for the production of proteins are called
   a. mutons.
   b. exons.
   c. introns.
   d. exposons.

30. Many biologists believe that having the genes of eukaryotic cells interrupted by non-coding sequences of nucleotides
   a. prevents the code from being copied.
   b. ensures that replication occurs correctly.
   c. causes severely damaging mutations.
   d. provides evolutionary flexibility to the genetic material.

31. Many thousands of proteins may have arisen from only a few thousand exons because
   a. an exon may be used by many different genes.
   b. there really is no difference between one protein and another.
   c. an exon does not actually code for any meaningful information.
   d. one gene can code for hundreds of different proteins.

32. Transposons can
   a. "jump" from one chromosomal location to another.
   b. inactivate genes in which they become located.
   c. bring together different combinations of genes.
   d. All of the above are correct.

33. Barbara McClintock received a Nobel prize for her work on
   a. introns.
   b. DNA polymerase.
   c. fruit fly mutations.
   d. transposons in corn.

34. Genes that can move from one location on a chromosome to another are called
   a. exons.
   b. introns.
   c. jumpons.
   d. transposons.

The following question(s) refer to the molecular diagram.

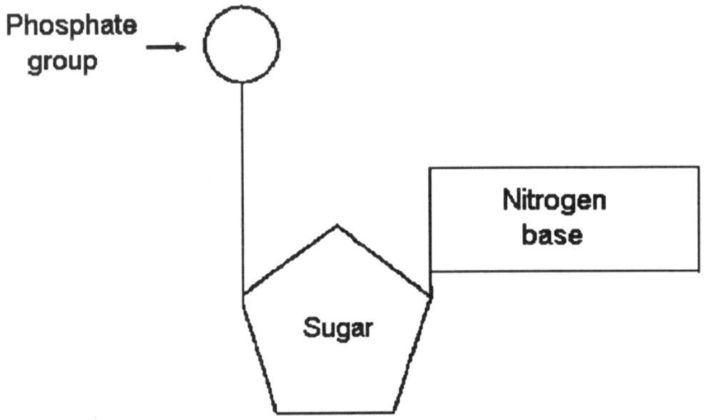

35. The entire molecule shown in the diagram is called a(n)
 a. amino acid.
 b. nucleotide.
 c. polysaccharide.
 d. pyrimidine.

36. In DNA, four forms of this molecule each have a different type of
 a. phosphate group.
 b. sugar.
 c. nitrogen base.
 d. None of the above

37. The part of the molecule for which deoxyribonucleic acid is named is the
 a. phosphate group.
 b. sugar.
 c. nitrogen base.
 d. None of the above; DNA is not named after part of the molecule.

SHORT ANSWER Complete each statement by writing the correct term or phrase in the space provided.

38. A(n) ________________________ is a preparation of killed or weakened microorganisms that is introduced into the body to produce immunity.

39. The ability of a virus to cause disease is referred to as its ________________________.

40. In Griffith's experiments, R bacteria were mixed with killed S bacteria and as a result the harmless R bacteria became deadly S bacteria. This changing of the nature of the organisms is known as ________________________.

41. Avery's prevention of transformation using DNA-destroying enzymes provided evidence that ________________________ molecules function as the hereditary material.

42. Viruses that infect bacteria are called ________________________.

43. A DNA subunit composed of a phosphate group, a five-carbon sugar, and a nitrogen-containing base is called a(n) ________________________.

44. The name of the five-carbon sugar that makes up a part of the backbone of molecules of DNA is ________________________.

45. Nitrogen bases containing a double ring of carbon and nitrogen atoms are called ________________________ .

46. Due to the strict pairing of nitrogen base pairs in DNA molecules, the two strands are said to be ______________________ to each other.

47. According to Chargaff's rule, adenine pairs with __________________ and guanine pairs with ______________________.

48. Watson and Crick determined that DNA molecules have the shape of a(n) ______________________.

49. The process by which DNA copies itself is called __________________.

50. Enzymes called ______________________ are responsible for unwinding the DNA double helix by breaking the hydrogen bonds that hold the complementary strands together.

51. The enzyme that is responsible for replicating molecules of DNA by attaching complementary bases in the correct sequence is ______________________.

52. Nucleotide segments of a DNA molecule that make up genes and are actually expressed in the phenotype of an organism are called ______________________.

53. Genes that can move from one chromosomal location to another are called ______________________.

54. A gene known as ______________________ has been found to code for a protein that seems to protect humans against the development of tumors.

ESSAY Read and answer each question or statement on a separate sheet of paper.

55. Briefly summarize the highlights of the experiments performed by Hershey and Chase that indicated that DNA was probably the genetic material.

56. Describe how a molecule of DNA is replicated.

57. How does p53 protect the human body from cancer?

ANALOGIES An analogy is a relationship between two pairs of words or phrases written as: a:b::c:d. The symbol : is read "is to," and the symbol :: is read "as." Write the letter of the pair of terms that best completes the analogy shown.

58. heat-killed S bacteria : mice to live::
  a. mice : S bacteria
  b. non-heat-killed S bacteria : mice to live
  c. S and R bacteria : mice to die
  d. R bacteria : mice to die
  e. mice : R bacteria

____

59. mutation : transposon::
  a. bacteria : vaccine
  b. tumor suppression : p53
  c. interruption : exon
  d. DNA : transformation
  e. phage : bacteria

____

60. multigene family : genes::
   a. monohybrid cross : phenotypes
   b. DNA : nucleotides
   c. homozygous : alleles
   d. heterozygous : alleles
   e. a genotype : phenotypes

---

61. Watson and Crick : DNA::
   a. Avery : nucleotides
   b. Hershey and Chase : protein
   c. Wilkins and Franklin : transformation
   d. McClintock : transposons
   e. Griffith : tumor suppressors

---

62. ATTG : TAAC::
   a. AAAT : TTTG
   b. TCGG : AGAT
   c. GTCC : CAGG
   d. CGAA : TGCG
   e. CCTT : GGCC

---

63. Wilkins and Franklin : Watson and Crick::
   a. snow boards : skate boards
   b. ink pen : quill
   c. automobile : horse-drawn wagon
   d. electric light bulb : candle light
   e. radio : television

TEST ANSWERS

TRUE/FALSE

1. F

2. T

3. T

4. F

5. F

6. T

7. T

8. F

9. T

10. T

MULTIPLE CHOICE

11. b

12. d

13. c

14. d

15. b

16. d

17. d

18. d

19. c

20. d

21. d

22. c

23. a

24. d

25. d

26. b

27. c

28. a

29. b

30. d

31. a

32. d

33. d

34. d

35. b

36. c

37. b

SHORT ANSWER

38. vaccine

39. virulence

40. transformation

41. DNA

42. bacteriophages

43. nucleotide

44. deoxyribose

45. purines

46. complementary

47. thymine; cytosine

48. double helix

49. replication

50. helicases

51. DNA polymerase

52. exons

53. transposons

54. p53

ESSAY

55. Hershey and Chase used radioactive labeling methods to tag bacteriophage DNA with ^{32}P and bacteriophage coat proteins with ^{35}S. They found that the bacteriophages injected the DNA into their host bacterial cells, but that the protein remained on the outside of the cell. It was observed that the genetic material—the DNA—was incorporated into the DNA of the bacteria and provided the code for the production of more phage particles, while the protein was not involved in phage reproduction.

56. To begin the replication process, enzymes called helicases break the hydrogen bonds that hold the two complementary strands of the DNA double helix together, allowing the helix to unwind. The complementary strands are held apart by additional protein molecules. At the replication fork, the point at which the double helix separates, a molecule of DNA polymerase attaches and begins to add nucleotides to the exposed bases according to the base-pairing rules. This continues until the DNA polymerase reaches a nucleotide sequence that signals it to detach, having completed the replication of the DNA strand.

57. The gene known as p53 codes for a protein that may act as a tumor suppressor. The product of the p53 gene coordinates a cell's response to DNA that has been damaged by mutation. It can cause a cell to shut down its growth and division processes, or in some cases to self-destruct, thus preventing the uncontrolled growth and division that is a hallmark of tumor development and cancer. If the p53 gene is itself inactivated by mutation, its protective function is lost and a tumor may develop.

ANALOGIES

58. C (caused)

59. B (is caused by a)

60. B (has many)

61. D (studied)

62. C (is complementary to)

63. E (came before)

Chapter 9

TRUE/FALSE Read each statement. If the statement is true, write T in the space provided. If the statement is false, write F in the space provided.

1.______ Proteins are manufactured directly from DNA molecules.

2.______ There are four steps in the process of gene expression.

3.______ During transcription, the information on a DNA molecule is "rewritten" into an mRNA molecule.

4.______ When mRNA leaves the nucleus and enters the cytoplasm, it has a complete set of both introns and exons.

5.______ A codon signifies either a specific amino acid or a stop signal.

6.______ It has been discovered that each species of organism has its own unique genetic code for synthesis of its proteins.

7.______ A functional ribosome forms when mRNA, two rRNA subunits, and a molecule of tRNA bind together.

8.______ The operator portion of an operon controls RNA polymerase's access to structural genes.

9.______ A chromosome puff is a portion of a chromosome that contains genes that have burst.

10.______ Mutations may be random, spontaneous mistakes or they may be caused by exposure to environmental mutagens.

11.______ Mutations may prevent evolution from occurring because they introduce genetic variation into a population.

MULTIPLE CHOICE Circle the letter of the term or phrase that best completes each statement or best answers each question.

12. RNA differs from DNA, in that RNA
   a. is single-stranded.
   b. contains a different sugar molecule.
   c. contains the nitrogen base uracil.
   d. All of the above are correct.

13. In RNA molecules, adenine is complementary to
   a. cytosine. c. thymine.
   b. guanine. d. uracil.

14. Each of the following is a type of RNA, *except*
   a. carrier RNA. c. ribosomal RNA.
   b. messenger RNA. d. transfer RNA.

15. During transcription
   a. proteins are synthesized. c. RNA is produced.
   b. DNA is replicated. d. translation occurs.

16. rRNA has a function in
a. synthesizing DNA.
b. synthesizing mRNA.
c. forming ribosomes.
d. transferring amino acids to ribosomes.

17. After mRNA has been transcribed
a. its introns are cut out.
b. its exons are joined together.
c. it leaves the nucleus by way of pores.
d. All of the above are true.

18. The fact that the genetic code is universal suggests that
a. all organisms evolved from separate original ancestors.
b. life forms have a common evolutionary ancestry.
c. proteins make up the genetic material.
d. evolution does not really occur.

19. Transfer RNA acts as an "interpreter" because it
a. carries an amino acid to its correct codon.
b. synthesizes amino acids as they are needed.
c. produces codons to match the correct anticodons.
d. converts DNA into mRNA.

20. A ribosome has
a. one binding site for DNA.
b. three binding sites used during translation.
c. four binding sites for tRNA.
d. no binding sites since the proteins must detach.

21. At the very beginning of translation, the first tRNA molecule
a. binds to the ribosome's A site.
b. attaches directly to the DNA codon.
c. connects an amino acid to its anticodon.
d. attaches to the P site of the ribosome.

22. After the primary structure of a protein has been completed
a. the codons and anticodons unite.
b. an enzyme attaches adjacent amino acids to each other to form a chain.
c. the protein folds into the secondary and tertiary structures.
d. the tRNA molecules remain attached until the protein is secreted from the cell.

23. In prokaryotic cells, a group of genes that code for functionally related proteins is a(n)
a. exon.
b. intron.
c. operon.
d. riboson.

24. The function of an operator is to
a. regulate access of RNA polymerase to structural genes.
b. turn on and off the molecules of tRNA.
c. control the process of transcription within the nucleus.
d. generate amino acids for protein synthesis.

25. An inducer molecule functions by
a. causing DNA replication.
b. binding the rRNA subunits of a ribosome.
c. removing a repressor molecule from an operator.
d. digesting lactose molecules in bacterial cells.

26. A portion of intense transcription formed by a piece of DNA that has looped out from a eukaryotic chromosome
a. does not contain useful genetic material.
b. is called a chromosome puff.
c. represents a gene that has been repressed.
d. becomes part of the nuclear envelope of the cell.

27. A sequence of nucleotides in a DNA molecule that aids in exposing RNA polymerase binding sites of specific genes is called a(n)
a. repressor.
b. operon.
c. operator.
d. enhancer.

28. Mutations that occur in gametes
a. may be passed on to future generations.
b. can be important in the evolution of a species.
c. affect the functioning of the gene in which they occur.
d. All of the above are correct.

29. A frameshift mutation can result from
a. addition of a nucleotide.
b. substitution of a nucleotide.
c. two nucleotides switching places.
d. None of the above

30. Which of the following may act as mutagens?
a. ultraviolet light
b. ionizing radiation
c. carcinogens
d. All of the above may cause mutations.

31. Metastasis is characterized by
a. the growth of a benign tumor.
b. spread of malignant cells beyond their original site.
c. localization of a tumor to a particular location.
d. conversion of a malignant tumor to a benign tumor.

32. An oncogene
a. metastasizes to other chromosomes.
b. prevents a cell from reproducing.
c. induces numerous mutations of the chromosome on which it is found.
d. may cause a cell to become a cancer cell.

The following question(s) refer(s) to the strand of mRNA CUCAAGUGCUUC and the chart showing the genetic code.

	U	C	A	G	
U	Phe	Ser	Tyr	Cys	U
	Phe	Ser	Tyr	Cys	C
	Leu	Ser	stop	stop	A
	Leu	Ser	stop	Trp	G
C	Leu	Pro	His	Arg	U
	Leu	Pro	His	Arg	C
	Leu	Pro	Gln	Arg	A
	Leu	Pro	Gln	Arg	G
A	Ile	Thr	Asn	Ser	U
	Ile	Thr	Asn	Ser	C
	Ile	Thr	Lys	Arg	A
	Met	Thr	Lys	Arg	G
G	Val	Ala	Asp	Gly	U
	Val	Ala	Asp	Gly	C
	Val	Ala	Glu	Gly	A
	Val	Ala	Glu	Gly	G

33. Which of the following would represent the strand of DNA from which the mRNA strand was made?
a. CUCAAGUGCUUC c. GAGTTCACGAAG
b. GAGUUCACGAAG d. AGACCTGTAGGA

34. The anticodons for the codons in the above mRNA are
a. GAG - UUC - ACG - AAG. c. CUC - GAA - CGU - CUU.
b. GAG - TTC - ACG - AAG. d. CUU - CGU - GAA - CUC.

35. What is the portion of the protein molecule coded for by the piece of mRNA shown?
a. Ser - Tyr - Arg - Gly c. Leu - Lys - Cys - Phe
b. Val - Asp - Pro - His d. Pro - Glu - Leu - Val

SHORT ANSWER Complete each statement by writing the correct term or phrase in the space provided.

36. Molecules of ______________________ carry instructions for protein synthesis from the nucleus to the cytoplasm.

37. The nitrogen-containing base that is only found in RNA is ______________________.

38. Transcription and translation are stages in the process of ______________________.

39. Messenger RNA is produced during the process of ___________________.

40. The information contained in a molecule of messenger RNA is used to make protein during the process of ______________________.

41. The enzyme responsible for making RNA is called ___________________.

42. A ______________________ is a sequence of DNA at the beginning of a gene that signals RNA polymerase to begin transcription.

43. The form of ribonucleic acid that carries genetic information from the DNA to the ribosomes is ______________________.

44. The sequence of three nucleotides that code for specific amino acids or stop signals in the synthesis of protein is called ________.

45. Nucleotide sequences of tRNA that are complementary to codons on mRNA are called ________.

46. A(n) ________ is a cluster of genes in a bacterial cell that codes for proteins with related functions.

47. Giant chromosomes found in salivary gland cells of flies such as *Drosophila* are called ________ chromosomes.

48. ________ are regions of DNA that stimulate transcription of certain genes.

49. A change in the DNA of a gene is called a(n) ________.

50. Mutations that change one nucleotide or just a few nucleotides in a gene are called ________ mutations.

51. Environmental agents that can cause defects in genetic material are called ________.

52. Agents that may cause mutations that may result in the development of cancer are called ________.

53. Diseases characterized by abnormal cell growth in which cells undergo repeated uncontrollable divisions are called ________.

54. A gene that, when mutated, can cause a cell to become cancerous is called a(n) ________.

ESSAY Read and answer each question or statement on a separate sheet of paper.

55. Identify the three types of RNA and briefly describe the function of each.

56. Generally describe the operon model of gene regulation as proposed by Jacob and Monod.

57. Distinguish among the different types of point mutations.

58. Distinguish between benign and malignant tumors.

ANALOGIES An analogy is a relationship between two pairs of words or phrases written as: a:b::c:d. The symbol : is read "is to," and the symbol :: is read "as." Write the letter of the pair of terms that best completes the analogy shown.

59. a site : tRNA::
   a. codon : mRNA
   b. mRNA : amino acid
   c. tRNA : amino acid
   d. mRNA : P site
   e. codon : A site

____

60. mRNA : nucleus::
   a. nucleus : protein
   b. protein : cytoplasm
   c. nucleus : ribosomes
   d. protein : nucleus
   e. ribosomes : nucleus

---

61. inducer : repressor::
   a. promoter : DNA
   b. deletion mutation : nucleotide
   c. DNA : terminator
   d. estrogen : receptor protein
   e. RNA : nucleus

---

62. UV light : mutations::
   a. cut : mend
   b. clatter : silence
   c. hot : flame
   d. flame : heat
   e. film : dense

---

63. cell division : RAS genes::
   a. gene regulation : operons
   b. operator : structural genes
   c. repressor : operators
   d. inducer : operons
   e. inducer : repressors

---

64. codon : nucleotides::
   a. P site : RNA molecules
   b. ribosome : DNA molecules
   c. RNA : bases
   d. DNA : bases
   e. ribosome : binding sites

---

65. repressor : operator::
   a. promoter : RNA polymerase
   b. promoter : DNA
   c. terminator : RNA polymerase
   d. terminator : DNA
   e. RNA polymerase : mRNA

---

66. insertion of nucleotide : genes::
   a. gene : deletion nucleotides
   b. point mutation : membranes
   c. benign tumor : tissues
   d. gene : substitution nucleotides
   e. malignant tumor : tissues

TEST ANSWERS

TRUE/FALSE

1. F
2. F
3. T
4. F
5. T
6. F
7. T
8. T
9. F
10. T
11. F

MULTIPLE CHOICE

12. d
13. d
14. a
15. c
16. c
17. d
18. b
19. a
20. b
21. d
22. c
23. c
24. a

25.c

26.b

27.d

28.d

29.a

30.d

31.b

32.d

33.c

34.a

35.c

SHORT ANSWER

36.RNA; ribonucleic acid

37.uracil

38.gene expression

39.transcription

40.translation

41.RNA polymerase

42.promoter

43.mRNA; messenger RNA.

44.a codon

45.anticodons

46.operon

47.polytene

48.Enhancers

49.mutation

50.point

51. mutagens

52. carcinogens

53. cancers

54. oncogene

ESSAY

55. There are three types of RNA: messenger RNA (mRNA), transfer RNA (tRNA), and ribosomal RNA (rRNA). Messenger RNA carries hereditary information from the DNA in the nucleus to the site of translation on the ribosomes; tRNA carries amino acids to the ribosomes for assembly into proteins; rRNA is a structural molecule, becoming part of the ribosomes upon which translation occurs.

56. An operon is a cluster of genes that code for functionally related proteins. A portion of the operon called an operator acts as a switch, controlling access of RNA polymerase to the structural genes of the operon. When the operator does not have a repressor molecule attached to it, the RNA polymerase catalyzes transcription of the structural genes of the operon. However, when a repressor molecule is attached to the operator, RNA polymerase is blocked from accessing the portions of the operon that code for a particular protein or group of proteins. An inducer molecule can remove the repressor molecule, thus allowing the RNA polymerase to attach to the operon, permitting transcription.

57. There are three types of point mutations: substitutions, insertions, and deletions. Substitutions involve the replacement of one nucleotide with another. These mutations may not have a serious effect if the new codon codes for the same amino acid as the original. Insertions and deletions, known as frameshift mutations, involve the addition or omission of a single nucleotide, resulting in a shifting of the reading of the triplet code. The result of a shift in the triplet code may result in nonsense codons that will not translate into a real protein.

58. A benign tumor is one that does not invade surrounding tissues. Such tumors can usually be surgically removed and cause relatively few problems. Malignant tumors are invasive, spreading into nearby tissues and interfering with organ functions. Cells of this latter type of tumor may break free of the original tumor and metastasize to other locations in the body.

ANALOGIES

59. C (holds)

60. B (is made in the)

61. B (removes)

62. D (causes)

63. A (is controlled by)

64. E (has 3)

65. C (blocks or stops the action of)

66.E (goes into or affects)

Chapter 10

TRUE/FALSE Read each statement. If the statement is true, write T in the space provided. If the statement is false, write F in the space provided.

1.______ Scientists have used genetic engineering to produce bacteria capable of synthesizing human proteins.

2.______ Plasmids are pieces of viral DNA that commonly infect human cells.

3.______ The "sticky ends" of a DNA fragment can combine with any other DNA fragment cut by the same restriction enzyme.

4.______ DNA ligase can seal the "sticky ends" of a DNA fragment.

5.______ The plasmid vectors prepared by Cohen and Boyer contained only one gene.

6.______ Injection of a particular vaccine can cause the body to produce antibodies that protect against the possibility of future infection by a particular pathogen.

7.______ Technological advances have now made it possible to transfer normal genes into the cells of a person who has a genetic disorder.

8.______ If a crop is made glyphosate-resistant, treating it with glyphosate would seriously reduce its yield.

9.______ Genetic engineers have developed a method of infecting cows with milk-producing bacteria to increase the amount of milk produced by the cows.

MULTIPLE CHOICE Circle the letter of the term or phrase that best completes each statement or best answers each question.

10. Cohen and Boyer transferred a gene from a frog chromosome into the genetic material of a
   a. different frog.
   b. different chromosome of the same frog.
   c. virus taken from the same frog.
   d. bacterial cell.

11. The use of genetic engineering to transfer human genes into bacteria
   a. is impossible with current technology.
   b. causes the human genes to manufacture bacterial proteins.
   c. results in the formation of a new species of organism.
   d. allows the bacteria to produce human proteins.

12. Recombinant DNA is formed by joining DNA molecules
   a. from two different species.
   b. from two chromosomes of the same organism.
   c. with RNA molecules.
   d. with a proteins from a different species.

13. Which of the following procedures is *not* a usual step in a genetic engineering experiment?
   a. inducing a mutation in a source chromosome
   b. cleaving DNA with a restriction enzyme
   c. recombining pieces of DNA from different species
   d. cloning and screening cells containing the gene of interest

14. Plasmids
   a. are circular pieces of bacterial DNA.
   b. can replicate independently of the organism's main chromosome.
   c. are often used as vectors in genetic engineering experiments.
   d. All of the above are true.

15. Fragments of DNA having complementary "sticky ends"
   a. are found only in bacterial cells.
   b. can join with each other.
   c. can only join with complementary fragments of the same species.
   d. are immediately digested by enzymes in the cytoplasm of the cell.

16. Cohen and Boyer used tetracycline in their genetic engineering experiments to
   a. prevent their cell cultures from becoming infected.
   b. kill the cells that possessed recombinant DNA.
   c. remove cells from the culture that had not taken up the vector DNA.
   d. None of the above; tetracycline is an antibiotic not to be used in experiments.

17. A technique that uses radioactively labeled RNA to identify specific genes in a piece of DNA that is being studied is called the
   a. Northern blot.
   b. Southern vector.
   c. Northern lights.
   d. Southern blot.

18. Restriction Fragment Length Polymorphism (RFLP) analysis is the procedure used for
   a. making recombinant DNA.
   b. DNA fingerprinting.
   c. creating plasmids in bacterial cells.
   d. cloning cells that might contain specific genes of interest.

19. The goal of the Human Genome Project is to
   a. collect samples of genes from humans around the world.
   b. determine the evolutionary origin of each human gene.
   c. perfect the genes of humans so as to eliminate disease.
   d. determine the nucleotide sequence of each human gene.

20. Although controversial, DNA fingerprinting has been used in criminal investigations because
   a. criminals leave DNA samples behind them when they touch an object at a crime scene.
   b. DNA analysis is believed to allow investigators to distinguish body cells of different individuals, who are unlikely to have the same DNA.
   c. bacterial DNA on the hands of criminals may provide a clue as to where that person was when the crime was committed.
   d. DNA found on murder weapons is easy to identify.

21. A medical condition that can be treated using proteins produced through genetic engineering is
   a. diabetes mellitus type I.
   b. heart attacks (dissolving blood clots).
   c. hemophilia (promoting blood clotting).
   d. All of the above may be treated with such products.

22. A vaccine is
   a. a substance that kills bacteria or viruses.
   b. a plasmid that contains disease-causing genes.
   c. an antibody.
   d. a harmless version of a disease-causing microorganism or its toxins.

23. The risk associated with vaccines prepared by injecting killed or weakened pathogenic microbes is that
   a. a few remaining live or unweakened microbes could still cause the disease.
   b. the antibodies that result may not work.
   c. the vaccine protects only against other diseases.
   d. None of the above; there is no risk with this type of vaccine.

24. Tumor necrosis factor (TNF)
   a. produces antibodies.
   b. is produced by cancer cells.
   c. attacks and kills cancer cells.
   d. destroys white blood cells.

25. Transferring normal human genes into human cells that lack them
   a. is impossible at this time.
   b. would cause cancer.
   c. will cause antibodies to kill those cells.
   d. is called gene therapy.

26. Glyphosate kills most actively growing plants by
   a. activating restriction enzymes.
   b. destroying a protein-producing enzyme.
   c. inducing production of antibodies.
   d. None of the above; glyphosate is a fertilizer.

27. In nature, the Ti plasmid
   a. contains a gene that causes tumors in plants.
   b. causes tumors in bacterial cultures.
   c. causes uncontrolled reproduction of bacterial cells.
   d. transfers restriction enzymes that cause DNA damage in plants.

28. A gene that codes for resistance to glyphosate has been added to the genome of certain plants. These plants will
   a. produce chemicals that kill weeds growing near them.
   b. die when exposed to glyphosate.
   c. convert glyphosate to fertilizer.
   d. survive when glyphosate is applied to the field.

29. Genetic engineers are developing approaches for improving agriculture in all of the following ways *except*
   a. making plants resistant to insects.
   b. making certain plants susceptible to Ti plasmids.
   c. enabling crops to fix nitrogen more economically.
   d. improving the productivity of livestock.

30. Legumes are plants that have nodules on their roots that contain bacteria, which convert N_2 gas into a form of nitrogen that plants can use. To accomplish this, the bacteria carry out a process called
   a. genetic fixation.
   b. nitrogen plasmid formation.
   c. nitrogen fixation.
   d. nitrogenous decomposition.

31. If major cash crops could be genetically engineered to carry out nitrogen fixation,
   a. the need for nitrogen fertilizers would be greatly reduced.
   b. plants would be able to absorb N_2 from the air and convert it to a useful form.
   c. the cost of cultivating the crops would be considerably reduced.
   d. All of the above are true.

32. Through genetic engineering, bacteria have been produced that are capable of all of the following *except*
a. producing insulin.
b. resisting the effects of plant-damaging insects.
c. producing milk, reducing the need for raising cows.
d. synthesizing frog rRNA.

33. "Genetic engineering" refers to the process of
a. creating new DNA molecules from nucleotide sequences.
b. rearranging nucleotides in a gene of an organism so that new traits appear in the development of an embryo.
c. moving genes from a chromosome of one organism to a chromosome of a different organism.
d. building a new species by combining genes of different organisms.

The following question(s) refer(s) to the diagram of a recombinant DNA plasmid.

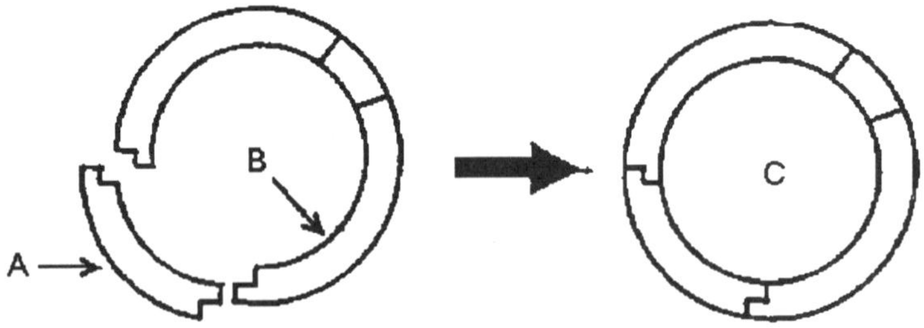

34. In the diagram above, the pieces of DNA labeled *A* and *B* were formed by the action of
a. planned mutations.
b. restriction enzymes.
c. viral injections.
d. digestive enzymes.

35. The structure labeled *C* is
a. bacterial DNA.
b. recombinant DNA.
c. EcoR1.
d. None of the above

SHORT ANSWER Complete each statement by writing the correct term or phrase in the space provided.

36. The process used to isolate a gene from the DNA of one organism and transfer the gene into the DNA of another is called ______________.

37. A DNA molecule that is produced from pieces of DNA from different organisms is called ______________.

38. Small circular forms of bacterial DNA are called ______________.

39. A(n) ______________ is an agent that is used to carry a DNA fragment isolated from one cell into another cell.

40. A large number of genetically identical cells grown from a single cell are called ______________.

41. Enzymes that cleave DNA at specific sequences, generating a set of small fragments of DNA, are called ______________.

42. A technique known as ______________ can be used to separate molecules in a mixture by subjecting them to an electrical field within a gel.

43. Unlimited copies of a gene or segment of a DNA molecule can be produced by a process known as ______________________.

44. The name of the scientific program that has the goals of constructing physical maps of human chromosomes and determining the DNA sequences of those chromosomes is the ______________________.

45. A(n) ______________________ is a solution that contains a harmless version of a pathogen or its toxins.

46. Proteins that kill infectious microbes in the body of an organism before they can cause disease are called ______________________.

47. A process by which legumes, such as soybeans, peanuts, and clover, convert N_2 gas in the air into chemical forms that plants can use is known as ______________________.

ESSAY Read and answer each question or statement on a separate sheet of paper.

48. Summarize the four steps required to conduct a genetic engineering experiment.

49. Some diseases are now being treated with medical products manufactured through genetic engineering. Using the treatment of diabetes mellitus, briefly explain how such products can be made.

50. The Ti plasmid causes tumors to develop in certain types of plants. Using an example, explain how it is possible for this plasmid to be used in a beneficial way in agriculture?

ANALOGIES An analogy is a relationship between two pairs of words or phrases written as: a:b::c:d. The symbol : is read "is to," and the symbol :: is read "as." Write the letter of the pair of terms that best completes the analogy shown.

51. foreign DNA fragment : recombinant DNA::
   a. heat : ice
   b. sugar : salt
   c. basketball : basketball court
   d. false tooth : complete set of teeth
   e. flag pole : building

____

52. genetic engineering : human health::
   a. vaccine : anticoagulant
   b. human insulin : diabetes patients
   c. anticoagulant : hemophilia
   d. diabetes : insulin
   e. tumors : TNF

____

53. factor VIII : hemophilia::
   a. bone marrow gene : diabetes
   b. anticoagulant : diabetes
   c. TNF : viruses
   d. insulin : diabetes
   e. vaccine : cancer

____

54. nitrogen-fixing bacteria : air::
   a. fertilizers : low-nitrogen fertilizer
   b. crops : air
   c. crops : high-nitrogen fertilizer
   d. high-nitrogen fertilizers : low-nitrogen fertilizers
   e. genetically engineered crops : fertilizer

____

55. erosion : glyphosate::
   a. success : hard work
   b. sickness : medicine
   c. knowledge : education
   d. music : tape player
   e. cold : freezing

____

56. PCR : DNA segments::
   a. genetically engineered bacterium : human proteins
   b. genetic engineering : new genes
   c. RNA polymerase : restriction enzymes
   d. DNA plasmid : gel electrophoresis
   e. RNA : PCR

____

57. RFLP : DNA fingerprints::
   a. screening cells : electrophoresis
   b. gel electrophoresis : chromosomes
   c. DNA fingerprints : fingerprints
   d. recombinant DNA : electrophoresis
   e. Human Genome Project : human genetic maps

____

58. "sticky ends" : DNA ligase::
   a. cancer : TNF
   b. nucleotides : DNA polymerase
   c. pathogens : antibodies
   d. blood clots : anticoagulants
   e. diabetes : insulin

____

59. plasmid : DNA segment coding for enzyme::
   a. DNA ligase : double-stranded DNA
   b. vector : restriction enzyme radioactively labeled
   c. cloned cell : DNA ligase
   d. recombinant DNA : DNA from another organism
   e. bacterial culture : inability to clone

____

TEST ANSWERS

TRUE/FALSE

1. T
2. F
3. T
4. T
5. F
6. T
7. T
8. F
9. F

MULTIPLE CHOICE

10. d
11. d
12. a
13. a
14. d
15. b
16. c
17. d
18. b
19. d
20. b
21. d
22. c
23. a
24. c

25. d

26. b

27. a

28. d

29. b

30. c

31. d

32. c

33. c

34. b

35. b

SHORT ANSWER

36. genetic engineering

37. recombinant DNA

38. plasmids

39. vector

40. clones

41. restriction enzymes

42. gel electrophoresis

43. polymerase chain reaction (PCR)

44. Human Genome Project

45. vaccine

46. antibodies

47. nitrogen fixation

ESSAY

48. First a source chromosome must be cleaved, producing a fragment that contains the gene to be transferred. Then the DNA fragment is put into a vector that will carry the gene into a bacterial cell. Next the recipient bacterial cells are cloned, and finally, the bacterial clone is screened to identify and isolate those specific bacterial cells that contain the transferred gene and that are producing the protein associated with it.

49. The human gene that controls the production of insulin is removed from human chromosome 11 using a restriction enzyme and is inserted into a bacterial plasmid. This plasmid is then introduced into a bacterial culture, and those bacteria containing the insulin gene are permitted to reproduce extensively. These bacteria produce large amounts of insulin that can be collected, purified, and used to treat diabetes patients.

50. Scientists have used genetic engineering techniques to remove the tumor-causing gene from the Ti plasmid. The remaining plasmid can then be used as a vector. A desirable gene, such as a gene for herbicide resistance, can be spliced into the plasmid. Infecting a plant with this altered plasmid results in the injection of the desired gene into the genome of the plant. Thus, the plant now has a new beneficial gene that can be expressed. For example, a plant receiving a gene that provides resistance to the herbicide glyphosate would survive when the weeds around it are treated with the herbicide.

ANALOGIES

51. D (is added to)

52. B (can benefit)

53. D (is genetically engineered to treat)

54. C (get nitrogen from)

55. B (is reduced with the use of)

56. A (makes)

57. E (analyzes)

58. B (are joined together with)

59. D (contains)

Chapter 11

TRUE/FALSE Read each statement. If the statement is true, write T in the space provided. If the statement is false, write F in the space provided.

1.______ Oparin based his theory of the origin of life on a lifetime of experimentation.

2.______ Miller and Urey succeeded in producing primitive living organisms from simple chemical compounds.

3.______ One of the first gases found in the Earth's atmosphere was oxygen.

4.______ One of the sources of energy for the origin of life on Earth was lightning.

5.______ The creation theory states that life evolved on a planet in another galaxy.

6.______ It is impossible to test the creation theory by scientific methods.

7.______ According to Oparin, the energy for the formation of molecules in the early oceans came from photosynthesis.

8.______ If life exists on another planet, heredity must depend on DNA.

9.______ RNA is a nucleic acid; it never behaves like an enzyme.

10.______ It is impossible to test any assumptions about the origin of life.

11.______ Scientists know that the Earth is the only planet in the universe that can sustain carbon-based life.

MULTIPLE CHOICE Circle the letter of the term or phrase that best completes each statement or best answers each question.

12.Which event in the Earth's history occurred first?
a. formation of the atmosphere
b. formation of simple organic compounds
c. formation of the oceans
d. formation of proteins

13.The surface of the Earth is protected from damaging ultraviolet light by
a. oxygen. c. hydrogen.
b. ozone. d. nitrogen.

14.A theory about the extraterrestrial origin of life refers to
a. the "little extra" that life on land needs.
b. the extra molecules needed for life to begin.
c. the origin of life outside the earth.
d. the terrestrial origin of life.

15.Oparin believed that proteins first appeared
a. in volcanoes. c. in water.
b. in the atmosphere. d. on iron pyrite and clay.

16. Which of the following gases was thought by Oparin to be part of the Earth's early atmosphere?
   a. oxygen
   b. ozone
   c. ammonia
   d. carbon dioxide

17. Louis Lerman suggested that the chemical reactions that led to the origin of life took place
   a. in the atmosphere.
   b. on another planet.
   c. in the primordial soup.
   d. in bubbles.

18. RNA
   a. has a three-dimensional structure.
   b. is a nucleic acid.
   c. can act like an enzyme.
   d. All of the above

19. Proteins and nucleotides
   a. form long chains of DNA.
   b. form spontaneously in water.
   c. have a half-life.
   d. do not form spontaneously in water.

20. When placed in water, proteins and lipids
   a. repel each other.
   b. spontaneously disintegrate.
   c. form plasma membranes.
   d. aggregate.

21. The theory that life was formed on Earth by divine forces is known as the
   a. creation theory.
   b. extraterrestrial origin theory.
   c. thermodynamics theory.
   d. evolution theory.

22. Isotopes are forms of the same element that differ in
   a. atomic number.
   b. number of electrons.
   c. atomic mass.
   d. number of protons.

23. The half-life of a radioisotope
   a. does not change.
   b. varies with the seasons.
   c. increases as the radioisotope ages.
   d. fluctuates.

24. If the half-life of a radioisotope is 5,000 years, how much of the radioisotope in a specimen will be left after 10,000 years?
   a. all of it
   b. one-half of the original amount.
   c. one-quarter of the original amount.
   d. none of it

25. The second law of thermodynamics
   a. applies only to photosynthetic plants.
   b. states that all systems are open.
   c. does not apply to processes involving organisms.
   d. states that all systems are closed.

26. Natural selection
   a. guides evolution.
   b. is random.
   c. is directionless.
   d. does not exist.

27. The second law of thermodynamics states that
   a. the universe never changes.
   b. disorder tends to increase in the universe.
   c. the universe is constantly becoming more orderly.
   d. None of the above

28. The first macromolecules on Earth were
a. probably composed of DNA.
b. proteins.
c. probably composed of RNA.
d. forms of ATP.

29. The energy necessary to support chemical reactions on the early Earth may have come from
a. volcanic activity.
b. lightning.
c. ultraviolet light.
d. All of the above

30. A search for extraterrestrial intelligence
a. has been carried out for several decades.
b. will be started in the twenty-first century.
c. is always conducted by the astronauts on the space shuttle.
d. is entirely dependent on the position of the moon in the sky.

31. Theories about the extraterrestrial origin of life
a. must necessarily conflict with creation theory.
b. explain how life evolved.
c. examine only life in our galaxy.
d. are not testable.

32. A layer of ozone in the atmosphere was critical to the formation of life on land because
a. land plants need ozone for photosynthesis.
b. there is a high concentration of ozone in the oceans.
c. ozone is needed to produce oxygen.
d. ozone absorbs ultraviolet light.

33. Miller and Urey did not put oxygen in their apparatus because oxygen
a. is not essential to most forms of life as we know it.
b. does not react with ammonia, methane, or hydrogen.
c. would have led to the formation of microorganisms.
d. was not believed to have been present in the Earth's early atmosphere.

34. In their experiment, Miller and Urey produced
a. proteins.
b. nucleic acids.
c. carbohydrates.
d. amino acids.

The following question(s) refer(s) to the apparatus used by scientists in the 1950s to re-create the conditions of the early Earth.

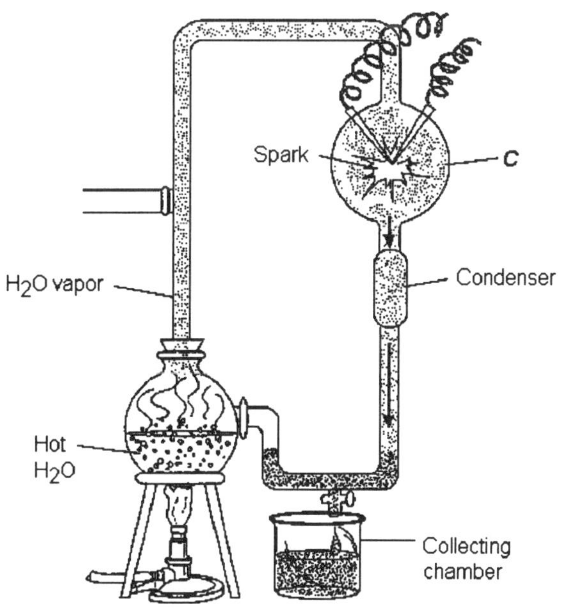

35. This apparatus was designed to demonstrate that life on Earth might have originated from
a. radioactive decay. c. simple organic molecules.
b. divine forces. d. extraterrestrial life.

36. Water vapor in the flasks was mixed with all of the following except
a. ammonia. c. oxygen.
b. hydrogen. d. methane.

37. Gases circulated through the apparatus. When the mixture reached the large flask labeled *C*, the gases were sparked so that
a. the gases could be removed for analysis.
b. extra nitrogen could be added.
c. lightning discharge through the gases could be simulated.
d. excess carbon monoxide could be removed.

38. Using this type of apparatus and different mixtures of gases, all of the following have been produced except
a. amino acids. c. lipids.
b. RNA. d. nucleotides.

SHORT ANSWER Complete each statement by writing the correct term or phrase in the space provided.

39. Miller and Urey based their experiment in 1953 on the work of ____________.

40. Most scientists think life had a(n) ____________ origin, developing by itself through natural processes.

41. Oparin's hypothesis about the origin of life suggested that the Earth's oceans were once a(n) ____________ soup containing organic molecules.

42. Scientists now think that ____________ was the first information-storing molecule to form on Earth.

43. Short chains of amino acids produced in the laboratory form tiny vesicles called ____________ .

44. Louis Lerman suggested that the key chemical processes leading to the origin of life took place in ____________ on the ocean's surface.

45. Carbon and silicon both need ____________ electrons to fill their outer energy level.

46. The oldest fossils that have been found are about ____________ billion years old.

47. The period of time it takes for one-half of a radioisotope to decay is called its ____________.

The following question(s) refer(s) to the diagram illustrating the radioactive decay of potassium-40.

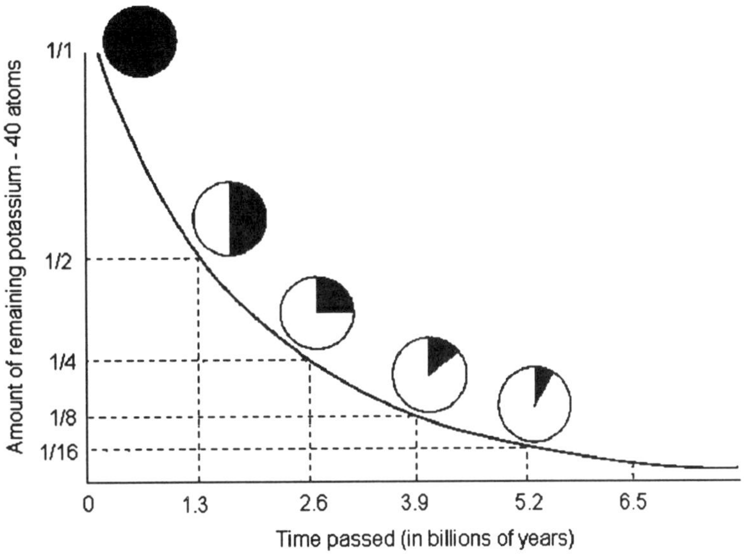

48. The half-life of potassium-40 is about ________________________ billion years.

49. Scientists have estimated the age of the Earth using similar ________________________ dating techniques.

50. The study of radioactive decay in rocks indicates that the Earth is about ________________________ billion years old.

51. Because of its ________________________ Earth has the right amount of gravitational pull to hold the gases found in our atmosphere.

ESSAY Read and answer each question or statement on a separate sheet of paper.

52. Why is Oparin's hypothesis about the origin of life in a primordial soup being reevaluated?

53. Explain how the use of radioisotopes helps scientists determine the age of a rock.

54. Why do scientists now think that proteins were not the first macromolecules to form?

55. The ozone layer in the Earth's atmosphere was discovered about 40 years ago. Scientists are not certain if the thinning of this layer is part of a natural atmospheric cycle or if the thinning has been caused by the activities of people in industrialized countries. Why do you think many biologists are urging industrialized nations to take steps that may prevent further destruction of the ozone layer?

ANALOGIES An analogy is a relationship between two pairs of words or phrases written as: a:b::c:d. The symbol : is read "is to," and the symbol :: is read "as." Write the letter of the pair of terms that best completes the analogy shown.

56. divine creation : spontaneous generation::
   a. nitrogen : gas
   b. billion : 1,000 million
   c. apple : orange
   d. extraterrestrial : outside of the Earth
   e. century : 100 years

____

57. salad dressing : oil globules::
   a. atmosphere : photons
   b. phospholipids : microspheres
   c. primordial soup : gases
   d. methane : membranes
   e. proteins : amino acid

____

58. complexity and diversity : DNA variation and time::
   a. randomness : design
   b. environment : natural selection
   c. environmental change : DNA
   d. natural selection : environment change
   e. time : complexity

____

59. closed systems : disorder::
   a. living things : closed systems
   b. Earth processes : closed systems
   c. open systems : organization
   d. open systems : simplicity
   e. open systems : energy loss

____

60. oxygen : electrons::
   a. carbon dioxide : oxygen
   b. nitrogen : protons
   c. methane : microspheres
   d. ozone : ultraviolet light
   e. water : DNA

____

TEST ANSWERS

TRUE/FALSE

1.F

2.F

3.F

4.T

5.F

6.T

7.F

8.F

9.F

10.F

11.F

MULTIPLE CHOICE

12.a

13.b

14.c

15.c

16.c

17.d

18.d

19.d

20.d

21.a

22.c

23.a

24.c

25. c

26. a

27. b

28. c

29. d

30. a

31. d

32. d

33. d

34. d

35. c

36. c

37. c

38. d

SHORT ANSWER

39. Oparin

40. spontaneous

41. primordial

42. RNA

43. microspheres

44. bubbles

45. four

46. 3.5

47. half-life

48. 1.3

49. radiometric

50. 4.5

51.mass

ESSAY

52.The primordial soup model is being reevaluated for two reasons (1) The discovery of 3.5-billion-year old fossils means that life might have originated faster than previously assumed. (2) Ozone was probably not present in the early atmosphere. Without ozone, ultraviolet light would have destroyed any ammonia and methane present in the atmosphere. Ammonia and methane are two key ingredients in the primordial soup suggested by Oparin.

53.Radioisotopes change to more stable isotopes through a process called radioactive decay. The period of time it takes a radioisotope to decay is called its half-life. By estimating how many half-lives of a radioisotope in a rock have passed, scientists can approximate the rock's age.

54.Scientists have not been able to cause amino acids dissolved in water to join together to form proteins. The energy-requiring reactions that link amino acids are reversible and do not occur on their own in water. Scientists now think that the initial macromolecules were composed of RNA and that RNA later catalyzed the formation of proteins.

55.Ozone shields the Earth from most of the damaging ultraviolet light from the sun.

ANALOGIES

56.C (is different from)

57.B (forms into)

58.D (is dependent on)

59.C (have more)

60.D (absorbs)

Chapter 12

TRUE/FALSE Read each statement. If the statement is true, write T in the space provided. If the statement is false, write F in the space provided.

1.______ Species that have evolved from a common ancestor should have certain characteristics in common.

2.______ In his "Essay on the Principle of Population" Malthus said humans were the only population that could continue to grow in size indefinitely.

3.______ Natural selection can cause the spread of an advantageous adaptation throughout a population.

4.______ The replacement of the dinosaurs by mammals as the dominant land animals is an example of macroevolution.

5.______ Evidence for evolution occurs only in the fossil record.

6.______ The way an embryo develops is not important in determining the evolutionary history of a species.

7.______ Punctuated gradualism refers to the theory that evolution occurs only in short periods of time.

8.______ The environment dictates only the direction and extent of evolution.

9.______ Within populations, divergence leads to speciation.

MULTIPLE CHOICE Circle the letter of the term or phrase that best completes each statement or best answers each question.

10. Which of the following ideas did Darwin accept before he began his voyage on the *Beagle*?
   a. The Earth was about 6,000 years old.
   b. The Earth had remained unchanged except for the effects of floods and other catastrophes.
   c. Organisms had appeared on the Earth in their present forms.
   d. All of the choices are correct.

11. Prior to his voyage on the *Beagle*, Darwin accepted the view of
   a. chromosomal theory. c. creationism.
   b. natural selection. d. radioactivity.

12. The major idea that Darwin presented in his book *The Origin of Species* was that
   a. species changed over time and never competed with each other.
   b. animals changed, but plants remained the same.
   c. giraffes and peppered moths changed constantly.
   d. species changed over time by natural selection.

13. Darwin conducted much of his research on
   a. the Samoan Islands. c. the Hawaiian Islands.
   b. Manhattan Island. d. the Galapagos Islands.

14. According to Darwin, evolution occurs
   a. by chance. c. by natural selection.
   b. over a half-life of 5,730 years. d. rapidly.

15. The change that results in a species being better suited to its environment is known as
   a. reproduction.
   b. variation.
   c. adaptation.
   d. selection.

16. Which of the following describes a population?
   a. Dogs and cats living in Austin, Texas
   b. Four species of fish living in a pond
   c. Dogwood trees in Middletown, Connecticut
   d. Roses and tulips in a garden

17. The hypothesis that evolution occurs at an irregular rate
   a. was supported by Darwin.
   b. is known as punctuated equilibrium.
   c. is supported by many transitional forms in the fossil record.
   d. can be demonstrated by examining the necks of giraffes.

18. The hypothesis that evolution occurs at an irregular rate through time is known as
   a. directional evolution.
   b. directional equilibrium.
   c. punctuated equilibrium.
   d. punctuated evolution.

19. A change in the frequency of a particular gene in one direction in a population is called
   a. directional selection.
   b. acquired variation.
   c. chromosome drift.
   d. balancing selection.

20. New species form
   a. when ecological races diverge more and more.
   b. because of natural selection.
   c. when members of the same species become adapted to new environments.
   d. All of the above

21. Populations of the same species that differ genetically because they have adapted to different living conditions are known as
   a. selected populations.
   b. ecological races.
   c. genetic populations.
   d. genetic races.

22. A situation in which two opposing forces affect the frequency of an allele in a population is known as
   a. mutation.
   b. directional variation.
   c. acquired variation.
   d. balancing selection.

23. Natural selection causes
   a. changes in the environment.
   b. plants and animals to produce more offspring than can survive.
   c. changes in the frequency of certain alleles in a population.
   d. All of the above

24. In his experiments with peppered moths, Kettlewell found that
   a. the color of the moths was not important.
   b. birds preferred the flavor of light-colored moths.
   c. moths that matched the color of the tree trunks were more likely to survive.
   d. coloration was an inherited trait.

25. Populations of the same species living in different places
   a. do not vary.
   b. always show balancing selection.
   c. have a half-life related to the size of the population.
   d. become increasingly different as each becomes adapted to its environment.

26.People heterozygous for the sickle cell allele
   a. are less susceptible to malaria.
   b. have an advantage if they live in Africa.
   c. are not favored by selection if they live in the United States.
   d. All of the above

27.Organisms well suited to the environment
   a. reproduce at a greater rate than those less suited to the environment.
   b. are always larger than organisms less suited to the environment.
   c. always live longer than species less suited to the environment.
   d. need less food than organisms less suited to the environment.

28.Natural selection is the process by which
   a. the age of the Earth is calculated.
   b. organisms with traits well suited to the environment survive and reproduce at a greater rate than other organisms.
   c. acquired traits are passed from one generation to the next.
   d. All of the above

29.The theory of evolution predicts that
   a. closely related species will show similarities in nucleotide sequences.
   b. if species have changed over time, their genes should have changed.
   c. closely related species will show similarities in amino acid sequences.
   d. All of the above

30.Homologous structures in organisms suggest that the organisms
   a. have a common ancestor.
   b. lived at different times.
   c. have a backbone.
   d. are extinct.

The following question(s) refer(s) to the diagram of bones in the limbs of fossil and modern horses.

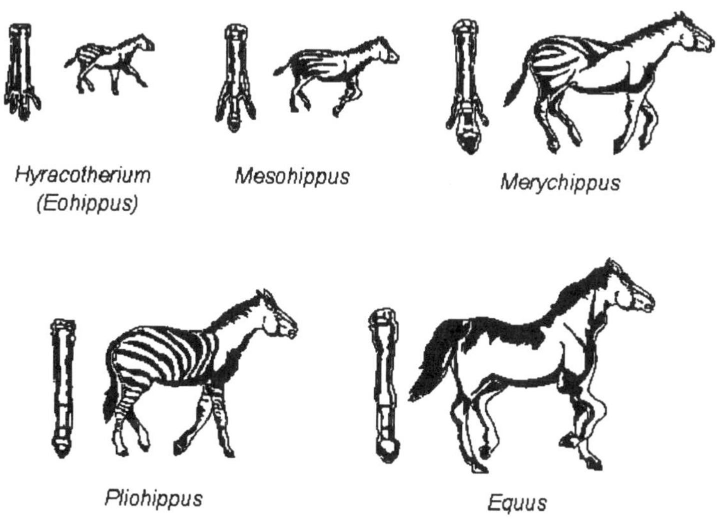

31.These fossils indicate that horse evolution probably has taken place
   a. rapidly.
   b. five times by the process of punctuated equilibrium.
   c. gradually.
   d. in only one place on the Earth.

The following question(s) refer(s) to the diagrams of the bones of a penguin, alligator, human, and bat.

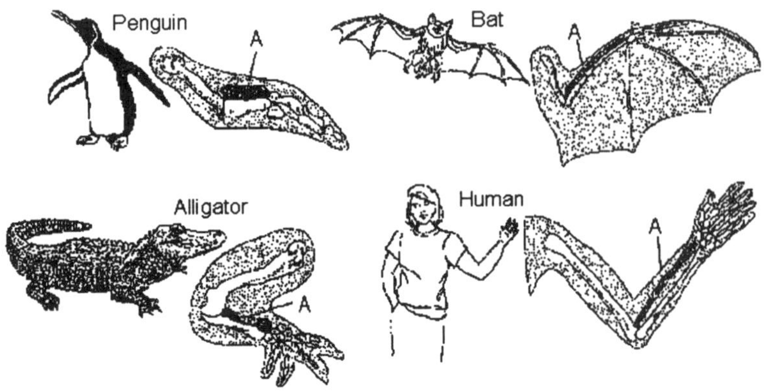

32. The similarity of these structures suggests that the organisms
   a. have a common ancestor.
   b. all grow at different rates.
   c. evolved slowly.
   d. live for a long time.

33. The bones labeled A are known as
   a. vestigial structures.
   b. sequential structures.
   c. homologous structures.
   d. fossil structures.

34. An analysis of the DNA from these organisms would indicate that
   a. their DNA is identical.
   b. they all have gill pouches.
   c. their nucleotide sequences show many similarities.
   d. they all have the same number of chromosomes.

SHORT ANSWER Complete each statement by writing the correct term or phrase in the space provided.

35. A change in a species over time is called ______________________.

36. A species that has disappeared permanently is said to be ______________________.

37. The most direct evidence that evolution has occurred comes from ______________________.

38. Closely related species show more similarities in nucleotide sequences than ______________________ related species.

39. Structures that are present in an organism but are reduced in size and have no function are called ______________________.

40. Homologous structures are similar because they are inherited from a common ______________________.

41. Scientists who study fossils are known as ______________________.

42. ______________________ is the gradual darkening of a population in response to pollution.

43. People who are heterozygous for the sickle cell allele are more resistant to ______________________ than people who do not have the allele.

44. A process by which organisms with traits well suited to the environment survive and reproduce at a greater rate than organisms less suited for the environment is ______________________.

45. Organisms produce more offspring than can ______________________.

ESSAY Read and answer each question or statement on a separate sheet of paper.

46. An agricultural plot of land is sprayed with a very powerful insecticide to destroy harmful insects. Nevertheless, many of the same species of insect are present on the land the following year. How might Darwin's theory account for this phenomenon?

47. You are a zoologist studying birds on a group of islands. You have just discovered four species of birds that have never before been seen. Each species lives on a different island. The birds are identical to each other except for the shape of their beaks. How can you explain their similarities and their differences?

48. What role does the environment play in natural selection?

ANALOGIES An analogy is a relationship between two pairs of words or phrases written as: a:b::c:d. The symbol : is read "is to," and the symbol :: is read "as." Write the letter of the pair of terms that best completes the analogy shown.

49. fossil : living animal::
   a. genes : population
   b. organisms on a family tree : each other
   c. scientists : fossils
   d. evolution : lack of change
   e. evolution : population

____

50. genes : variations::
   a. variations : genes
   b. isolations : extinction
   c. different habitats : different populations
   d. different adaptations : fossils
   e. habitats : different mutations

____

51. best suited individuals : most offspring::
   a. best adapted individuals : most extinctions
   b. most isolated individuals : most fossils
   c. most offspring : least fossils
   d. most mineralized remains : most fossils
   e. most genes : least adapted traits

____

52. vestigial structures : macroevolution::
   a. homologous structures : common ancestry
   b. common ancestry : fossils
   c. common ancestry : rock
   d. homologous structures : unrelated species
   e. living animals : fossils

____

53. struggle for survival : competition::
   a. time : environment
   b. survival of the fittest : best traits
   c. trait : time
   d. environment : traits
   e. environment : best traits

---

54. dark peppered moth : industrialized areas::
   a. light peppered moth : unindustrialized areas
   b. dark peppered moth : light tree trunks
   c. light peppered moth : industrialized areas
   d. lack of soot : industrialized areas
   e. light peppered moth : dark tree trunks

---

55. presence of malaria : heterozygous sickle cell condition::
   a. presence of malaria : malaria
   b. presence of malaria : normal hemoglobin
   c. presence of malaria : no sickle cells
   d. absence of malaria : sickle cell anemia
   e. absence of malaria : normal hemoglobin

---

56. Galapagos Island organisms : South America::
   a. Lyell : South America
   b. Darwin : England
   c. England : Darwin
   d. South American organisms : Galapagos Islands
   e. islands : organisms

---

57. microevolution : within a species::
   a. speciation : in a gene
   b. fossil : to a family tree
   c. geometric progression : within a food supply
   d. macroevolution : among species
   e. extinction : by speciation

TEST ANSWERS

TRUE/FALSE

1. T
2. F
3. T
4. T
5. F
6. F
7. F
8. T
9. T

MULTIPLE CHOICE

10. c
11. c
12. d
13. d
14. c
15. c
16. c
17. b
18. c
19. a
20. d
21. b
22. d
23. c
24. c

25. d

26. d

27. a

28. b

29. d

30. a

31. c

32. a

33. c

34. c

SHORT ANSWER

35. microevolution

36. extinct

37. fossils

38. distantly

39. vestigial

40. ancestor

41. paleontologists

42. Industrial melanism

43. malaria

44. natural selection

45. survive

ESSAY

46. A part of Darwin's theory states that variations exist within a species. A small percentage of the insects exposed to the insecticides were not affected by the insecticide. They survived and produced offspring that were also resistant to the insecticide.

47. The four species evolved from a common ancestor, each species adapting to the conditions on its island. The difference in beak shape may be the result of differences in available food among the islands. Each bird species adapted to the food that was available on its island.

48. According to Darwin's theory, those organisms with traits that are best suited to the environment survive to reproduce.

ANALOGIES

49. b (resemble(s))

50. c (contain)

51. d (will produce)

52. a (is/are evidence of)

53. b (is based on)

54. a (escapes predation in)

55. e (favors)

56. b (came from)

57. d (involves change)

Chapter 13

TRUE/FALSE Read each statement. If the statement is true, write T in the space provided. If the statement is false, write F in the space provided.

1._______ The first living things to appear on the Earth were prokaryotes.

2._______ On the early Earth, oxygen was found in the atmosphere millions of years before it was present in the oceans.

3._______ All protists are single-celled.

4._______ Some protists are photosynthetic.

5._______ Mass extinctions occur with regularity every 125 million years.

6._______ The first multicellular creatures to populate the land were animals that came from the oceans.

7._______ Arthropods evolved from insects.

8._______ Bony fishes became dominant during the Devonian period.

9._______ The first fish to evolve had small, powerful jaws.

10._______ Amphibians can lay their eggs on dry land because the eggs are surrounded by a shell that prevents water loss.

MULTIPLE CHOICE Circle the letter of the term or phrase that best completes each statement or best answers each question.

11. Protists evolved from
    a. archaebacteria. c. eubacteria.
    b. protobacteria. d. pseudobacteria.

12. Cyanobacteria are believed to be the ancestors of
    a. mitochondria. c. ribosomes.
    b. nuclei. d. chloroplasts.

13. The eukaryotic kingdom that exhibits the greatest diversity of organisms is the
    a. plants. c. fungi.
    b. animals. d. protists.

14. Cyanobacteria changed the early Earth's atmosphere by giving off
    a. carbon dioxide. c. hydrogen.
    b. ammonia. d. oxygen.

15. One major problem faced by amphibians moving onto land was
    a. not enough oxygen. c. body structures that constantly lose water.
    b. too many insects. d. All of the above

16. All of the living things on the Earth today can be grouped into
    a. three kingdoms. c. five kingdoms.
    b. four kingdoms. d. six kingdoms.

17. Eukaryotes may have descended from
   a. eubacteria.
   b. archaebacteria.
   c. cyanobacteria.
   d. None of the above

18. Multicellularity
   a. is a very recent evolutionary event.
   b. means having more than one cell.
   c. is found in some protists.
   d. All of the above

19. Archaebacteria
   a. cause most of the diseases found in people today.
   b. often cause decay.
   c. have ways of producing energy without using oxygen.
   d. are important in the production of cheese and other dairy products.

20. All of the major phyla of animals on the Earth today evolved
   a. in the Burgess Shale.
   b. after the mass extinctions that have taken place.
   c. during the Cambrian period.
   d. prior to the appearance of the kingdom Protista.

21. Although there was no soil present, plants were able to invade the surface of the early Earth because they
   a. were supported by arthropods.
   b. extracted minerals from protists.
   c. could live off dead material.
   d. formed a partnership with fungi.

22. Arthropods have a hard outer skeleton and
   a. a backbone.
   b. hair.
   c. a four-chambered heart.
   d. jointed appendages.

23. A layer of ozone in the atmosphere was critical to the formation of life on land because
   a. land plants need ozone to live.
   b. there is a high concentration of ozone in the oceans.
   c. ozone is necessary to produce oxygen.
   d. ozone absorbs ultraviolet radiation.

24. Insects have been very successful because
   a. they reproduce in large numbers.
   b. they evolved the ability to fly.
   c. some of them can feed on flower nectar.
   d. All of the above

25. Lobsters, insects, and spiders are all examples of
   a. amphibians.
   b. vertebrates.
   c. arthropods.
   d. monerans.

26. The first animals to invade the land were the
   a. amphibians.
   b. arthropods.
   c. dinosaurs.
   d. mycorrhizae.

27. The most diverse group of animals on Earth are the
   a. reptiles.
   b. mammals.
   c. insects.
   d. vertebrates.

28. The first vertebrates
   a. had skeletons made of cartilage.
   b. evolved on the land.
   c. were jawless fishes.
   d. resembled amphibians.

29. Vertebrates that are adapted to life both on land and in water are
a. reptiles.
b. birds.
c. bony fish.
d. amphibians.

30. Lampreys and hagfishes do not have
a. a backbone.
b. jaws.
c. fins.
d. bony skeletons.

31. The most diverse eukaryotic kingdom is
a. Protista.
b. Plantae.
c. Animalia.
d. Fungi.

32. The first organisms to populate the surface of the land were
a. bacteria and plants.
b. plants and animals.
c. plants and fungi.
d. bacteria and fungi.

The following question(s) refer(s) to the drawing of the organisms.

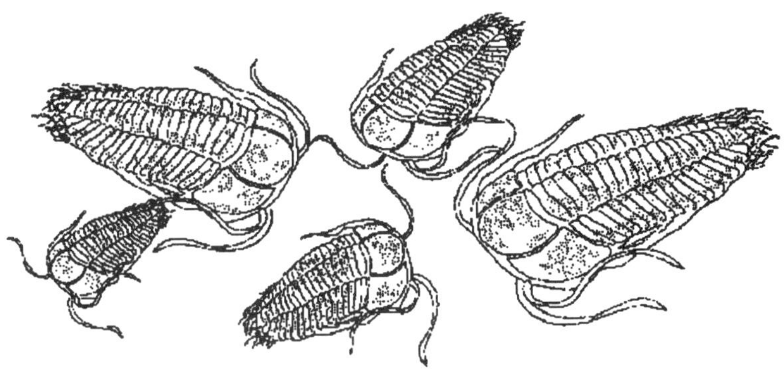

33. The organisms in the diagram are
a. a type of marine arthropod.
b. all extinct.
c. trilobites.
d. All of the above

The following question(s) refer(s) to the drawing of the plant roots.

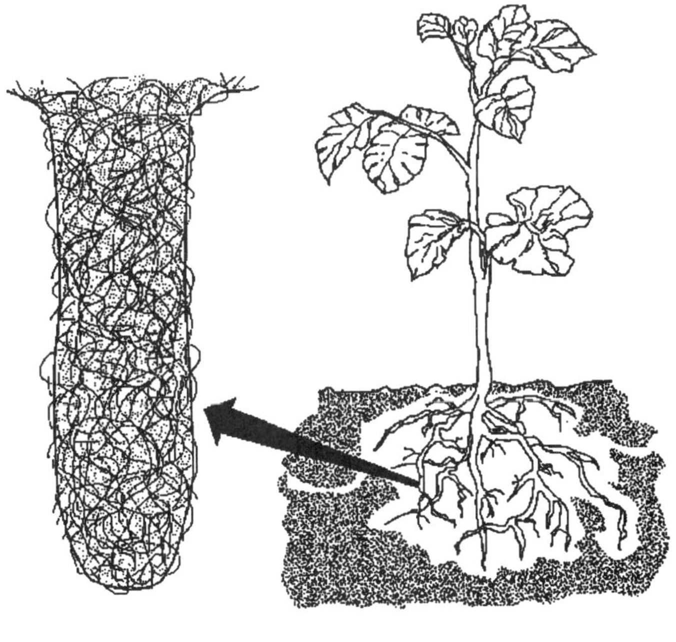

34. The associations between plants and fungi illustrated in the diagram are known as
a. powdery mildew. c. lichens.
b. mycorrhizae. d. mitochondria.

35. In this association the fungi provide
a. food to the plant. c. minerals to the plant.
b. energy to the plant. d. all of the above.

36. This type of partnership, in which two different species help each other, is known as
a. multicellularity. c. catalysis.
b. symbiosis. d. parasitism.

SHORT ANSWER Complete each statement by writing the correct term or phrase in the space provided.

37. Having more than one cell is known as ______________________.

38. The ______________________ are the most common bacteria found on Earth today.

39. ______________________ were the first eukaryotes.

40. Primitive ______________________ produced the first oxygen in the Earth's atmosphere.

41. At the end of the ______________________ period 250 million years ago, about 96 percent of all species of animals became extinct.

42. The first terrestrial organisms that were able to fly were the ______________________.

43. Three atoms of oxygen can combine together to make up a molecule of ______________________.

44. Ozone forms a protective layer in the atmosphere and absorbs ______________________ radiation.

45. Reptiles evolved from ______________________.

46. ______________________ evolved from early, heavily armored jawed fishes.

47. The large number of marsupials found in Australia and South America can be explained by the movement of the Earth's land masses. This movement is known as ______________________.

48. Birds and mammals became the dominant vertebrates on the land after the ______________________ extinction.

49. The ______________________ probably were the ancestors of all eukaryotic cells.

50. All of the major phyla on Earth today evolved during the ______________________ period.

ESSAY Read and answer each question or statement on a separate sheet of paper.

51. Most cells found in organisms on the Earth today are aerobic; they need oxygen. Explain why the first cells could not have been aerobic bacteria.

52. Why has there been a burst of evolution after each of the great mass extinctions?

53. Describe the structural innovations that helped amphibians adapt to life on land.

ANALOGIES An analogy is a relationship between two pairs of words or phrases written as: a:b::c:d. The symbol : is read "is to," and the symbol :: is read "as." Write the letter of the pair of terms that best completes the analogy shown.

54. archaebacteria : eukaryotes::
   a. eubacteria : cyanobacteria
   b. archaebacteria : eubacteria
   c. eubacteria : archaebacteria
   d. eukaryotes : prokaryotes
   e. cyanobacteria : eubacteria

____

55. sun's rays on O_2 : ozone::
   a. fungi : minerals
   b. fungi : food
   c. plants : minerals
   d. ozone : O_2
   e. cyanobacteria : O_2

____

56. arthropods : live on land::
   a. plants : absorb minerals from rocks
   b. fungi : make food
   c. scorpions : fly
   d. plants : give rise to fungi
   e. insects : fly

____

57. mycorrhizae : plant roots::
   a. mitochondria : cells
   b. Golgi apparatus : nucleus
   c. mitochondria : nucleus
   d. nucleus : mitochondria
   e. cells : mitochondria

____

58. lungs : amphibians::
   a. bony skeleton : reptiles
   b. water-tight skin : reptiles
   c. land animal design : protists
   d. circulatory system : reptiles
   e. walking : fish

____

59. birds and mammals : the whole Earth::
   a. marsupials : Australia
   b. reptiles : ocean
   c. fish : land
   d. amphibians : ocean
   e. birds : ocean

____

60. many new reptiles : Permian extinction::
  a. fish and amphibians : Permian extinction
  b. fish : Cretaceous extinction
  c. Permian extinction : beginning of continental drift
  d. mammals, birds, and small reptiles : Cretaceous extinction
  e. dinosaurs and reptiles : Cretaceous extinction

TEST ANSWERS

TRUE/FALSE

1.T

2.F

3.F

4.T

5.F

6.F

7.F

8.T

9.F

10.F

MULTIPLE CHOICE

11.a

12.d

13.d

14.d

15.c

16.d

17.b

18.d

19.c

20.c

21.d

22.d

23.d

24.d

25. c

26. b

27. c

28. c

29. d

30. b

31. a

32. c

33. d

34. b

35. c

36. b

SHORT ANSWER

37. multicellularity

38. eubacteria

39. Protists

40. cyanobacteria

41. Permian

42. insects

43. ozone

44. ultraviolet

45. amphibians

46. Sharks

47. continental drift

48. Cretaceous

49. archaebacteria

50. Cambrian

ESSAY

51. There was no oxygen in the atmosphere of the primitive Earth. Oxygen was found in the atmosphere only after the evolution of the photosynthetic cyanobacteria.

52. The organisms that survive the mass extinctions find themselves in a world of opportunity—an Earth full of food and space that is no longer used by others.

53. A number of adaptations evolved in amphibians that allowed them to live on land. These structural changes included: lungs and changes in the circulatory system, including the development of a new blood vessel and extra heart chamber. In addition, amphibians evolved limbs and a skeletal system that allowed them to walk on land.

ANALOGIES

54. e (are ancestors of)

55. e (make)

56. e (were the first to)

57. a (are found in)

58. b (developed in)

59. a (dominate)

60. d (appeared after)

Chapter 14

TRUE/FALSE Read each statement. If the statement is true, write T in the space provided. If the statement is false, write F in the space provided.

1.______ Today, prosimians can be found in South America, Asia, and Africa.

2.______ Fossils of apes have never been found in North and South America.

3.______ Fossil evidence indicates that most extinct primate species lived in trees.

4.______ Depth perception is a characteristic of most primates.

5.______ The oldest hominid skeleton unearthed to date is of the species *Australopithecus afarensis*.

6.______ Apes and humans are classified as hominids.

7.______ The arms of a gorilla are longer than its legs.

8.______ The Dutch physician Eugene Dubois thought that orangutans resembled what a missing link should look like.

9.______ Scientists think that early *Homo sapiens* had language capability.

10.______ Neanderthals are members of the species *Homo sapiens*.

MULTIPLE CHOICE Circle the letter of the term or phrase that best completes each statement or best answers each question.

11. The large eye sockets of early primates indicate that these animals probably
   a. had color vision.
   b. had poor eyesight.
   c. had large eyes to compensate for their inability to judge distance.
   d. were active at night.

12. Monkeys
   a. have smaller eyes than prosimians have.
   b. have larger brains than prosimians have.
   c. and prosimians have binocular vision.
   d. All of the above

13. Apes can be distinguished from monkeys because only
   a. monkeys can climb trees.
   b. apes can be found in Africa.
   c. monkeys have an opposable toe.
   d. monkeys have a tail.

14. Since DNA sequences in humans and chimpanzees are very similar,
   a. humans must have evolved from chimpanzees.
   b. chimpanzees must have single-stranded DNA.
   c. humans and chimpanzees must have a common ancestor in very recent geologic history.
   d. humans and chimpanzees are the same species.

15. A biologist analyzes the DNA sequences in three different organisms. The biologist finds that organisms A and B have almost exactly the same DNA sequences. The DNA sequences in organism C are very different. All three organisms are primates. This information allows the biologist to infer that
   a. organisms A and B are more closely related to each other than either is to organism C.
   b. all three organisms appeared on Earth at the same time.
   c. either organism A or B must be the direct ancestor of organism C.
   d. organism C could not have been the ancestor of A or B.

16. Fossil evidence indicates that primates evolved from
   a. the great apes.
   b. arthropods.
   c. small, insect-eating, rodentlike mammals.
   d. small hominids.

17. In order to accurately judge distance
   a. the field of vision of both eyes must be identical.
   b. an animal must be able to distinguish colors.
   c. an animal must be able to walk in an upright position.
   d. the field of vision of both eyes must overlap slightly.

18. Two distinctive features of primates are
   a. binocular vision and grasping hands.
   b. grasping hands and feet.
   c. binocular vision and a large skull.
   d. large eye sockets and binocular vision.

19. *A. boisei* was nicknamed "nutcracker man" because of its
   a. massive jaws.
   b. large skull.
   c. strong hands.
   d. None of the above

20. Most scientists agree that the oldest known hominid was
   a. *Homo habilis*.
   b. *Australopithecus afarensis*.
   c. *Homo sapiens*.
   d. *Homo erectus*.

21. An examination of australopithecine fossils indicates that australopithecines
   a. were shorter than modern humans.
   b. were bipedal.
   c. probably were active during the day.
   d. All of the above

22. Lucy's skeleton revealed that she was bipedal. This means that she
   a. walked on all four limbs.
   b. lived in a tree.
   c. walked on two legs.
   d. crawled on all fours.

23. The name *Homo habilis* translates roughly as
   a. hairy ape.
   b. human habitat.
   c. handy man.
   d. apelike human.

24. *Homo erectus* had a
   a. pair of large eye sockets.
   b. lighter skeleton than other hominids.
   c. written language.
   d. brain almost as large as that of modern humans.

25. *Homo erectus*
   a. had a large brain.
   b. walked erect.
   c. was larger than *Homo habilis*.
   d. All of the above

26. *Homo sapiens* appeared in Africa about
   a. 50,000,000 years ago. c. 500,000 years ago.
   b. 5,000,000 years ago. d. 50,000 years ago.

27. Hunting large animals and butchering them using flint and bone tools, as well as living in caves or wooden shelters, are associated with
   a. *H. habilis*. c. *A. robustus*.
   b. *A. africanus*. d. *H. erectus*.

28. The hypothesis that *Homo sapiens* evolved in Africa is supported by the finding that
   a. all *Homo sapiens* have the same number of chromosomes.
   b. mitochondrial DNA is inherited only from the mother.
   c. the greatest variety of mitochondrial DNA sequences are found among modern Africans.
   d. the fewest number of mitochondrial DNA mutations are found among modern Africans.

29. The first primates were
   a. australopithecines. c. gorillas.
   b. prosimians. d. humans.

30. Hominid fossils have been found by each of the following except
   a. Raymond Dart. c. Donald Johanson.
   b. Richard Leakey. d. Charles Darwin.

31. Cultural evolution refers to
   a. conceptual thought.
   b. social interactions in a population.
   c. controlling evolution.
   d. accumulating and transmitting knowledge.

32. Modern humans are most closely related to
   a. *A. boisei*. c. *A. africanus*.
   b. Neanderthals. d. *H. habilis*.

33. Changes in nucleotide sequences in mitochondrial DNA are caused by
   a. mating. c. crossing-over.
   b. meiosis. d. mutations.

34. Evidence of the intelligence of *Homo erectus* includes their
   a. large eye sockets. c. tools.
   b. written logs. d. use of agriculture.

35. The Neanderthals are noted for
   a. burying their dead.
   b. using weapons.
   c. having a brain capacity larger than that of modern humans.
   d. All of the above

36. *Homo erectus* evolved into
   a. *Homo habilis*. c. *Homo sapiens*.
   b. *Homo erectus*. d. the genus *Homo*.

The following question(s) refer(s) to the diagram illustrating the phylogenetic tree of primates.

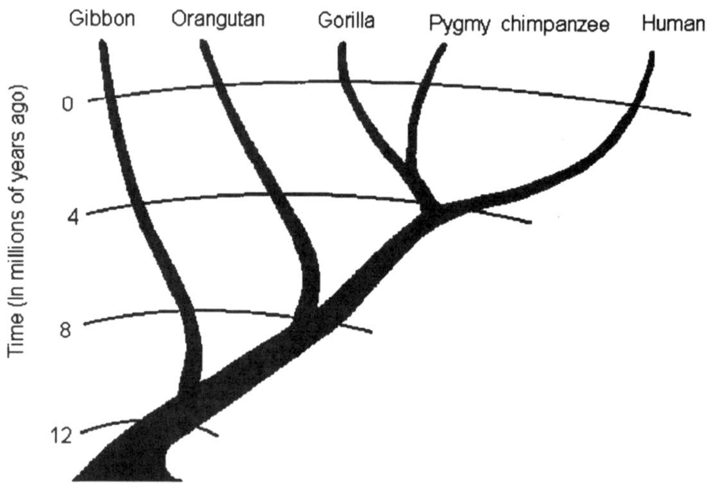

37. According to the diagram, gorillas evolved
   a. about 3 million years ago.
   b. more than 10 million years ago.
   c. about the same time as modern humans.
   d. after chimpanzees.

38. The diagram indicates that
   a. gorillas always walk on all fours.
   b. the gibbons were one of the first groups of apes to evolve.
   c. gibbons spend their whole lives in trees.
   d. humans have larger skulls than gorillas.

The following question(s) refer(s) to the diagram of the skeletons of two primates.

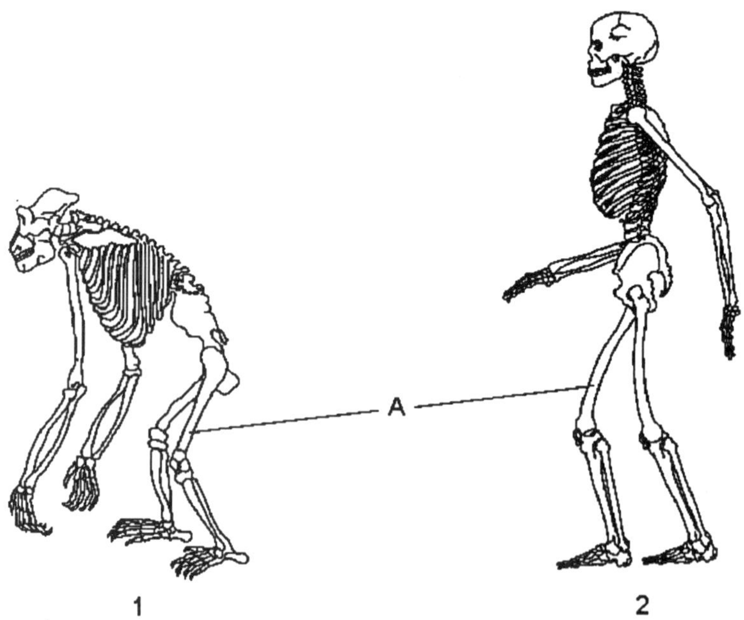

39. The bones labeled A are the
   a. pelvis.
   b. femurs.
   c. molars.
   d. spine.

40. Diagram 2 is probably the skeleton of a
   a. chimpanzee.
   b. gorilla.
   c. bonobo.
   d. hominid.

41. By examining the skeletons, scientists conclude that both organisms
   a. could walk upright on two legs.
   b. had grasping fingers.
   c. had rather large brains.
   d. All of the above

SHORT ANSWER Complete each statement by writing the correct term or phrase in the space provided.

42. Primates have ______________________ hands and feet.

43. The position of the eyes in primates allows them to have ______________________ vision.

44. The first primates were ______________________.

45. Organisms that are active during the day and sleep at night are said to be ______________________.

46. A(n) ______________________ thumb can be bent towards the other fingers.

47. Humans and their closest fossil relatives are known as ______________________.

48. The most complete hominid fossil ever found was that of a female nicknamed ______________________.

49. Analysis of ______________________ DNA indicates that *Homo sapiens* evolved in Africa.

50. The first *Homo sapiens* fossils in Europe were found in the ______________________ Valley of Germany.

51. A human ancestor that made simple tools was ______________________.

52. Java man and ______________________ man belong to the species *Homo erectus*.

53. Modern humans all over the Earth belong to the species ______________________.

ESSAY Read and answer each question or statement on a separate sheet of paper.

54. Summarize the evidence that all *Homo sapiens* evolved in one place.

55. How were the Neanderthals similar to modern humans?

56. Why was the evolution of a large brain important in human evolution?

ANALOGIES **An analogy is a relationship between two pairs of words or phrases written as: a:b::c:d. The symbol : is read "is to," and the symbol :: is read "as." Write the letter of the pair of terms that best completes the analogy shown.**

57. prosimians : primates::
   a. primates : hominids
   b. *Australopithecus* : hominids
   c. *Homo habilis* : hominids
   d. apes : hominids
   e. hominids : apes

____

58. orangutan : ape::
   a. *Homo erectus* : hominid
   b. *Australopithecus* : *Homo*
   c. monkey : ape
   d. hominid : ape
   e. *H. erectus* : ape

____

59. prosimians : big::
   a. lemurs : small
   b. lorises : small
   c. tarsiers : able to see in daylight
   d. monkeys : able to see color
   e. apes : able to see at night

____

60. prosimians : night::
   a. monkeys : night
   b. apes : night
   c. lemurs : day
   d. lorises : day
   e. monkeys : day

____

61. ape arm : *Australopithecus* arm::
   a. length of a driveway : length of a highway
   b. 1 yard : football field
   c. 1 meter : 1 centimeter
   d. distance across U.S. : distance around the Earth
   e. distance to the moon : distance to the sun

____

62. *robustus* and *boisei* : *Australopithecus* ::
   a. *afarensis* and *africanus* : *Homo*
   b. *erectus* and *habilis* : *Australopithecus*
   c. *erectus* and *habilis* : *Homo*
   d. *robustus* and *boisei* : *Homo*
   e. Neanderthal and *sapiens* : *Australopithecus*

____

63. human DNA : chimpanzee DNA::
   a. human hemoglobin : chimp hemoglobin
   b. *n* in humans : 2*n* in humans
   c. cell : centromere
   d. chromosome number in haploid cell : chromosome number in diploid cell
   e. 2*n* in duck : 2*n* in frog

64. 640 cu. cm brain : *H. habilis*::
   a. 700 cu. cm brain : prosimians
   b. daytime activity : prosimians
   c. tail : apes
   d. 200 cu. cm brain : *Australopithecus*
   e. 400–500 cu. cm brain : *Australopithecus*

TEST ANSWERS

TRUE/FALSE

1. F

2. T

3. T

4. T

5. T

6. F

7. T

8. T

9. T

10. T

MULTIPLE CHOICE

11. d

12. d

13. d

14. c

15. a

16. c

17. d

18. a

19. a

20. b

21. d

22. c

23. c

24. d

25. d

26. c

27. d

28. c

29. b

30. d

31. d

32. b

33. d

34. c

35. d

36. c

37. a

38. b

39. b

40. d

41. d

SHORT ANSWER

42. grasping

43. binocular

44. prosimians

45. diurnal

46. opposable

47. hominids

48. Lucy

49. mitchondrial

50. Neander

51. *Homo habilis*

52. Peking

53. *Homo sapiens*

ESSAY

54. Mitochondrial DNA is inherited only from the mother. Changes in mitochondrial DNA are caused only by mutations. The oldest mitochondrial DNA should show the most mutations. Since the greatest number of differences in mitochondrial DNA sequences occurs in modern Africans, scientists conclude that *Homo sapiens* evolved in Africa.

55. The bodies of Neanderthals were similar to the bodies of modern humans, and their brains were at least as large as those of modern humans. In addition, Neanderthals made complex stone tools and shelters and buried their dead.

56. Increased brain size allowed humans to take advantage of their environments and to develop tools, art, symbolic communication, and language.

ANALOGIES

57. b (were the first)

58. a (is a type of)

59. d (have eyes that are)

60. e (are active during the)

61. c (is longer than)

62. c (are species of)

63. a (is similar to)

64. e (is found in)

Chapter 15

TRUE/FALSE Read each statement. If the statement is true, write T in the space provided. If the statement is false, write F in the space provided.

1._______ The study of demographics helps predict changes in the size of a population.

2._______ Very small populations are less likely to become extinct than larger populations.

3._______ Clumped population distributions result from organisms seeking particular microhabitats.

4._______ Toxic wastes tend to accumulate in the environment as a population reaches the carrying capacity.

5._______ Populations of *K*-strategists grow rapidly, while *r*-strategist populations grow slowly.

6._______ *K*-strategists exhibit a high degree of specialization within their habitats.

7._______ The Hardy-Weinberg Principle states that the proportions of recessive and dominant alleles in a population fluctuate randomly from generation to generation.

8._______ Mutations are so common that they are the major cause of changes in allele frequencies within a population.

9._______ Natural selection acts on phenotypes, not genotypes.

10._______ Natural selection always eliminates any genetic disorders from a population, regardless of the frequency of the gene that is responsible for it.

MULTIPLE CHOICE Circle the letter of the term or phrase that best completes each statement or best answers each question.

11. Which of the following does not represent a population?
    a. all the robins in Austin, Texas
    b. all the grass frogs in the pond of Central Park, New York City
    c. all the birds in Chicago, Illinois
    d. all the earthworms in Yosemite National Park

12. Since individuals in a population usually tend to produce more than one offspring,
    a. populations tend to increase in size.
    b. populations remain stable in size.
    c. individuals tend to die quickly.
    d. the number of individuals declines rapidly.

13. All of the following are problems arising from inbreeding, *except*
    a. production of a genetically uniform population.
    b. increases in the diversity within a population.
    c. increased chance of homozygous recessive alleles occurring.
    d. reduction of a population's ability to adapt to environmental changes.

14. Demographic studies of populations must take into consideration
    a. population size.
    c. population dispersion.
    b. population density.
    d. All of the above

15. Regarding population dispersion patterns, which of the following is an *improper* pairing?
   a. randomly spaced — chance
   b. evenly spaced — regular intervals
   c. clumped — clusters
   d. dispersive — randomly distributed

16. The equation $N = rN\ [(K-N)\ /\ K]$ represents
   a. the logistic model of population growth.
   b. the number of organisms in a population at a particular moment in time.
   c. exponential growth of bacterial populations.
   d. the population density in a given habitat.

17. As a population reaches its carrying capacity, there is an increase in competition for
   a. food.
   b. shelter.
   c. mates.
   d. All of the above

18. Environments that are unpredictable and rapidly changing tend to support populations of
   a. *Q*-strategists.
   b. *K*-strategists.
   c. *N*-strategists.
   d. *r*-strategists.

19. All of the following are true of *r*-strategists, *except*
   a. early maturation and reproduction.
   b. little parental care.
   c. few offspring.
   d. offspring are small.

20. Which of the following is an *r*-strategist?
   a. redwoods
   b. dandelions
   c. whales
   d. humans

21. Which of the following is *improperly* paired?
   a. *K*-strategists — reproduce late in life
   b. *K*-strategists — minimal parental care
   c. *r*-strategists — reproduce early in life
   d. *r*-strategists — mature quickly

22. In 1908, Hardy and Weinberg independently demonstrated that
   a. *r*- and *K*-strategist populations are actually the same.
   b. recessive alleles replace dominant alleles in a population over long periods of time.
   c. dominant alleles do not replace recessive alleles in a population.
   d. recessive alleles are usually more common than dominant alleles.

23. The equation: $p^2 + 2pq + q^2 = 1$ represents the
   a. population growth rate.
   b. carrying capacity.
   c. logistic growth curve.
   d. Hardy-Weinberg principle.

24. Actual proportions of homozygotes and heterozygotes can differ from Hardy-Weinberg predictions because of
   a. the occurrence of mutations.
   b. nonrandom mating among individuals.
   c. genetic drift within the population.
   d. All of the above

25. The movement of alleles into or out of a population due to migration is called
   a. mutation.
   b. gene flow.
   c. nonrandom mating.
   d. natural selection.

26. Inbreeding
a. is a form of random mating.
b. causes mutations to occur.
c. increases the proportion of heterozygotes.
d. increases the proportion of homozygotes.

27. Natural selection acts
a. on heterozygous genotypes.
b. only on recessive alleles.
c. on phenotypes that are expressed.
d. on all mutations.

28. A characteristic that is controlled by several genes is called a _______ trait.
a. directional
b. polymorphic
c. polygenic
d. stabilizing

29. The type of selection that may eliminate intermediate phenotypes is
a. direction selection.
b. disruptive selection.
c. polygenic selection.
d. stabilizing selection.

30. Directional selection tends to eliminate
a. both extremes in a range of phenotypes.
b. one extreme in a range of phenotypes.
c. intermediate phenotypes.
d. None of the above; it causes new phenotypes to form.

The following question(s) refer(s) to the graph showing population growth over time.

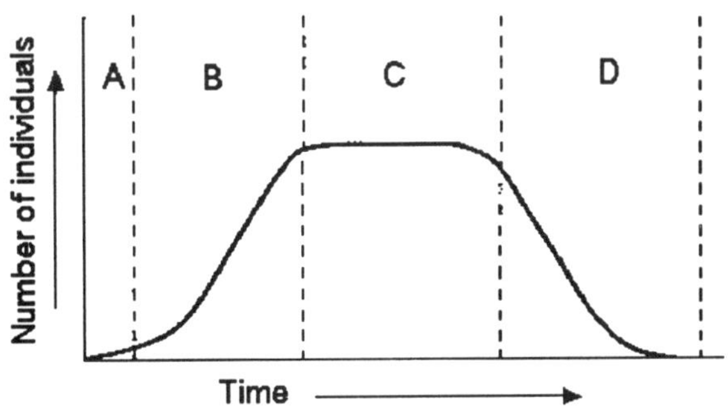

31. Which time period shows exponential growth of the population?
a. period "A"
b. period "B"
c. period "C"
d. period "D"

32. During which time period are the birth rate and death rate equal?
a. period "A"
b. period "B"
c. period "C"
d. period "D"

33. The rate of growth of a population is represented by *r*. During which time period will $r = 0$?
a. period "A"
b. period "B"
c. period "C"
d. period "D"

34. The time period during which *r* would have a negative value is
a. period "A."
b. period "B."
c. period "C."
d. period "D."

SHORT ANSWER Complete each statement by writing the correct term or phrase in the space provided.

35. A(n) ______________________ consists of all the individuals of a particular species in a particular place.

36. ______________________ is the measurement of the factors that determine how populations grow.

37. Population density refers to how many ______________________ are present in a particular location.

38. The way in which members of a population are arranged in a given area is referred to as ______________________.

39. Clumped distributions result when individuals in a population seek out particular sets of environmental conditions called ________________.

40. A population ______________________ is a hypothetical population that has key characteristics of the real population being studied.

41. ______________________ is defined as the difference between the birth rate and death rate of a population.

42. The ______________________ is the population size that can be sustained by an environment.

43. Species that are ______________________ tend to have periods of exponential growth followed by sudden crashes in population size.

44. Small population sizes and slow population growth are typical of organisms that are ______________________.

45. Stable and predictable environments tend to be inhabited by ______________________ populations.

46. Alternative versions of genes are called ______________________.

47. In the Hardy-Weinberg equation, the symbols *p* and *q* represent ______________________.

48. The movement of individuals from one population to another is called ______________________.

49. Genetic variation that results from a population having more than one allele of a gene at a significant frequency is called ______________.

50. A characteristic influenced by several genes is called a(n) ______________________ trait.

51. ______________________ selection causes the range of phenotypes to become narrower, increasing the number of individuals with characteristics near the middle of the range.

ESSAY Read and answer each question or statement on a separate sheet of paper.

52. Explain the difference between *r*-strategist and *K*-strategist populations.

53. In a population of birds, 16 of 100 individuals suffer from a recessive genetic disorder that causes the feathers to fall off their wings. What proportion of the bird population are heterozygous carriers of the disorder?

54. Distinguish between the three types of natural selection acting on polygenic traits.

ANALOGIES An analogy is a relationship between two pairs of words or phrases written as: a:b::c:d. The symbol : is read "is to," and the symbol :: is read "as." Write the letter of the pair of terms that best completes the analogy shown.

55. population density : number of individuals in a given area ::
a. population : an area where organisms live
b. logistic model : how populations grow in nature
c. logistic growth curve : rate of growth
d. population size : population density
e. carrying capacity : birth rate minus death rate

____

56. $\frac{1}{4}$: 0.25 ::
a. p^2 : 1
b. q^2 : 1
c. $p^2 + 2pq + q^2$: 1
d. $2pq$: 1
e. $p^2 + 2pq$: 1

____

57. nonrandom mating : increasing proportion of homozygotes ::
a. migration of individuals : gene flow
b. mutation : major change in allele frequencies
c. Hardy-Weinberg equation : natural selection
d. inbreeding : frequency of alleles
e. mutation : migration

____

58. RR : homozygous dominant ::
a. Rr : heterozygous
b. rr : heterozygous recessive
c. Yy : homozygous
d. yy : heterozygous dominant
e. Ry : homozygous recessive

____

59. natural selection : mutations ::
a. efficient natural selection : rare alleles
b. genetic change : homozygous recessive alleles
c. mutation : homozygous dominant alleles
d. heterozygous offspring : common alleles coming together
e. homozygous offspring : different alleles coming together

____

60. homozygous : heterozygous ::
   a. heterozygous : Bb
   b. probability : predicting chances
   c. dominant : recessive
   d. factor : gene
   e. PKU : inability to break down phenylalanine

____

61. randomly spaced individuals : chance ::
   a. clumped individuals : regular intervals
   b. clumped individuals : chance
   c. evenly spaced individuals : chance
   d. evenly spaced individuals : clusters
   e. evenly spaced individuals : regular intervals

____

62. birth and death rates : constant on exponential growth curve ::
   a. birth rates : equal to death rates
   b. *r*-strategists : equal to *k*-strategists
   c. birth and death rates : not constant on logistic growth curve
   d. exponential models : same as logistic models
   e. population growths : equal for all species

____

63. bacteria : *r*-strategists ::
   a. gorillas : *k*-strategists
   b. insects : *k*-strategists
   c. annual plants : *k*-strategists
   d. rhinoceroses : *r*-strategists
   e. tigers : *r*-strategists

____

64. recessive allele frequency : 0.02 ::
   a. dominant allele frequency : 0.01
   b. dominant allele frequency : 0.04
   c. dominant allele frequency : 0.98
   d. dominant allele frequency : 0.08
   e. dominant allele frequency : 1.0

____

65. natural selection : genetic changes ::
   a. nonrandom mating : mutations
   b. mutation : gene flow
   c. gene flow : Hardy-Weinberg principle
   d. loss of an individual : genetic drift
   e. Hardy-Weinberg principle : changes in allele frequencies

____

TEST ANSWERS

TRUE/FALSE

1. T
2. F
3. T
4. T
5. F
6. T
7. F
8. F
9. T
10. F

MULTIPLE CHOICE

11. c
12. a
13. b
14. d
15. d
16. a
17. d
18. d
19. c
20. b
21. b
22. c
23. d
24. d

25. b

26. d

27. c

28. c

29. b

30. b

31. b

32. c

33. c

34. d

SHORT ANSWER

35. population

36. Demography

37. individuals; members of a species

38. dispersion

39. microhabitats

40. model

41. Growth rate

42. carrying capacity

43. *r*-strategists

44. *K*-strategists

45. *K*-strategist

46. alleles

47. allele frequencies

48. migration

49. genetic polymorphism

50. polygenic

51. Stabilizing

ESSAY

52. These two types of populations differ in their rates of maturation and reproduction, amount of parental care of offspring, and the type of environments that they inhabit. Species that are *r*-strategists mature quickly, reproduce in large numbers, and invest little energy in care of their offspring. This adapts them for unpredictable, rapidly changing habitats. *K*-strategists tend to be larger organisms that mature more gradually and exhibit parental care of a smaller number of offspring. They are found in more stable environments.

53. Given that 16 of 100 birds express the disorder, the proportion of the population that is homozygous recessive (q^2) is 0.16. Based on this, $q = 0.4$. Since $p + q = 1$, then $p = 1 - q = 1 - 0.4 = 0.6$. The proportion of birds that are heterozygous carriers of the disorder: $2pq = 2\ (0.6)\ (0.4) = 0.48$.

54. Directional selection tends to eliminate individuals in a population that are at one or the other extreme of the range of phenotypes. Stabilizing selection narrows the range of individuals to those in the middle of the range. Finally, disruptive selection divides a population into two different groups with varying phenotypes by selecting against the individuals with intermediate phenotypes.

ANALOGIES

55. b (describes)

56. c (is equal to)

57. a (results in)

58. a (is allele symbol for)

59. b (depends upon)

60. c (is the opposite of)

61. e (are located by)

62. c (are assumed to be)

63. a (are an example of)

64. c (is equal to)

65. d (causes)

Chapter 16

TRUE/FALSE Read each statement. If the statement is true, write T in the space provided. If the statement is false, write F in the space provided.

1.______ All organisms in an ecosystem are part of the food web of that ecosystem.

2.______ The roots of trees are usually surrounded by fungi known as lichens.

3.______ Different locations on the Earth are ecologically isolated from other places.

4.______ The source of energy of an organism determines its trophic level.

5.______ Decomposers absorb energy from organisms by breaking down living tissue.

6.______ The lowest trophic level of any ecosystem is occupied by the producers.

7.______ Producers are at the bottom of ecological pyramids.

8.______ A food chain is made up of interrelated food webs.

9.______ Biomass pyramids can point up or down depending on the ecosystem.

10.______ Omnivores feed only on primary producers.

11.______ The number of organisms in a trophic level is always directly proportional to the amount of energy at that level.

12.______ Nitrogen gas makes up about 79 percent of the Earth's atmosphere.

13.______ During nitrification, decomposers break down the roots of plants to produce nitrates.

14.______ Phosphorus is often found in detergents and fertilizers.

MULTIPLE CHOICE Circle the letter of the term or phrase that best completes each statement or best answers each question.

15. All the organisms that live in a particular place and their physical location make up a(n)
    a. ecosystem. c. community.
    b. habitat. d. food chain.

16. An ecosystem consists of
    a. a community of organisms. c. the soil, water, and weather.
    b. energy. d. All of the above

17. Ecology is the study of the interaction of living organisms
    a. with each other and their habitat.
    b. and their communities.
    c. with each other and their physical environment.
    d. and the food they eat.

18. When an organism dies, the nitrogen in its body
   a. can never be reused by other living things.
   b. is immediately released into the atmosphere.
   c. is released by the action of decomposers.
   d. None of the above

19. A relationship between a producer and consumer is best illustrated by
   a. a snake eating a bird.
   b. a fox eating a mouse.
   c. a lion eating a zebra.
   d. a zebra eating grass.

20. In a food web, which type of organism receives energy from every other type?
   a. producer
   b. carnivore
   c. decomposer
   d. herbivore

21. In an ecosystem, the producers may include
   a. lions, zebras, and bacteria.
   b. trees, grass, and shrubs.
   c. vultures, deer, and eagles.
   d. bacteria, fungi, and algae.

22. The most abundant organisms in a food chain are the
   a. producers.
   b. herbivores.
   c. carnivores.
   d. omnivores.

23. Animals that feed on plants are at least in the
   a. first trophic level.
   b. second trophic level.
   c. third trophic level.
   d. fourth trophic level.

24. The number of trophic levels in an ecological pyramid
   a. is limitless.
   b. is limited by the amount of energy that is lost at each trophic level.
   c. never exceeds four.
   d. never exceeds three.

25. Food webs are more commonplace than food chains because
   a. the animals that comprise the links in the food chain are mobile, hence the chains become intertwined.
   b. organisms almost always eat, and are eaten by, many different organisms.
   c. over time food chains always disappear.
   d. None of the above

26. In going from one trophic level to the next higher level
   a. the number of organisms increases.
   b. the amount of usable energy increases.
   c. the amount of usable energy decreases.
   d. diversity of organisms increases.

27. Which of the following is part of the nitrogen cycle?
   a. the conversion of atmospheric nitrogen into usable organic compounds by bacteria
   b. the conversion of decaying organisms into ammonia
   c. nitrogen fixation
   d. All of the above

28. Nitrogen fixation is best described as
   a. the conversion of atmospheric nitrogen into a form usable by living organisms.
   b. the conversion of nitrogen into nitrogen ions.
   c. the removal of nitrogen from animal wastes.
   d. the recycling of nitrogen.

29. Nitrogen is a component of
a. proteins.
b. fats.
c. carbohydrates.
d. water.

30. The continuous movement of nutrients and other chemicals throughout an ecosystem is called
a. restructuring.
b. cycling.
c. modeling.
d. rebuilding.

31. Materials such as water, nitrogen, and carbon
a. are constantly replenished from deep within the Earth.
b. cycle within ecosystems.
c. are only needed in small quantities by living things.
d. are not found in the deserts of the world.

32. Precipitation and evaporation are important components of the
a. nitrogen cycle.
b. water cycle.
c. carbon cycle.
d. All of the above

33. The level of phosphorus in freshwater lake ecosystems
a. never varies.
b. may change as a result of human activity.
c. is dependent on phosphorylation.
d. None of the above

34. Succession is
a. an organism's ability to survive in its environment.
b. the number of species living in an ecosystem.
c. the regular progression of species replacement in an environment.
d. the transfer of energy through a food chain.

35. Humans are affecting the carbon cycle by
a. burning fossil fuels.
b. destroying vegetation that absorbs carbon dioxide.
c. Both a and b
d. None of the above

The following question(s) refer(s) to the diagram of organisms in an ocean ecosystem.

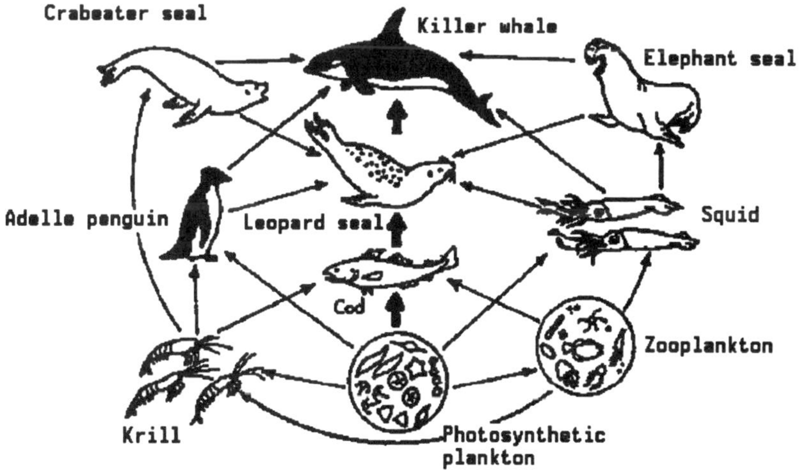

36. The diagram, which shows how energy moves through an ecosystem, is known as a
a. trophic net.
b. food chain.
c. food net.
d. food web.

37. The photosynthetic algae are
 a. producers.
 b. consumers.
 c. parasites.
 d. decomposers.

38. Leopard seals are
 a. producers.
 b. omnivores.
 c. herbivores.
 d. carnivores.

39. Killer whales feed at the
 a. first and second trophic levels.
 b. second trophic level only.
 c. second and third trophic levels.
 d. third and fourth trophic levels.

The following question(s) refer(s) to the diagram showing levels of an ecosystem.

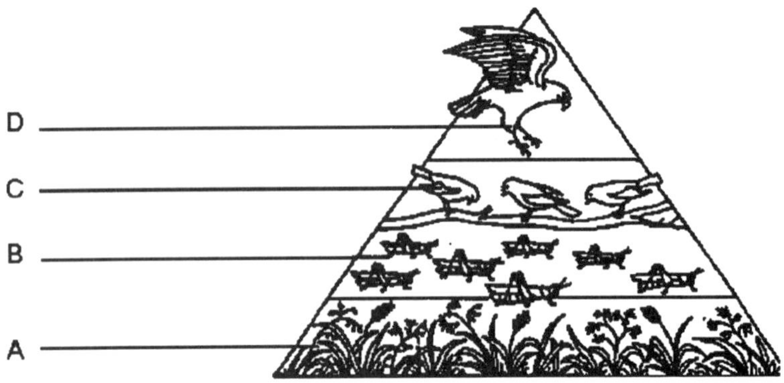

40. The diagram illustrates a(n)
 a. food chain.
 b. community.
 c. food web.
 d. ecological pyramid.

41. Level "A" is made up of
 a. carnivores.
 b. herbivores.
 c. producers.
 d. omnivores.

42. How much energy is available to the organisms at level "C"?
 a. all of the energy at level "A" plus the energy at level "B"
 b. all of the energy at level "A" minus the energy at level "B"
 c. 10 percent of the energy at level "B"
 d. 90 percent of the energy at level "B"

43. The diagram represents the decrease
 a. in biomass from lower to higher trophic levels.
 b. in available energy from lower to higher trophic levels.
 c. in diversity of organisms from lower to higher trophic levels.
 d. All of the above

SHORT ANSWER Complete each statement by writing the correct term or phrase in the space provided.

44. An ecosystem consists of the living and ________________________ environment.

45. Organisms that obtain their energy from the organic wastes and dead bodies at all the energy levels in an ecosystem are called
________________________.

46. A path of energy through the energy levels of an ecosystem is called a ________________________.

47. Every time energy is transferred in an ecosystem, potential energy is lost as ________________________.

48. The total dry weight of the organisms in an ecosystem is called the ________________________.

49. The primary productivity of an ecosystem is a measure of the amount of organic material that the ________________________ organisms in the ecosystem produce.

50. An energy pyramid shows the amount of energy contained in the bodies of organisms at each ________________________ level.

51. The sequence of one organism feeding upon another is called a food chain. A network of intertwined food chains is called a ________________________.

52. Water that seeps into the soil is called ________________________ water.

53. Plants lose water from their leaves in a process known as ________________________.

54. In the nitrogen cycle, plants use nitrates and nitrites to form ________________________.

55. The conversion of nitrogen gas to ammonia by bacteria is called ________________________.

56. Nitrogen ________________________ is the absorption and incorporation of nitrogen into plant and animal components.

57. The conversion of nitrate to nitrogen gas is called ________________.

ESSAY Read and answer each question or statement on a separate sheet of paper.

58. Why is it cheaper for a farmer to produce a pound of grain than a pound of meat?

59. Describe how energy is transferred from one trophic level to another.

60. Why are decomposers necessary for the continuation of life on Earth?

61. Compare the flow of matter and energy in ecosystems.

ANALOGIES An analogy is a relationship between two pairs of words or phrases written as: a:b::c:d. The symbol : is read "is to," and the symbol :: is read "as." Write the letter of the pair of terms that best completes the analogy shown.

62. secondary succession : cleared forest ::
a. pile of rock and gravel : secondary succession
b. secondary succession : bare soil
c. primary succession : new volcanic island
d. succession : the absence of plants
e. exposed area : a climax community

____

63. cows : herbivores ::
  a. horses : carnivores
  b. plants : producers
  c. algae : consumers
  d. caterpillars : producers
  e. lowest trophic level organisms : consumers

---

64. pond : evaporation ::
  a. leaf : absorption
  b. root : absorption
  c. plant : transpiration
  d. lake : transpiration
  e. river : transpiration

---

65. pumping a bicycle tire : the tire air pressure ::
  a. more rain : transformation of rain forests
  b. more transpiration : arid weather
  c. burning fossil fuels : the release of carbon to atmosphere
  d. slowing of evaporation : transpiration
  e. combustion : ground water

---

66. carbon : organic molecules ::
  a. nitrogen : photosynthesis
  b. phosphate : ATP and DNA
  c. nitrogen-fixation : water
  d. photosynthesis : carbon
  e. carbon : nitrogen

---

67. primary succession : areas of no previous plant growth ::
  a. new habitat : a climax community
  b. rain forest : a desert
  c. tundra : a desert
  d. secondary succession : bare rock
  e. secondary succession : abandoned farm fields

---

68. energy in third trophic level : energy in second trophic level ::
  a. $1 : $5
  b. diamond ring : toy car
  c. hurricane : drizzle
  d. superhighway : dirt road
  e. soccer ball : cheese

TEST ANSWERS

TRUE/FALSE

1.T

2.F

3.F

4.T

5.F

6.T

7.T

8.F

9.F

10.F

11.F

12.T

13.F

14.T

MULTIPLE CHOICE

15.a

16.d

17.c

18.c

19.d

20.c

21.d

22.a

23.b

24.b

25. b

26. c

27. d

28. a

29. a

30. b

31. b

32. b

33. b

34. c

35. c

36. d

37. d

38. d

39. d

40. d

41. c

42. c

43. d

SHORT ANSWER

44. nonliving

45. detritivores

46. food chain

47. heat

48. biomass

49. photosynthetic

50. trophic

51. food web

52. ground

53. transpiration

54. amino acids

55. nitrogen fixation

56. assimilation

57. denitrification

ESSAY

58. Animals are on higher trophic levels than plants. It takes several pounds of grain to produce one pound of meat.

59. A portion of the energy available to organisms at each level of the food chain is stored in the chemical bonds of nutrients that are not used by the organism to sustain life. When that organism is eaten by another, the stored chemical energy is transferred to the new organism and used to sustain its life.

60. Matter cycles through ecosystems. Decomposers release matter from waste materials and dead organisms. Were it not for the action of decomposers, the Earth would eventually run out of essential matter such as carbon and nitrogen that organisms need. Decomposers are essential to the recycling of matter. Without decomposers we would not have cycles of matter.

61. Energy goes one way through ecosystems. It enters from the process of photosynthesis and is lost from one trophic level to the next. Matter goes in cycles and is reused again and again.

ANALOGIES

62. c (occurs following)

63. b (are examples of)

64. c (releases water by)

65. c (increases)

66. b (is used to make)

67. e (occurs in)

68. a (is less than)

Chapter 17

TRUE/FALSE Read each statement. If the statement is true, write T in the space provided. If the statement is false, write F in the space provided.

1.______ A long-term relationship in which both participating species benefit is known as parasitism.

2.______ The space that an organism uses has nothing to do with its niche.

3.______ The seasons are caused by the daily rotation of the Earth.

4.______ Warm air can hold more moisture than cold air.

5.______ Patterns of circulation in the ocean are determined by the circulation of the wind.

6.______ All planktonic organisms are photosynthetic.

7.______ Freshwater habitats are independent of terrestrial habitats.

8.______ Tropical rain forests have the most fertile soil on Earth.

9.______ Deserts are most extensive in the interiors of continents.

10.______ Permafrost is a characteristic of the taiga.

11.______ When two species compete for limited resources, competitive exclusion is sure to take place.

MULTIPLE CHOICE Circle the letter of the term or phrase that best completes each statement or best answers each question.

12. The evolution of flowers and the insects that feed on them is known as
 a. parasitism. c. coevolution.
 b. secondary succession. d. stability.

13. A tick feeding on a human is an example of
 a. parasitism. c. symbiosis.
 b. mutualism. d. predation.

14. Urushiol, which causes the irritation from contact with poison ivy, is an example of a
 a. primary compound. c. secondary compound.
 b. complex carbohydrate. d. selective nucleic acid.

15. The relationship between a clown fish and a sea anemone is known as
 a. parasitism. c. mutualism.
 b. competition. d. commensalism.

16. An organism's niche includes
 a. what it eats. c. how it reproduces.
 b. where it eats. d. All of the above

17. Most ecosystems are complex because
   a. they are found all over the globe.
   b. potential competitors in the ecosystem occupy slightly different niches.
   c. they all contain producers.
   d. of decomposition.

18. If the niches of two organisms overlap,
   a. the organisms may have to compete directly.
   b. the two organisms will form a symbiotic relationship.
   c. both organisms will die.
   d. one organism will die.

19. Which of the following would not be included in a description of an organism's niche?
   a. its trophic level
   b. the humidity it prefers
   c. its color
   d. when it reproduces

20. Competitive exclusion occurs when
   a. a species is eliminated locally within an ecosystem.
   b. new species enter an ecosystem.
   c. species reproduce.
   d. a species occupies a fundamental niche.

21. An evolutionary adjustment that reduces competition among species is known as
   a. competitive exclusion.
   b. establishing a niche.
   c. competitive coexistence.
   d. character displacement.

22. The closer an ecosystem is to the equator
   a. the longer its growing season.
   b. the greater its diversity.
   c. the warmer its temperature.
   d. All of the above

23. The different amounts of moisture found on opposite sides of a mountain is referred to as the
   a. seasonal adjustment.
   b. atmospheric alignment.
   c. rain-shadow effect.
   d. alpine effect.

24. Organisms with light-producing body parts would most likely be found in
   a. the deep-water zone of lakes.
   b. shallow ocean waters.
   c. open ocean surfaces.
   d. deep ocean waters.

25. Major ecosystems that occur over wide areas of land are called
   a. communities.
   b. habitats.
   c. biomes.
   d. food chains.

26. The most diverse terrestrial ecosystems on the Earth are
   a. deciduous forests.
   b. coniferous forests.
   c. grasslands.
   d. tropical rain forests.

27. Tropical ecosystems are more diverse than temperate-zone ecosystems because
   a. the growing season in tropical ecosystems never stops.
   b. the climate in tropical ecosystems does not vary much from year to year.
   c. a greater amount of food is produced in tropical ecosystems.
   d. All of the above

28. Herds of grazing animals are most likely to be found in a
   a. savanna.
   b. tropical rain forest.
   c. deciduous forest.
   d. taiga.

29.The greatest diversity of life in the ocean is found in
a. shallow ocean waters.
b. the ocean surface.
c. deep ocean waters.
d. tidal areas.

30.Almost all of the Earth's surface water is contained in
a. ocean ecosystems.
b. freshwater biomes.
c. tropical rain forests.
d. ponds and lakes.

31.The biome that makes up most of the central part of the United States is
a. rain forest.
b. temperate grassland.
c. tundra.
d. coniferous forest.

32.A biome characterized by ground that is always frozen is a
a. savanna.
b. coniferous forest.
c. tundra.
d. deciduous forest.

33.Sea stars are fierce competitors of marine organisms such as clams and mussels. An ecologist studying an ocean ecosystem performed an experiment in which the sea stars were removed from the ecosystem. After the removal of the sea stars,
a. the ecosystem became more diverse.
b. the size of the ecosystem was reduced.
c. food webs in the ecosystem became more complex.
d. the number of species in the ecosystem was reduced.

The following question(s) refer(s) to the diagrams illustrating experiments performed with two species of barnacles that live in the same area.

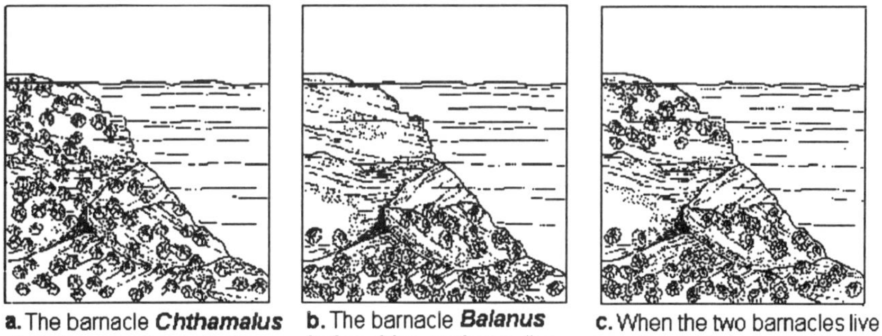

a. The barnacle ***Chthamalus stellatus*** can live in both shallow and deep water on a rocky coast.

b. The barnacle ***Balanus balanoides*** prefers to live in deep water.

c. When the two barnacles live together, ***Chthamalus*** is restricted to shallow water.

34.Diagram "A" indicates that the barnacle *Chthamulus stellatus* can live in both shallow and deep water on a rocky coast. This is the barnacle's
a. competitive niche.
b. realized niche.
c. fundamental niche.
d. exclusive niche.

35.Diagram "B" indicates that the barnacle *Balanus balanoides* prefers to live in deep water. Deep water is the barnacle's
a. competitive niche.
b. realized niche.
c. fundamental niche.
d. exclusive niche.

36.Diagram "C" indicates that when the two barnacles live together, *Chthamalus* is restricted to shallow water. Shallow water is the barnacle's
a. competitive niche.
b. realized niche.
c. fundamental niche.
d. exclusive niche.

37. Since the two species of barnacles attempt to use the same resources, they are
a. parasitic.
b. in competition with one another.
c. mutualistic.
d. symbiotic.

The following question(s) refer(s) to the diagram of the Earth.

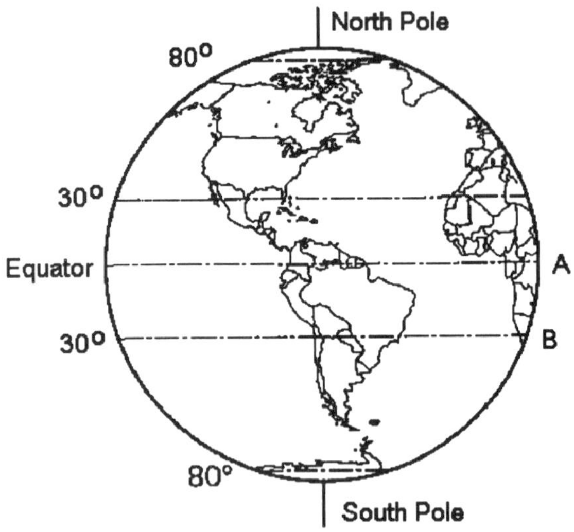

38. An ecosystem located along latitude "A" would
a. have a shorter growing season than an ecosystem on latitude "B."
b. probably contain fewer species than an ecosystem at latitude "B."
c. probably be more diverse than an ecosystem at latitude "B."
d. probably have less rainfall than an ecosystem at latitude "B."

SHORT ANSWER Complete each statement by writing the correct term or phrase in the space provided.

39. The struggle among organisms for limited natural resources is called ________________________.

40. When one organism feeds on another, ________________________ occurs.

41. When two or more species evolve in response to each other, ________________________ occurs.

42. A relationship in which one species benefits and the other is neither harmed nor helped is known as ________________________.

43. The niche of a deer in a forest is that of a herbivore. The niche of a cougar that eats the deer is that of a ________________________.

44. An organism's ________________________ is the sum of its interactions with other organisms and the environment.

45. Local elimination of one competing species is known as ____________.

46. The total niche that an organism is potentially able to use within an ecosystem is called that organism's ________________________.

47. The circulation of the Earth's atmosphere is caused by ________________________ heating.

48. A region of rising air is one of ______________________ pressure.

49. A dry grassland found in a tropical area is called a ______________.

50. Coniferous trees are predominantly found in the __________________ biome.

ESSAY Read and answer each question or statement on a separate sheet of paper.

51. Can two species occupy the same niche? Explain.

52. Why are regions on the Earth near the equator warmer than regions near the poles?

53. Why is the climate on the windward side of a mountain usually different from the climate on the leeward side?

ANALOGIES An analogy is a relationship between two pairs of words or phrases written as: a:b::c:d. The symbol : is read "is to," and the symbol :: is read "as." Write the letter of the pair of terms that best completes the analogy shown.

54. coevolution : prey to improve at eluding predator ::
- a. predation : prey to fail at eluding a predator
- b. coevolution : predator to improve at catching prey
- c. coevolution : predator to improve at catching parasites
- d. parasitism : parasite to fail at finding a host
- e. parasitism : prey to get a host

____

55. commensalism : one organism ::
- a. parasitism : both organisms
- b. predation : neither organism
- c. competition : one organism
- d. mutualism : both organisms
- e. mutualism : one organism

____

56. competitive exclusion : one of the species ::
- a. physical establishment : adaptation purposes
- b. individual balancing : number of species
- c. fundamental competition : specialization
- d. resource stimulation : predation
- e. character displacement : competition

____

57. Earth's surface : different amounts of solar energy ::
- a. equator : least solar energy
- b. equator : less perpendicular rays than other locations
- c. temperate area : perpendicular rays all year long
- d. tropical area : least solar energy
- e. each pole : less energy per unit area than the equator

____

58. moisture-holding capacity of air : when it is cooled ::
a. precipitation : leeward side of a mountain
b. air temperature : leeward side of a mountain
c. precipitation : windward side of a mountain
d. rain-shadow effect : on leeward side of a mountain
e. rain-shadow effect : great plains of United States

____

59. atmospheric air circulation : ocean circulation patterns ::
a. leeward mountainside : windward weather
b. warm air on leeward mountainside : cold air on windward side
c. desert : mountain weather
d. landmass : ocean circulation patterns
e. landmass : latitude

____

60. many fish : shallow ocean water habitat ::
a. nutrients : deep-sea waters
b. plankton : deep-sea-water habitat
c. plankton : open-sea surface habitat
d. animals producing own light : shallow-ocean-water habitat
e. animals producing own light : open-sea surface habitat

____

61. tropical rain forest : infertile soil ::
a. tundra : warm weather
b. temperate deciduous forest : little rain
c. taiga : coniferous forest
d. temperate grassland : nonproductive soil for agriculture
e. desert : high annual precipitation

____

62. rotation of Earth : daily alternation of warming and cooling ::
a. solar energy : Earth's shape
b. solar energy : high temperatures at poles
c. solar energy : low temperatures at equator
d. Earth rotation : extreme buildup of heat
e. 23.5° Earth axis tilt : seasons at higher latitudes

____

TEST ANSWERS

TRUE/FALSE

1. F
2. F
3. F
4. T
5. T
6. F
7. F
8. F
9. T
10. F
11. F

MULTIPLE CHOICE

12. c
13. a
14. c
15. c
16. d
17. b
18. b
19. c
20. b
21. d
22. d
23. c
24. d

25. c

26. d

27. d

28. a

29. a

30. a

31. b

32. c

33. d

34. c

35. c

36. b

37. b

38. c

SHORT ANSWER

39. competition

40. predation

41. coevolution

42. commensalism

43. carnivore

44. niche

45. competitive exclusion

46. fundamental niche

47. solar

48. low

49. savanna

50. taiga

ESSAY

51. No two species can have the exact same niche. The principle of competitive exclusion states that if two species are competing for the same resource, the species that uses the resource more efficiently will eventually eliminate the other.

52. Different places on the Earth receive different amounts of solar energy. At regions near the equator, the sun's rays arrive almost perpendicularly. Nearer the poles, the angle at which the sun's rays hit the Earth spreads them out over a much greater area, providing less energy per unit area.

53. When mountains force incoming air upward, the air's moisture-holding capacity decreases as the air cools, resulting in increased precipitation on the windward side. As the air descends on the leeward side, it is warmed and its moisture holding capacity increases. It tends to draw up moisture from the surface. The leeward side of a mountain is often much drier than the windward side.

ANALOGIES

54. b (causes)

55. d (is beneficial to)

56. e (eliminates)

57. e (receives)

58. a (decreases)

59. d (determines)

60. c (are found in the)

61. c (has)

62. e (causes)

Chapter 18

TRUE/FALSE Read each statement. If the statement is true, write T in the space provided. If the statement is false, write F in the space provided.

1.______ Sulfur-rich smoke produced by burning fossil fuels causes limestone to change color.

2.______ Acid rain usually has a pH between 8.3 and 9.7.

3.______ Devices known as scrubbers are routinely used worldwide to clean the inside surfaces of tall smokestacks.

4.______ In the United States, the number of cases of malignant melanoma almost doubled between 1980 and 1996.

5.______ CFCs are the only chemicals that destroy ozone in the atmosphere.

6.______ About one-half of the Earth's tropical rain forests have been destroyed.

7.______ The human birth rate has remained about the same for the last 350 years.

8.______ The population of every country on the Earth is steadily increasing.

9.______ The population of the Earth has not changed very much over the last 350 years.

10.______ The last stage of addressing any environmental problem is risk analysis.

11.______ One approach to reducing pollution is to have government subsidize industries that cause pollution.

12.______ Secondary sewage treatment plants remove chemicals and bacteria from sewage.

MULTIPLE CHOICE Circle the letter of the term or phrase that best completes each statement or best answers each question.

13. CFCs were once
   a. thought to be chemically inert.
   b. used as refrigerants.
   c. used as aerosol propellants.
   d. All of the above

14. CFCs in the atmosphere
   a. stick to frozen water vapor.
   b. change oxygen into ozone.
   c. convert sunlight into ozone.
   d. convert ozone into methane.

15. Tall smokestacks were placed on power plants because the smoke they produced from the burning of coal contained high concentrations of
   a. ozone.
   b. sulfur.
   c. oxygen.
   d. nitrogen.

16. The decrease in species diversity of some lakes in the northeastern United States during this century may best be explained by
   a. global warming.
   b. evolutionary trends.
   c. the destruction of the ozone layer.
   d. acid precipitation.

17. Ozone shields the Earth from the harmful effects of

a. radioactivity. c. cosmic rays.
b. infrared rays. d. ultraviolet rays.

18. The destruction of the ozone layer may be responsible for an increase in

a. cataracts. c. cancer of the retina.
b. melanoma. d. All of the above

19. Halon, carbon tetrachloride, and methyl bromide are examples of chemicals that

a. cause acid rain.
b. contain high concentrations of sulfur.
c. contribute to the destruction of the ozone layer.
d. have not been manufactured in the United States since 1944.

20. Solar energy can be trapped in the atmosphere by

a. the chemical bonds of carbon dioxide. c. radiation.
b. ozone. d. sunlight.

21. The heat-trapping ability of some gases in the atmosphere can be compared to

a. the melting of snow.
b. the way glass traps heat in a greenhouse.
c. condensation due to heating.
d. heating water on a stove.

22. As a result of the discovery of the ozone hole,

a. tall smokestacks were placed on power plants.
b. the production of most CFCs will end during the 1990s.
c. methane has been substituted for nitrous oxides in some chemicals.
d. large greenhouses were built in Europe, the United States, and Canada.

23. Topsoil

a. is a renewable resource.
b. is formed from the remains of plants and animals.
c. forms at the rate of 1 cm each growing season.
d. forms by the action of water and wind.

24. The oil spill from the *Exxon Valdez*

a. polluted many kilometers of coastline.
b. was a minor problem.
c. only affected fish that were near the ship.
d. was easily contained.

25. Topsoil and ground water

a. exist in unlimited quantities in aquifers throughout the world.
b. are found only on the prairie.
c. are renewable resources.
d. are nonreplaceable resources.

26. Most of the human food supply

a. comes from tropical rain forests. c. contains dioxin.
b. comes from about 20 species. d. is in danger of extinction.

27. The extinction of species

a. is a problem limited to the tropics.
b. has been speeded up by the activities of people.
c. is a problem only where topsoil and ground water are limited.
d. is not a problem in the twentieth century.

28. The human population increases in size when the
  a. death rate equals the birth rate.
  b. death rate exceeds the birth rate.
  c. birth rate equals the death rate.
  d. birth rate exceeds the death rate.

29. Human population growth is most rapid in
  a. Europe.
  b. the United States.
  c. Japan.
  d. developing countries.

30. At the current rate of growth, the human population will double in approximately
  a. 10 years.
  b. 20 years.
  c. 40 years.
  d. 80 years.

31. Pollution is an example of
  a. a direct cost of production.
  b. an indirect cost of production.
  c. a direct and indirect cost of production.
  d. None of the above

32. In order to reduce emissions, cars are required to have
  a. nitrogen scrubbers.
  b. mufflers.
  c. carburetors.
  d. catalytic converters.

33. Aluminum cans
  a. can be recycled.
  b. contain dioxin.
  c. cause carcinogens to be released into the atmosphere during their manufacture.
  d. do not change in price due to demand.

34. The first stage in addressing an environmental problem is
  a. an analysis of supply and demand.
  b. political action.
  c. a public outcry.
  d. assessment about what is happening.

35. Risk analysis involves
  a. supplying a population with information.
  b. predicting the consequences of environmental intervention.
  c. political action.
  d. education.

The following question(s) refer(s) to the graph.

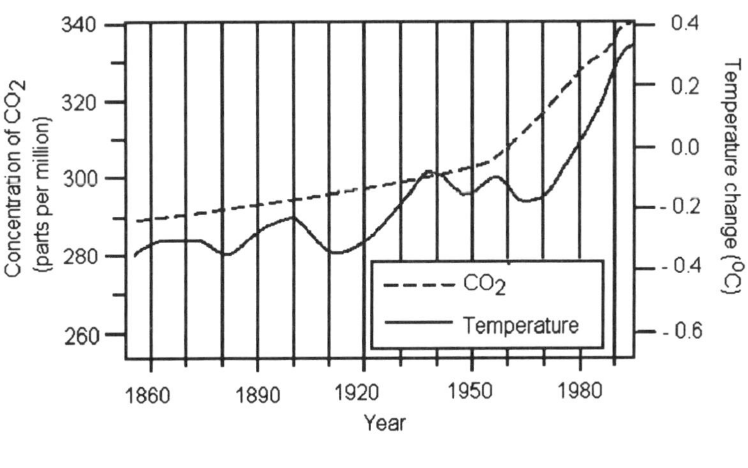

36. The graph shows
  a. the concentration of CO_2 in the atmosphere from 1860 to the present.
  b. the average global temperature from 1860 to present.
  c. that the concentration of oxygen in the atmosphere has increased since 1860.
  d. All of the above

37. According to the graph,
  a. from 1900 to 1950 the average global temperature constantly increased.
  b. the concentration of CO_2 in the atmosphere has increased at the same steady rate for the past 10 years.
  c. the concentration of CO_2 and the temperature were the same in 1940.
  d. None of the above

The following question(s) refer(s) to the graph showing human population growth.

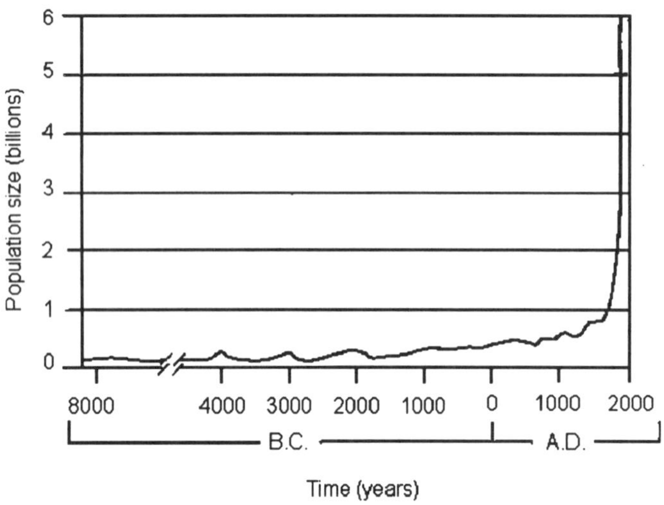

38. According to the graph, the human population
  a. remained essentially unchanged for thousands of years.
  b. doubled in size from 2000 BC to 1000 BC.
  c. reached 1 billion in 1492.
  d. will stop growing in the year 2000.

39. According to the graph,
  a. there were no humans on the Earth around 6000 BC.
  b. the human population has never decreased in size.
  c. an increase in the food supply was responsible for the increase in the population.
  d. the human population is currently rising dramatically.

SHORT ANSWER Complete each statement by writing the correct term or phrase in the space provided.

40. Ozone in the atmosphere blocks harmful ____________________ radiation from reaching the surface of the Earth.

41. The heat-trapping ability of carbon dioxide, methane, and nitrous oxide in the atmosphere is known as the ____________________.

42. CFCs catalyze the conversion of ozone into molecular ____________________.

43. Topsoil is an example of a ______________________ resource.

44. The main reason the Earth's human population has increased over the past 350 years is because of a decrease in the ____________________ rate.

45. Human beings first reached North America about ____________________ years ago.

46. A great deal of water is found beneath the soil within porous rock reservoirs called ______________________.

47. The population of the Earth is expected to exceed ________________ billion people by the year 2000.

48. The ______________________ countries on the Earth are experiencing the greatest increase in population growth.

49. The first step of addressing an environmental problem is __________.

50. A tax on pollution is called a ______________________.

51. Carbon monoxide, soot, and ______________________ are three pollutants produced by the burning of coal.

52. The economy of the United States is based on a system of ______________________ and demand.

53. Secondary sewage treatment facilities remove bacteria and ______________________ from sewage.

ESSAY Read and answer each question or statement on a separate sheet of paper.

54. Describe how acid precipitation is formed.

55. Why has our society made it profitable for industries to pollute the environment?

56. What are the five steps that must be followed in order to successfully solve an environmental problem?

ANALOGIES An analogy is a relationship between two pairs of words or phrases written as: a:b::c:d. The symbol : is read "is to," and the symbol :: is read "as." Write the letter of the pair of terms that best completes the analogy shown.

57. CFC : ozone ::
   a. ozone : carbon dioxide
   b. ozone : oxygen
   c. acid rain : fish and amphibians
   d. acid rain : carbon dioxide
   e. acid rain : mammals and birds

____

58. burning fossil fuels : atmospheric carbon dioxide ::
a. greenhouse effect : CFCs
b. atmospheric carbon dioxide : global warming
c. ground level ozone : CO_2
d. atmospheric carbon dioxide : ozone
e. ozone : global warming

---

59. 10,000 years ago : agriculture ::
a. 10,000 years ago : major pollution
b. 10,000 years ago : advanced medical techniques
c. recent times : continental ice sheets
d. recent times : increased death rate
e. recent times : sanitation and medical techniques

---

60. industrialized countries' populations : slowly ::
a. developing countries' populations : rapidly
b. developing countries' populations : slowly
c. populations of Asia and Latin America : slowly
d. populations of North America and Europe : rapidly
e. problems of pollution and consumption : more slowly

---

61. political action : solving an environmental problem ::
a. making jewelry : fishing
b. digging a hole : collecting stamps
c. dribbling : playing basketball
d. throwing a beach ball : repaving a driveway
e. landscaping : singing

---

62. frogs' disappearance : acid rain ::
a. pH : acidity
b. frog skin color : absorption
c. greenhouse effect : atmospheric CO_2
d. warm weather : snow melt
e. warm weather : cold weather

---

63. sulfur and water : acid rain ::
a. acid rain and ozone : carbon dioxide
b. CFCs and ozone : oxygen
c. oxygen and CFCs : ozone
d. ozone and carbon dioxide : acid rain
e. carbon dioxide and acid rain : ozone

---

64. greenhouse gases cause global warming : some scientists ::
   a. global warming is caused by weather fluctuations : other scientists
   b. global warming is decreasing : recent reports
   c. weather is the cause of carbon dioxide increase : some scientists
   d. CFCs increase CO_2 : some studies
   e. CFCs cause weather changes : some reports

____

65. world population : exceeding 6 billion by year 2000 ::
   a. industrialized countries : increasing population slowly
   b. industrialized countries : increasing population rapidly
   c. industrialized countries : decreasing population rapidly
   d. developing countries : increasing population very slowly
   e. world population in year 2000 : stabilizing

____

66. not paying indirect costs : consumer prices for goods ::
   a. an increase in humans : population
   b. Clean Air Act (1990) : pollution
   c. carbon dioxide : pollution
   d. sulfur : pollution
   e. CFC : ozone hole

____

TEST ANSWERS

TRUE/FALSE

1.F

2.F

3.F

4.T

5.F

6.T

7.T

8.F

9.F

10.F

11.F

12.T

MULTIPLE CHOICE

13.d

14.a

15.b

16.d

17.d

18.d

19.c

20.a

21.b

22.b

23.b

24.a

25. d

26. b

27. b

28. d

29. d

30. c

31. b

32. d

33. a

34. d

35. b

36. a

37. d

38. a

39. d

SHORT ANSWER

40. ultraviolet

41. greenhouse effect

42. oxygen

43. nonreplaceable

44. death

45. 12,000

46. aquifers

47. 6

48. developing

49. assessment

50. pollution permit

51. sulfur dioxide

52. supply

53. chemicals

ESSAY

54. Sulfur in the smoke from burning sulfur-rich coal is released into the atmosphere, condensation occurs, and the sulfuric acid is brought back to Earth in the form of acid rain or snow.

55. Since the indirect costs are not included in the price that a consumer pays for a product, far more is consumed than if they had been included. By not adding the indirect costs to the price of energy and manufactured goods, our society has made it profitable to pollute.

56. In order to solve the problem facing an ecosystem it is necessary to
   1. assess the problem by collecting data;
   2. perform a risk analysis to evaluate the various possible courses of action;
   3. educate the public about the most feasible course of action and its costs;
   4. implement a solution through political action; and
   5. follow through on any action taken, to verify that the problem is being effectively addressed and solved.

ANALOGIES

57. c (destroys)

58. b (increases)

59. e (there was development of)

60. a (are growing)

61. d (is part of)

62. a (is a measure of)

63. b (join together to make)

64. a (according to)

65. a (will be)

66. b (reduces)

Chapter 19

TRUE/FALSE Read each statement. If the statement is true, write T in the space provided. If the statement is false, write F in the space provided.

1.______ Eukaryotes are the most abundant inhabitants of the Earth.

2.______ The gene translation machinery of eubacteria is different from that of both archaebacteria and eukaryotes.

3.______ Obligate anaerobes are organisms that require oxygen.

4.______ Methanogens, which produce the poisonous gas methane, can be found in the digestive tracts of cows.

5.______ Mitosis distinguishes prokaryotic organisms from eukaryotes.

6.______ Genetic diversity is the raw material for evolution.

7.______ In plants, spores give rise to a haploid phase called the gametophyte.

8.______ The main advantage of multicellularity is that each cell in a multicellular organism requires less food and oxygen than a unicellular organism.

9.______ A colonial organism is a collection of cells that are permanently associated.

10.______ Each of the kingdoms of living things exhibit multicellularity.

11.______ The Greek philosopher Aristotle grouped animals according to their structural similarities.

12.______ Cladograms represent direct information about ancestors and descendants, showing who came from whom.

MULTIPLE CHOICE Circle the letter of the term or phrase that best completes each statement or best answers each question.

13. Simple, non-nucleated organisms that use hydrogen gas to produce methane belong to the kingdom
   a. Archaebacteria.   c. Protista.
   b. Eubacteria.   d. Fungi.

14. Chloroplasts in the cells of plants may have arisen due to
   a. circular genomes.   c. sporic meiosis.
   b. nondisjunction.   d. endosymbiosis.

15. Archaebacteria can be distinguished from eubacteria because of differences in their
   a. cell walls.   c. gene architecture.
   b. plasma membranes.   d. All of the above

16. Budding is an example of
   a. endosymbiosis.   c. meiosis.
   b. asexual reproduction.   d. sexual reproduction.

17. The essential act of sexual reproduction is
   a. zygotic meiosis.
   b. the formation of buds.
   c. the fusion of two haploid cells to form a zygote.
   d. The appearance of multicellularity and specialization.

18. The life cycle in which the zygote is the only diploid cell is known as
   a. gametic meiosis.
   b. sporic meiosis.
   c. zygotic meiosis.
   d. sexual meiosis.

19. The process in which a single diploid cell becomes a multicellular individual with many kinds of cells is called
   a. mitosis.
   b. differentiation.
   c. endocytosis.
   d. aggregation.

20. A characteristic of multicellular organisms in which the cells of the organism adjust in response to each other is known as
   a. sporic aggregation.
   b. multicellular meiosis.
   c. colonial reduction.
   d. intercell coordination.

21. Organisms in different genera
   a. may share the second word of their scientific names.
   b. may be in the same family.
   c. may be in different orders.
   d. All of the above

22. Which of the following scientists developed the system of classifying organisms with a genus and species name?
   a. Leakey
   b. Aristotle
   c. Darwin
   d. Linnaeus

23. If two structurally similar individuals produce fertile offspring in nature, they are considered to be members of the same
   a. class.
   b. genus.
   c. kingdom.
   d. species.

24. The red maple is also known as *Acer rubrum*. Its scientific name is
   a. red.
   b. *Acer*.
   c. maple.
   d. *Acer rubrum*.

25. The organism *Quercus phellos* is a member of the genus
   a. Plantae.
   b. *phellos*.
   c. *Quercus*.
   d. Protista.

26. Two organisms in the same class but different orders will
   a. be in different kingdoms.
   b. have the same genus name.
   c. be in the same phylum.
   d. be members of the same species.

27. As we move through the biological hierarchy from the kingdom to species level, organisms
   a. vary more and more.
   b. are less and less related to each other.
   c. become more similar in appearance.
   d. always are members of the same order.

28. Multicellular, nucleated organisms that carry on photosynthesis belong to the kingdom
   a. Animalia.
   b. Fungi.
   c. Monera.
   d. Plantae.

29. Nearly all single-celled eukaryotes that are either heterotrophs or photosynthetic belong to the kingdom
a. Eubacteria.
b. Fungi.
c. Plantae.
d. Protista.

30. The shark and the dolphin, which evolved similar features independently, are an example of
a. analogous evolution.
b. similar evolution.
c. divergent evolution.
d. convergent evolution.

31. The beak of a bird and the beak of a giant squid evolved independently and serve the same function. The beaks are
a. convergent structures.
b. phenetic structures.
c. analogous structures.
d. hybrid structures.

32. An organism that breaks down organic matter, which it then absorbs, is a member of the kingdom
a. Fungi.
b. Plantae.
c. Monera.
d. Protista.

33. It seems likely that sexual reproduction evolved as a mechanism to
a. keep cells alive longer.
b. repair chromosomal damage.
c. preserve the sporophyte generation.
d. produce many well-adapted offspring in a short period of time.

The following question(s) refer(s) to the branching diagram.

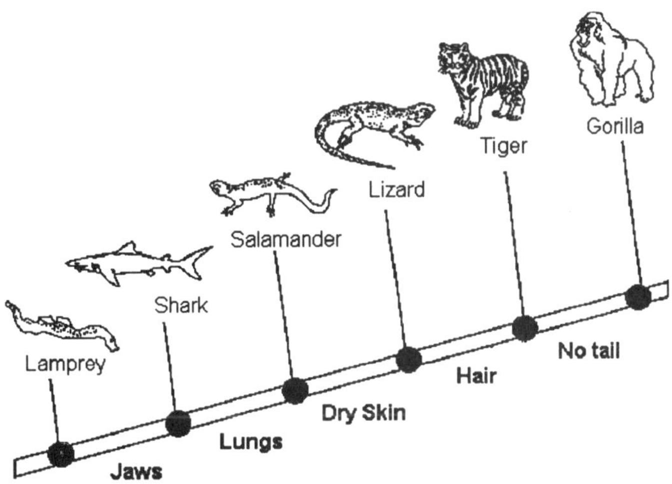

34. A branching diagram like the one illustrated is called a
a. phenetic tree.
b. cladogram.
c. family tree.
d. homology.

35. Each unique character, such as dry skin, that is used to assign an organism to a group is known as a(n)
a. special character.
b. analogous character.
c. derived character.
d. homologous character.

The following question(s) refer(s) to the diagram of the phylogenetic tree.

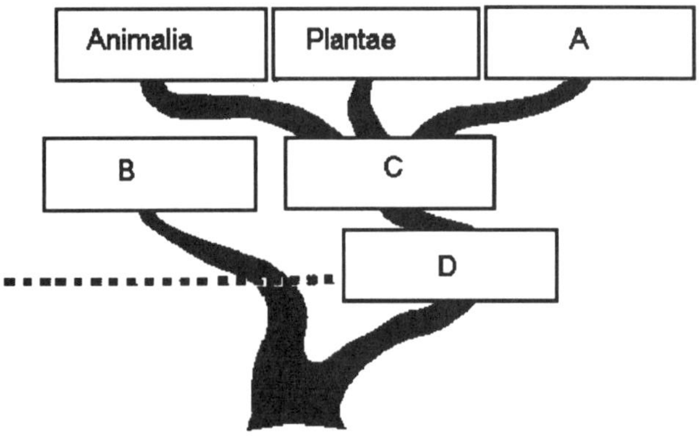

36. The kingdom represented in box A is
   a. Archaebacteria. c. Eubacteria.
   b. Protista. d. Fungi.

37. The kingdom represented in box B is
   a. Archaebacteria. c. Eubacteria.
   b. Protista. d. Fungi.

38. The kingdom represented in box C is
   a. Archaebacteria. c. Eubacteria.
   b. Protista. d. Fungi.

SHORT ANSWER Complete each statement by writing the correct term or phrase in the space provided.

39. The cell walls of eubacteria are made of ________________________.

40. In the life cycle that has ________________________ meiosis, a diploid reproductive cell undergoes meiosis and gives rise to haploid gametes.

41. The diploid phase in the life cycle of plants is called the ________________________.

42. A temporary collection of cells that come together for a period of time is called a(n) ________________________.

43. The key characteristic of complex multicellular organisms that distinguishes them from multicellular protists is cell ________________________.

44. All scientific names are made up of two words derived from the ________________________ language.

45. The science of naming and classifying organisms is called ________________________.

46. The offspring that result when individuals from two different species interbreed are called ________________________.

47. The term ________________________ is used in place of phylum when plants are classified.

48. The evolutionary history of a species is called its ________________.

The following question(s) refer(s) to the drawings of the representatives of the five kingdoms.

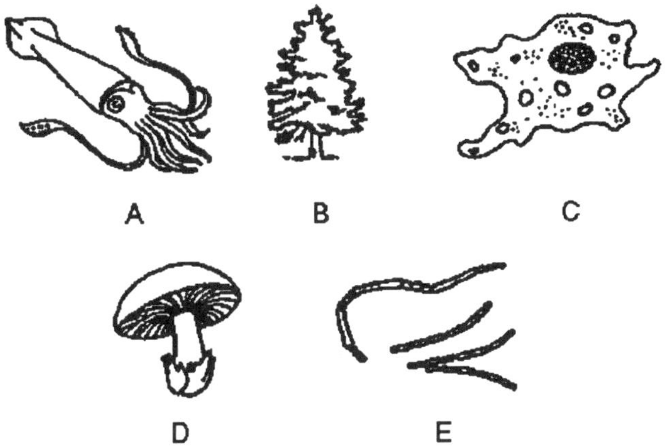

49. Organism A is classified in the kingdom ______________________.

50. Organism B is classified in the kingdom ______________________.

51. Organism C is classified in the kingdom ______________________.

52. Organism D is classified in the kingdom ______________________.

ESSAY Read and answer each question or statement on a separate sheet of paper.

53. What observations support the idea that mitochondria are descended from bacteria?

54. Congratulations! While on a biological expedition to a tropical rain forest you discover a new insect. Explain how you would determine the genus and species name of the insect.

55. A species is defined as a group of organisms that are similar in appearance and that can interbreed and produce fertile offspring in nature. Horses and donkeys can interbreed and produce mules. Mules cannot produce offspring. Is it possible that horse and donkeys belong to the same species? Explain.

56. Explain two reasons why scientists believe that *Pelomyxa palustris* may resemble an early stage in the evolution of the eukaryotic cell.

ANALOGIES An analogy is a relationship between two pairs of words or phrases written as: a:b::c:d. The symbol : is read "is to," and the symbol :: is read "as." Write the letter of the pair of terms that best completes the analogy shown.

57. *Pelomyxa*: prokaryotic cell::
   a. pair of tweezers : sweatshirt
   b. loaf of bread : hamburger bun
   c. feather : chair
   d. penny : spool of thread
   e. bouquet of flowers : diesel truck

____

58. gametic meiosis : gametes::
   a. meiosis : two replications
   b. reduction division : diploid cells
   c. animal : spores
   d. crossing-over : identical genetic material
   e. binary fission : identical offspring

____

59. non-sulfur purple bacteria : mitochondria::
   a. archaebacteria : eubacteria
   b. eubacteria : prokaryotes
   c. eubacteria : archaebacteria
   d. *Prochloron* : chloroplasts
   e. *Pelomyxa*: bacteria

____

60. simple multicellularity : complex multicellularity::
   a. alga : redwood tree
   b. eukaryote : prokaryote
   c. redwood tree : alga
   d. fungus : prokaryote
   e. multicellular organism : unicellular organism

____

61. simplicity : zygotic meiosis::
   a. multicellularity : bacteria
   b. alternation of generations : animals
   c. zygotic meiosis : most animals
   d. cell specialization : multicellular organisms
   e. presence of mitochondria : bacteria

____

62. Ray : definition of species::
   a. Mayr : polynomial naming system
   b. Mayr : two-word naming system
   c. Aristotle : scientific naming
   d. Linnaeus : binomial nomenclature
   e. Ray : biological species concept

____

63. *Quercus rubra* : *Quercus phellos*::
   a. *Anolis carolinensis* : *Parus carolinesis*
   b. *Erithacus rubicula* : *Turdus migratoria*
   c. *Aphis pomi* : *Aphis gossypii*
   d. carp : goldfish
   e. green anole lizard : Carolina chickadee

____

64. analogous characters : convergent evolution::
   a. two members of the same genus : same species
   b. cladogram : evolutionary relationships
   c. common names : universal identification
   d. cladograms : exact, direct information
   e. evolutionary systemics : small amounts of information

65. class : family::
   a. order : phylum
   b. genus : class
   c. species : genus
   d. phylum : order
   e. family : class

TEST ANSWERS

TRUE/FALSE

1. F

2. T

3. F

4. T

5. F

6. T

7. T

8. F

9. T

10. F

11. T

12. F

MULTIPLE CHOICE

13. a

14. d

15. d

16. b

17. c

18. c

19. b

20. d

21. d

22. d

23. d

24. d

25. c

26. c

27. c

28. d

29. d

30. d

31. c

32. a

33. d

34. b

35. c

36. d

37. c

38. b

SHORT ANSWER

39. peptidoglycan

40. gametic

41. sporophyte

42. aggregation

43. specialization

44. Latin

45. taxonomy

46. hybrids

47. division

48. phylogeny

49. Animalia

50. Plantae

51.Protista

52.Fungi

ESSAY

53.The following four major observations support the idea that mitochondria descended from bacteria,

1. the size and structure of mitochondria;
2. the mitochondrial genome;
3. the gene translation machinery of mitochondria;
4. the way mitochondria reproduce.

Mitochondria are similar in size and shape to eubacteria; their DNA is in the form of a circular molecule similar to that in most bacteria; their gene translation machinery is similar and mitochondria, like bacteria, reproduce by simple fission.

54.Based on the insect's physical characteristics you would determine if it belongs in a known genus. If it does, it has that genus name. If a new genus name is needed, it should be descriptive. Scientific names usually describe the appearance, location, or lifestyle of the organism. Scientific names must be Latin words or terms constructed according to the rules of Latin grammar.

55.Horses and donkeys cannot belong to the same species because their offspring, mules, are infertile.

56.*Pelomyxa palustris*, like early eukaryotes, is much larger than a prokaryotic cell and has a complex system of internal membranes. It also resembles some of the fossils of early prokaryotes.

ANALOGIES

57.b (is larger than)

58.e (produces)

59.d (evolved into)

60.a (is simpler than)

61.d (is a characteristic of)

62.d (devised)

63.c (is a member of the same genus as)

64.b (show evidence of)

65.d (is a larger unit than)

Chapter 20

TRUE/FALSE Read each statement. If the statement is true, write T in the space provided. If the statement is false, write F in the space provided.

1.______ Although viruses do not consist of cells, biologists consider them to be living because they are capable of reproduction.

2.______ Some viruses have a membranous envelope surrounding the protein coat that helps them gain entry into host cells.

3.______ Reverse transcriptase makes DNA from a template of RNA.

4.______ Bacteria are the oldest life-forms on Earth.

5.______ Gram-negative bacteria have a thick layer of peptidoglycan that stains purple.

6.______ Bacteria lack nuclei and therefore also lack genetic material.

7.______ Although there are some bacteria that are heterotrophic, the vast majority are autotrophic.

8.______ Photosynthetic bacteria are present in leguminous plants and convert atmospheric nitrogen into a form that is usable by the plants.

9.______ Certain antibiotics have been ineffective against certain strains of bacteria. These bacteria have developed a resistance, which may be passed on from one generation of bacteria to the next.

10.______ People can contract the influenza virus more than once because the virus tends to mutate rapidly, avoiding the actions of the immune system.

MULTIPLE CHOICE Circle the letter of the term or phrase that best completes each statement or best answers each question.

11. Tobacco mosaic virus
   a. is able to be crystallized.
   b. causes disease in tobacco plants.
   c. is smaller than a bacterium.
   d. All of the above are correct.

12. Biologists now know that viruses
   a. are the smallest organisms.
   b. consist of a protein surrounded by a nucleic acid coat.
   c. contain RNA or DNA in a protein coat.
   d. all form the same crystalline shape.

13. A membranous envelope surrounding some viruses may be composed of
   a. lipids.
   b. proteins.
   c. glycoproteins.
   d. All of the above

14. A typical virus consists of
   a. a protein coat and a cytoplasm core.
   b. a carbohydrate coat and a nucleic acid core.
   c. a protein coat and a nucleic acid core.
   d. a polysaccharide coat and a nucleic acid core.

15. Viruses
   a. are cellular organisms.
   b. reproduce only in living cells.
   c. have nuclei and organelles.
   d. are surrounded by a polysaccharide coat.

16. An animal virus enters its host cell by
   a. being injected into the cell.
   b. penetrating a rip in the cell wall.
   c. punching a hole in the cell membrane.
   d. endocytosis across the cell membrane.

17. The capsid of a virus is the
   a. protective outer coat.
   b. cell membrane.
   c. nucleus.
   d. cell wall and membrane complex.

18. Animal viruses often infect only specific host cells because
   a. they must have the same DNA as their host cell.
   b. the host cell has receptors for the glycoprotein on the virus.
   c. viruses have receptors for host cell glycoproteins.
   d. the enzymes of the virus can attach only to specific host cells.

19. A pathogen is an agent that is
   a. beneficial to humans.
   b. harmful only to plants.
   c. harmful to living organisms.
   d. nearly extinct.

20. Viruses that use reverse transcriptase to cause their host cells to transcribe DNA from an RNA template are called
   a. bacteriophages.
   b. antibodies.
   c. retroviruses.
   d. capsoviruses.

21. Bacteria are the only organisms characterized as
   a. unicellular.
   b. prokaryotic.
   c. eukaryotic.
   d. photosynthetic.

22. When tested with a Gram stain, gram-positive bacteria are stained
   a. green.
   b. yellow.
   c. pink.
   d. purple.

23. Gram-staining provides important medical information in that
   a. the results can provide a clue as to what type of antibiotics will be useful.
   b. it can tell the species of bacteria causing the infection.
   c. gram-negative bacteria never cause disease.
   d. the proper type of Gram-staining can prevent infections.

24. Bacteria cells
   a. have only a cell wall.
   b. have only a cell membrane.
   c. have both a cell membrane and an outer cell wall.
   d. have a cell wall inside of their cell membrane.

25. The bacteria that cause botulism may survive in canned food for a long time because
   a. the can was left open.
   b. some cans may contain viruses that protect the bacteria.
   c. the bacteria may form endospores.
   d. sterilized cans do not have enough oxygen to harm the bacteria.

26. Which of the following comparisons is *incorrect*?

	PROKARYOTES	EUKARYOTES
a.	smaller	larger
b.	circular chromosomes	linear chromosomes
c.	binary fission	mitosis
d.	chloroplasts	mitochondria

27. Which of the following might be found in the cytoplasm of a bacterial cell?
a. chloroplasts
b. Golgi apparatus
c. mitochondria
d. None of the above

28. The cytoplasm of bacteria
a. contains numerous types of organelles.
b. is divided into compartments.
c. has varying numbers of chromosomes, depending on the species of bacteria.
d. contains a single molecule of DNA.

29. Bacteria lack a true nucleus and membrane-bound organelles; therefore, they are classified as
a. prokaryotes.
b. aerobes.
c. anaerobes.
d. eukaryotes.

30. Bacterial endospores
a. occur where there is plenty of available food.
b. allow certain species to survive harsh environmental conditions.
c. are similar to human tumors.
d. can cause growth abnormalities in plants.

31. Bacteria *do not* exist
a. in the depths of the oceans.
b. under the ice of Antarctica.
c. in ordinary ground water.
d. None of the above is true; bacteria have been found in all of these places.

32. Bacteria can be classified according to their
a. type of cell walls.
b. methods of obtaining energy.
c. Gram-staining characteristics.
d. All of the above

33. A type of autotrophic bacteria found in aerobic environments is
a. cyanobacteria.
b. green sulfur bacteria.
c. purple sulfur bacteria.
d. None of the above is true; bacteria live in anaerobic environments only.

34. Nitrogen-fixing bacteria
a. repair nitrogen-damaged legume roots.
b. damage the environment by using atmospheric oxygen to produce toxic nitrogen.
c. convert atmospheric nitrogen into ammonia.
d. convert ammonia in the soil into nitrogen gas.

35. Some pathogens damage the body by the production of
a. toxins.
b. antibodies.
c. antigens.
d. DNA.

36. Alexander Fleming, a British bacteriologist, is credited with the discovery of
   a. photosynthetic bacteria.
   b. antibodies.
   c. tuberculosis.
   d. penicillin.

37. Antibiotics are ineffective against viral infections because
   a. viruses are protected inside their host cells.
   b. viruses have enzymes that inactivate the antibiotics.
   c. antibiotics interfere with metabolic processes that viruses do not perform.
   d. viral protein coats block the antibiotics from entering the virus.

38. *Escherichia coli* is an example of a bacterium that has short, thin, hairlike projections called
   a. pili.
   b. cilia.
   c. cocci.
   d. ribosomes.

39. Bacterial cells such as *E. coli* transfer pieces of genetic material in a process called
   a. binary fission.
   b. mitosis.
   c. conjugation.
   d. sexual reproduction.

40. Cell organelles that *E. coli* and other bacteria have in common with eukaryotes are
   a. chloroplasts.
   b. mitochondria.
   c. nuclei.
   d. ribosomes.

41. A bacterial disease carried from rodents to humans by fleas is
   a. tuberculosis.
   b. bubonic plague.
   c. cholera.
   d. Lyme disease.

42. A viral disease that causes painful swelling in the salivary glands is
   a. mumps.
   b. AIDS.
   c. polio.
   d. measles.

The following question(s) refer(s) to the illustration of Organisms A, B, and C.

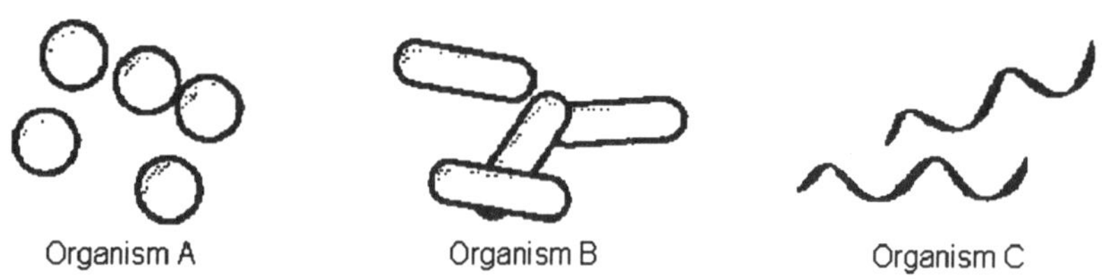

43. The shape represented by Organism A is called
   a. coccus.
   b. spirillum.
   c. bacillus.
   d. filamentous.

44. You would expect that the bacterial genus *Bacillus* has a shape like that shown by
   a. Organism A.
   b. Organism B.
   c. Organism C.
   d. None of the above

45. Which of the illustrations shows the organism that is the cause of AIDS?
   a. Organism A
   b. Organism B
   c. Organism C
   d. None of the above

SHORT ANSWER Complete each statement by writing the correct term or phrase in the space provided.

46. Segments of nucleic acid wrapped in a protein coat are called ______________________.

47. A disease of tobacco plants in which growth is stunted and leaves are blotchy is caused by the ______________________.

48. The protein coat of a virus is called a(n) ______________________.

49. ______________________ are bacterial viruses with a polyhedral head attached to a helical tail.

50. Agents that cause disease are called ______________________.

51. A(n) ______________________ is a protein secreted by cells in the mammalian immune system in response to a foreign substance in the body.

52. An enzyme called ______________________ manufactures DNA that is complementary to a virus's RNA.

53. A virus that transcribes DNA from an RNA template is called a(n) ______________________.

54. Rod-shaped bacteria are called ______________________.

55. Spherical bacteria are called ______________________.

56. Spiral bacteria are called ______________________.

57. The cell walls of eubacteria are composed of a combination of polysaccharide and polypeptide called ______________________.

58. Protective coverings that some bacteria may form under harsh conditions are ______________________.

59. In general, organisms that obtain their energy from sunlight are called ______________________.

60. Certain cyanobacteria, such as *Anabaena*, can fix nitrogen by using enzymes contained in specialized structures called ________________.

61. Bacteria that obtain their energy by removing electrons from inorganic molecules, rather than obtaining energy from the sun, are called ______________________ bacteria.

62. The oxidation of ammonia to nitrates that can be used by plants is called ______________________.

63. ______________________ is a disease of the respiratory tract caused by the bacterium *Mycobacterium tuberculosis*.

64. A(n) ______________________ is a substance obtained from bacteria or fungi that is used as a drug to fight pathogenic microorganisms.

65. A species of bacteria that lives in the intestines of many mammals is ______________________.

ESSAY Read and answer each question or statement on a separate sheet of paper.

66. Viruses are not considered living organisms, but they are still studied as part of biology. Explain.

67. Explain the difference between gram-positive and gram-negative bacteria.

68. It has been observed that antibiotics are generally effective against bacterial infections but cannot be used to treat viral infections. Why is this the case?

ANALOGIES An analogy is a relationship between two pairs of words or phrases written as: a:b::c:d. The symbol : is read "is to," and the symbol :: is read "as." Write the letter of the pair of terms that best completes the analogy shown.

69. plant virus : tiny rips::
   a. receptor protein : cell membrane
   b. virus : immune system
   c. hepatitis : mutation
   d. animal virus : endocytosis
   e. antibody : glycoprotein

____

70. viral capsid : nucleic acid core::
   a. DNA : virus
   b. envelope : capsid
   c. capsid : envelope
   d. capsid : glycoprotein
   e. RNA : plant

____

71. virus : chemicals::
   a. water : ice
   b. brick : clay
   c. simplicity : complexity
   d. a jigsaw puzzle : individual pieces
   e. virus : cell

____

72. influenza : respiratory tract::
   a. influenza : a rash
   b. polio : nerve tissue
   c. smallpox : contaminated food or water
   d. cold : smallpox
   e. paralysis : influenza

____

73. Lyme disease : ticks::
   a. tuberculosis : food and feces
   b. cholera : human tubercles
   c. typhus : fleas
   d. bubonic plague : fleas
   e. bubonic plague : drinking water

____

74. eukaryotes : nuclei
   a. prokaryotes : nuclei
   b. chromosomes : prokaryotes
   c. DNA fragments : proteins
   d. prokaryotes : microtubules
   e. prokaryotes : diverse metabolic abilities

___

75. photosynthetic bacteria : sunlight::
   a. chemotrophic bacteria : dead organisms
   b. chemoautotrophic bacteria : inorganic molecules
   c. photosynthesis : nitrification
   d. heterotrophic bacteria : inorganic molecules
   e. heterotrophic bacteria : photosynthesis

___

76. viruses : cells::
   a. photosynthetic bacteria : chemosynthetic bacteria
   b. bacteria : viruses
   c. antibiotics : bacteria
   d. cyanobacteria : chlorophyll
   e. heterocysts : nitrogen

___

77. viral reverse transcription : cellular RNA transcription::
   a. far : distant
   b. impossible : undoable
   c. winner : victor
   d. illness : health
   e. heavy : weighted

___

TEST ANSWERS

TRUE/FALSE

1. F
2. T
3. T
4. T
5. F
6. F
7. F
8. F
9. T
10. T

MULTIPLE CHOICE

11. d
12. c
13. d
14. c
15. b
16. d
17. a
18. b
19. c
20. c
21. b
22. d
23. a
24. c

25. c

26. d

27. d

28. d

29. a

30. b

31. d

32. d

33. a

34. c

35. b

36. d

37. c

38. a

39. c

40. d

41. b

42. a

43. a

44. b

45. d

SHORT ANSWER

46. viruses

47. tobacco mosaic virus

48. capsid

49. Bacteriophages

50. pathogens

51. antibody

52. reverse transcriptase

53. retrovirus

54. bacilli

55. cocci

56. spirilli

57. peptidoglycan

58. endospores

59. autotrophs

60. heterocysts

61. chemoautotrophic

62. nitrification

63. Tuberculosis

64. antibiotic

65. *E. coli*

ESSAY

66. Viruses are active inside living cells, making them an important part of the study of biology. Although viruses contain genetic material and can evolve as this material changes over time, they are not considered living because they are not cellular, cannot make their own proteins, and cannot use energy in metabolic processes.

67. Gram-positive and gram-negative bacteria are distinguished from each other through Gram-staining due to differences in their cell walls. Gram-positive bacteria retain a purple dye during the Gram-staining procedure because they have a thick peptidoglycan layer that prevents the dye from leaving the cell during the alcohol-rinse stage of the process. Gram-negative bacteria have only a thin layer of peptidoglycan and an outer layer that is dissolved by the alcohol-rinse, allowing the purple dye to leave the cell. These cells usually stain pink because of a pink dye that is applied after the alcohol-rinse.

68. Antibiotics work by interfering with cell wall formation or cellular processes such as protein synthesis. Some antibiotics can cause disruption of the cell membrane. All of these actions can negatively affect the survival of bacteria. However, since viruses do not undergo cellular processes, they are not influenced by the actions of antibiotics. Thus, antibiotic treatment is ineffective against viral infections.

ANALOGIES

69. d (enters a cell by way of)

70.b (surrounds)

71.d (is an assembly of)

72.b (is a virus that affects)

73.d (is spread by)

74.e (have)

75.b (receive energy from)

76.c (destroy)

77.d (is the opposite of)

Chapter 21

TRUE/FALSE Read each statement. If the statement is true, write T in the space provided. If the statement is false, write F in the space provided.

1.______ The first protists were prokaryotes.

2.______ In protists, flagella are composed of nine pairs of microtubules encircling a single microtubule.

3.______ The kingdom Protista contains all eukaryotes that cannot be classified as animals, plants, or fungi.

4.______ The major phyla of protists are, for the most part, closely related to each other.

5.______ While they can parasitize animals, protists cannot parasitize plants.

6.______ Sporangia produce zoospores by mitosis.

7.______ Amoebas move by means of pseudopods.

8.______ Diatoms are the only type of protists with shells.

9.______ Oomycetes are water molds.

10.______ The oocytes of sporozoans are diploid cells that are resistant to unfavorable environmental conditions.

11.______ Some strains of *Plasmodium* are resistant to all known anti-malarial drugs.

12.______ Amoebic dysentery is usually passed from person to person by coughing and sneezing.

MULTIPLE CHOICE Circle the letter of the term or phrase that best completes each statement or best answers each question.

13. All protists
    a. are parasites.
    b. have a true nucleus.
    c. move toward light.
    d. use flagella to move.

14. The kingdom Protista includes
    a. most of the single-celled eukaryotes.
    b. slime and water molds.
    c. the multicellular algae.
    d. All of the above

15. A protist may be
    a. unicellular and heterotrophic.
    b. unicellular and autotrophic.
    c. multicellular and autotrophic.
    d. All of the above

16. Zoospores are
    a. produced as a result of meiosis.
    b. diploid.
    c. produced as a result of mitosis.
    d. all parasitic.

17. Algae are
  a. heterotrophic.
  b. only microscopic in size.
  c. autotrophic protists.
  d. only found in fresh water.

18. A *Chlamydomonas* recognizes a different mating type by
  a. its size.
  b. the presence of cell-surface protein molecules.
  c. color.
  d. the type of zygospore it forms.

19. The haploid, gamete-producing phase in the life cycle of some protists is known as the
  a. zygospore generation.
  b. gametophyte.
  c. conjugation generation.
  d. sporophyte generation.

20. Zoomastigotes
  a. are unicellular.
  b. are heterotrophic.
  c. have a flagellum.
  d. All of the above

21. *Euglena* is an example of a protist that
  a. is both autotrophic and heterotrophic.
  b. is only a parasitic heterotroph.
  c. is always autotrophic.
  d. swims away from light.

22. Amoebas capture food by
  a. engulfing it.
  b. using cilia.
  c. trapping it with flagella.
  d. taking it into in a gullet.

23. Green algae are considered to be the ancestors of
  a. euglenoids.
  b. parasitic algae.
  c. the plant kingdom.
  d. heterotrophic slime molds.

24. A "mass of cytoplasm that can ooze around obstacles and even pass through cloth" is most likely a
  a. salamander.
  b. water mold.
  c. slime mold.
  d. amoeba.

25. Pseudopods are used for
  a. conjugation.
  b. movement.
  c. reproduction.
  d. mitosis.

26. The process in which two paramecia come together after meiosis to exchange parts of their genetic material is called
  a. mitosis.
  b. replication.
  c. pollination.
  d. conjugation.

27. A "red tide" that turns the ocean a reddish color and kills fish and other organisms is caused by a population explosion of
  a. red algae.
  b. *Didinium*.
  c. kelp.
  d. dinoflagellates.

28. Malaria is caused by
  a. *Toxoplasma*.
  b. *Phytophthora*.
  c. *Trypanosoma*.
  d. *Plasmodium*.

29. The symptoms of malaria
  a. include high fever, delirium, and sweating.
  b. follow a cycle.
  c. severe chills and fever.
  d. All of the above

30.One of the reasons that the number of cases of malaria has increased since the mid-1970s is

a. the mosquitoes that spread malaria have grown larger and stronger in the acid rain that has been produced since 1975.
b. the drugs that control malaria were found to cause cancer and taken off the market.
c. the parasite that causes malaria has evolved immunity to the drugs that control it.
d. All of the above

31.African sleeping sickness is spread by

a. tsetse flies.
b. mosquitoes.
c. contaminated food.
d. All of the above

32.The protozoan that causes malaria reproduces in the

a. intestine of a human.
b. red blood cells of a mosquito.
c. red blood cells of a human.
d. liver of a mosquito.

33.The stage in the live cycle of *Plasmodium* in which it lives in mosquitoes and is injected into humans is called the

a. gametophyte.
b. sporozoite.
c. sporophyte.
d. zoospore.

The following question(s) refer(s) to the diagram of *Paramecium*.

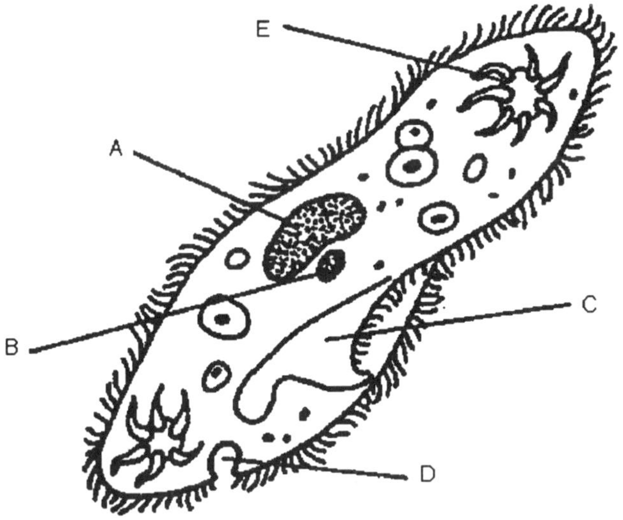

34.The structure that contains small pieces of DNA is labeled

a. A.
b. B.
c. C.
d. E.

35.Structure E is the

a. macronucleus.
b. contractile vacuole.
c. gullet.
d. micronucleus.

SHORT ANSWER Complete each statement by writing the correct term or phrase in the space provided.

36.Two important features that evolved in the protists were sexual reproduction and

________________________.

37.Some protists undergo sexual reproduction only at times of environmental

________________________.

38. Some protist have ______________________ that contain light-sensitive pigments.

39. Forams have porous shells called ______________________.

40. A flexible ______________________ allows a euglenoid to change its shape.

41. Poisonous "red tides" are caused by population explosions of ______________________.

42. Diatoms exhibit either radial or ______________________ symmetry.

43. The stage in the life cycle of *Plasmodium* that is transferred to humans when a mosquito bites is known as the ____________________.

44. Sleeping sickness is caused by a group of ______________________ called trypanosomes.

The following question(s) refer(s) to the diagrams of the three single-celled organisms.

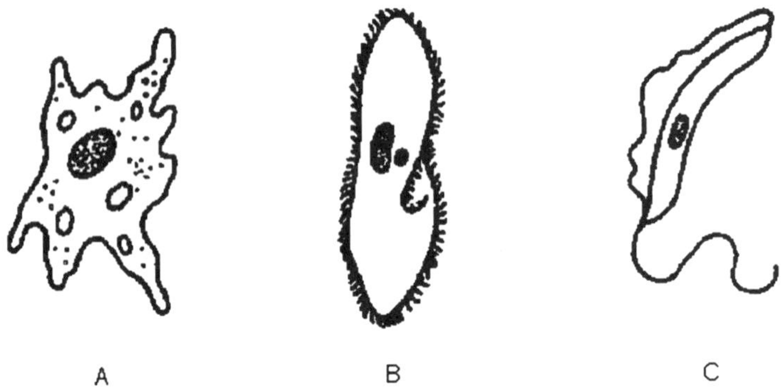

45. The organism shown in diagram A moves by means of ________________.

46. The organism shown in diagram B moves by means of ________________.

47. The organism shown in diagram C moves by means of a ______________.

The following question(s) refer(s) to the diagram of the life cycle of *Chlamydomonas*.

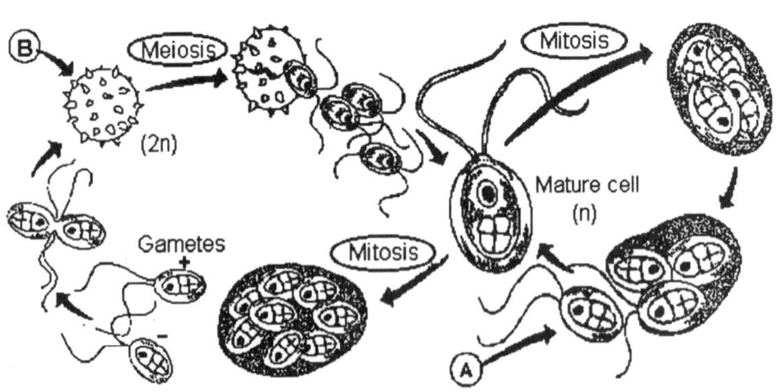

48. According to the diagram, *Chlamydomonas* reproduces both sexually and ______________________.

49. Arrow A is pointing to the ______________________ stage in the life cycle.

50. Arrow B is pointing to the ______________________ stage in the life cycle.

The following question(s) refer(s) to the diagram showing side-by-side filaments of *Spirogyra*.

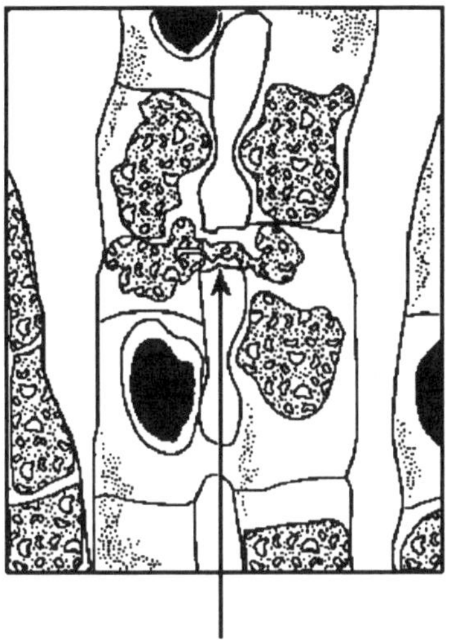

51. The diagram illustrates reproduction in a filamentous green algae. The arrow is pointing to a structure known as a(n)
______________________.

ESSAY Read and answer each question or statement on a separate sheet of paper.

52. How does the zoomastigote *Trichonympha* help termites?

53. How do euglenoids illustrate the problems of classifying protists as plants or animals?

54. After a vacation in the tropics you become seriously ill. Your symptoms include cycles of chills and fever every 48 hours. Why does your doctor suspect malaria?

ANALOGIES An analogy is a relationship between two pairs of words or phrases written as: a:b::c:d. The symbol : is read "is to," and the symbol :: is read "as." Write the letter of the pair of terms that best completes the analogy shown.

55. multicellular protists : a small degree of coordination among cells ::
   a. protists : a high degree of complexity
   b. eukaryotes : great similarity to prokaryotes
   c. *Pelomyxa* : mitosis
   d. flagella and cilia : "9 + 2" structure
   e. eukaryotes : quick evolution

____

56. protists : damp soil or sand ::
   a. cells : cilia
   b. protists : desert
   c. green algae : dry ground
   d. forams : freshwater
   e. parasitic protists : human tissues or blood

____

57. *Spirogyra* : thick-walled resting spore ::
   a. sexually reproducing *Chlamydomonas* : zoospore
   b. *Oedogonium* : thick-walled resting spore
   c. asexually reproducing *Chlamydomonas* : zygospore
   d. *Oedogonium* : thin-walled resting spore
   e. *Oedogonium* : sporangium

---

58. dinoflagellates : flagella ::
   a. amoebas : pseudopodia
   b. forams : flagella
   c. ciliates : pseudopodia
   d. amoebas : flagella
   e. forams : cilia

---

59. chloroplasts of green algae : chlorophyll a and b ::
   a. red algae : species called kelp
   b. red algae : similar pigments to diatoms
   c. brown algae : symbionts
   d. red algae : phycobilins
   e. brown algae : phycobilins

---

60. pseudopodia and cilia : locomotion structures ::
   a. nucleus and gullet : eyespots in euglenoids
   b. food and contractile : vacuoles in *Paramecia*
   c. cilia and pellicle : nuclei in amoebas
   d. "dino" and "zoo" : ciliates
   e. *Euglena* and *Paramecia* : forams

---

61. giardiasis : contaminated water ::
   a. amoebic dysentery : mosquito
   b. amoebic dysentery : giardiasis
   c. malaria : mosquito
   d. malaria : food contamination
   e. food contamination : mosquito

---

62. drug : malaria patients ::
   a. quinine derivative : drugs
   b. drug : cinchona tree
   c. drug : *Anopheles* mosquitoes
   d. DDT : *Anopheles* mosquitoes
   e. DDT : humans

TEST ANSWERS

TRUE/FALSE

1. F

2. F

3. F

4. F

5. F

6. F

7. T

8. F

9. T

10. T

11. T

12. F

MULTIPLE CHOICE

13. b

14. d

15. d

16. c

17. c

18. b

19. b

20. d

21. a

22. a

23. c

24. c

25. b

26. d

27. d

28. d

29. d

30. c

31. a

32. c

33. b

34. a (macronucleus)

35. b

SHORT ANSWER

36. multicellularity

37. stress

38. eyespots

39. tests

40. pellicle

41. dinoflagellates

42. bilateral

43. sporozoite

44. zoomastigotes

45. pseudopods

46. cilia

47. flagellum or flagella

48. asexually

49. zoospore

50. zygote or zygospore

51. conjugation tube

ESSAY

52. *Trichonympha* lives in the guts of termites, where it produces the enzymes that enable termites to digest wood.

53. Some species of euglenoids are photosynthetic; others lack chloroplasts and are heterotrophic. Some photosynthetic euglenoids may reduce the size of their chloroplasts and become heterotrophic if kept in a dark environment.

54. In the second stage of the life cycle of *Plasmodium*, merozoites reenter the host's bloodstream, invade red blood cells, and divide rapidly. In about 48 hours the blood cells rupture, releasing merozoites and toxic substances throughout the host's body, and initiating a cycle of chills and fever. The cycle repeats itself every 48 hours as new blood cells are infected.

ANALOGIES

55. d (display)

56. e (are found in)

57. b (forms a)

58. a (move by)

59. d (contain)

60. b (are types of)

61. c (is transmitted by)

62. d (is used on)

Chapter 22

TRUE/FALSE Read each statement. If the statement is true, write T in the space provided. If the statement is false, write F in the space provided.

1.______ Most fungi are heterotrophic.

2.______ The cell walls of fungi are made up of cellulose.

3.______ All fungi, except yeasts, have bodies composed of filaments.

4.______ Sporangia are reproductive structures in which spores form asexually.

5.______ Some fungi are predators.

6.______ The Fungi Imperfecti reproduce sexually and asexually.

7.______ Ringworm is caused by a small wormlike animal.

8.______ Mitosis in the tips of ascomycetes hyphae gives rise to conidiophores.

9.______ Some yeasts can cause diseases in humans.

10.______ Lichens have been found all over the world except in arid desert regions.

11.______ Some fungi aid in the transfer of minerals from the soil to the roots of plants.

MULTIPLE CHOICE Circle the letter of the term or phrase that best completes each statement or best answers each question.

12.Fungi obtain food by
a. photosynthesis. c. digesting food and absorbing it.
b. nitrogen fixation. d. None of the above

13.The individual filaments that make up the body of a fungus are called
a. vascular tissue. c. rhizoids.
b. hyphae. d. stems.

14.Chitin is found in fungi and in
a. clam shells. c. cytoplasm.
b. the outer shells of insects. d. snail shells.

15.Most fungi are spread throughout the world by
a. animals. c. birds.
b. water. d. wind.

16.Fungi
a. do not contain chloroplasts.
b. have cell walls that contain chitin.
c. do not produce their own food.
d. All of the above

17.Reproductive structures in which spores form sexually are known as
a. gametophytes. c. yeasts.
b. sporangia. d. None of the above

18. Fungi obtain energy
   a. directly from the sun.
   b. from inorganic material in their environment.
   c. by absorbing organic molecules.
   d. from nuclear fusion.

19. A hypha is a long string of cells divided by
   a. spindle fibers. c. mycorrhizae.
   b. an ascus. d. septa.

20. Fungi digest food
   a. through photosynthesis. c. inside their bodies.
   b. outside their bodies. d. All of the above

21. Reproductive structures in which spores form asexually are known as
   a. septa. c. mycorrhizae.
   b. cnidophores. d. sporangia.

22. Mushrooms and toadstools are members of the phylum
   a. Ascomycota. c. Zygomycota.
   b. Basidiomycota. d. Deuteromycota.

23. An example of a fungus is
   a. a mushroom. c. *Penicillium.*
   b. a bread mold. d. All of the above

24. The Fungi Imperfecti
   a. cause skin diseases in humans. c. are used to ferment soy sauce.
   b. do not reproduce sexually. d. All of the above

25. All of the following are true of ascomycetes except that
   a. they have saclike reproductive cells in which spores grow.
   b. some are truffles.
   c. they are often used in the baking of bread.
   d. they are imperfect fungi.

26. Molds that are used in the manufacture of cheese are
   a. club fungi. c. slime molds.
   b. sac fungi. d. imperfect fungi.

27. An economically important use of fungi is
   a. bread making.
   b. the production of antibiotics.
   c. the manufacture of alcoholic beverages.
   d. All of the above

28. Zygospores allow molds to
   a. survive unfavorable conditions. c. grow unusually large.
   b. change to algae. d. produce antibiotics.

29. Mushrooms and shelf fungi are examples of
   a. club fungi. c. molds.
   b. sac fungi. d. imperfect fungi.

30. A lichen is
   a. an organism consisting of a fungus and an alga.
   b. a sac fungus.
   c. a protozoan living as a parasite on a fungus.
   d. a mold found on certain rocks and sand.

31. Mycorrhizae
   a. aid in the transfer of minerals from the soil to a plant.
   b. cause a variety of plant diseases.
   c. aid in the transfer of minerals to fungi.
   d. are only found on aquatic fungi.

32. Mycorrhizae that wrap around the root of a plant are called
   a. endomycorrhizae.
   b. ascomycorrhizae.
   c. ectomycorrhizae.
   d. lichens.

The following question(s) refer(s) to the illustration of the fungus.

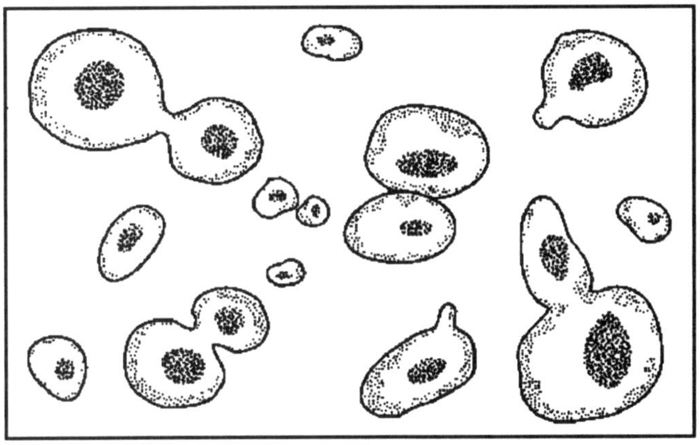

33. The organism illustrated is a(n)
   a. mycorrhizae.
   b. Zycomycota.
   c. Fungi Imperfecti.
   d. unicellular fungus.

34. This organism can reproduce
   a. by fission.
   b. by forming buds.
   c. sexually.
   d. All of the above

The following question(s) refer(s) to the diagram that illustrates the life cycle of a mold.

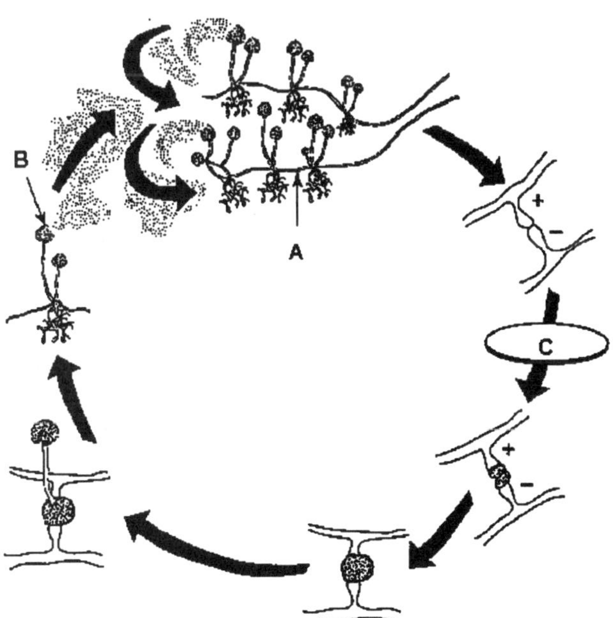

35. Structure A is

a. a rhizoid.
b. vascular tissue.
c. a hypha.
d. a stem.

36. Structure B is

a. commensal.
b. haploid.
c. embryonic.
d. diploid.

37. The process that takes place at structure C is known as

a. meiosis.
b. conjugation.
c. mitosis.
d. fusion.

SHORT ANSWER Complete each statement by writing the correct term or phrase in the space provided.

38. Fungi produce sexually and ______________________.

39. When hyphae grow, they form a mass called a(n) __________________.

40. Fungi obtain food by ______________________ organic matter.

41. Mildews and yeasts are examples of ______________________.

42. A fungal ______________________ is a haploid reproductive cell that is capable of developing into a new organism.

43. The common name of unicellular fungi is ______________________.

44. The common mushroom belongs to a group called the ________________.

45. Chestnut blight and Dutch elm disease are caused by ________________.

46. A(n) ______________________ is a saclike structure in which haploid spores are formed.

47. A lichen consists of a fungus and an ______________________ living together in a symbiotic relationship.

48. Lichens are sensitive to ______________________ such as sulfur dioxide in the atmosphere.

49. Mycorrhizae that do not physically penetrate a plant root are known as ______________________.

50. Lichens are able to carry out ______________________ fixation.

ESSAY Read and answer each question or statement on a separate sheet of paper.

51. How does mitosis in a mushroom differ from that in most other eukaryotic organisms?

52. Why are fungi well suited for absorbing food from the environment?

53. Why are lichens a good indictor of the environment's health?

54. How did mycorrhizae help plants invade the land?

ANALOGIES An analogy is a relationship between two pairs of words or phrases written as: a:b::c:d. The symbol : is read "is to," and the symbol :: is read "as." Write the letter of the pair of terms that best completes the analogy shown.

55. mushroom : long, slender filaments ::
   a. chitin : cellulose
   b. fungus cell wall : chitin
   c. pea plant cell wall : chitin
   d. fungus cell wall : cellulose
   e. insect exoskeleton : cellulose

___

56. fungal food : organic molecules ::
   a. pea plant cell wall : chitin
   b. insect exoskeleton : cellulose
   c. fungus cell wall : cellulose
   d. mycelium : hyphae
   e. chitin : cellulose

___

57. sporangia : reproductive structures ::
   a. gametangia : reproductive structures
   b. hyphae : reproductive structures
   c. septae : filaments
   d. mycelia : sporangia
   e. fungi : plants

___

58. Ascomycota : phylum that includes yeasts ::
   a. *Penicillium* : yeast that makes alcohol
   b. Deuteromycota : phylum that includes toadstools
   c. Zygomycota : phylum that includes mushrooms
   d. *Rhizopus* : common black bread mold
   e. Basidiomycota : phylum that includes athlete's foot fungus

___

59. commercial mushrooms : food for humans ::
   a. yeasts : fermenting cheese
   b. mushrooms : genetic engineering
   c. yeasts : genetic engineering
   d. black bread mold : producing alcoholic beverages
   e. puffballs : flavoring cheese

___

60. algae : layers of a lichen's hyphae ::
   a. bridges : rivers
   b. pages : covers of a book
   c. locks : key rings
   d. baseballs : baseball bats
   e. calculators : harmonicas

___

61. sperm : gamete ::

a. gamete : sex
b. gamete : sperm
c. spore : haploid cell
d. spore : diploid cell
e. chromosome : chromatid

---

62. fungi : algae in lichens ::

a. algae : plant roots
b. plant roots : lichens
c. mycorrhizae : plant roots
d. fungi : chlorophyll
e. chlorophylls : bare soil

---

63. plant roots : soil ::

a. lichens : plant roots
b. plant roots : lichens
c. endomycorrhizae : plant roots
d. ectomycorrhizae : plant roots
e. ectomycorrhizae : rocks

TEST ANSWERS

TRUE/FALSE

1. F

2. F

3. T

4. F

5. T

6. F

7. F

8. T

9. T

10. F

11. T

MULTIPLE CHOICE

12. c

13. b

14. b

15. d

16. d

17. b

18. c

19. d

20. b

21. b

22. b

23. d

24. d

25.d

26.d

27.d

28.a

29.a

30.a

31.a

32.c

33.d

34.d

35.c

36.b

37.d

SHORT ANSWER

38.asexually

39.mycelium

40.decomposing

41.ascomycetes or sac fungi

42.spore

43.yeasts

44.basidiomycetes or club fungi

45.Ascomycota

46.ascus

47.alga

48.pollutants

49.ectomycorrhizae

50.nitrogen

ESSAY

51. Fungi exhibit nuclear mitosis. In dividing cells, the nuclear envelope remains intact from prophase to anaphase. Spindle fibers form within the nucleus, dragging chromosomes to opposite poles of the cell. Mitosis is completed when the nuclear envelope pinches in two.

52. The mycelium of a fungus can be made of many meters of individual hyphae. This creates a high surface-area-to-volume ratio, which makes a fungus well suited for absorbing food from the environment.

53. Lichens are particularly sensitive to pollutants in the atmosphere because they readily absorb substances dissolved in rain and dew. Pollutants from cars and industrial activity can destroy a lichen's chlorophyll, decreasing its rate of photosynthesis and killing the lichen.

54. Fossil evidence indicates that the earliest land plants often had endomycorrhizae. The soil at that time was completely lacking in organic matter. Mycorrhizal plants are particularly successful in infertile soil.

ANALOGIES

55. b (is made of)

56. d (is made out of)

57. a (are types of)

58. d (is the name of the)

59. c (are used for)

60. b (are found between)

61. c (is a type of a)

62. c (have a symbiotic relationship with)

63. c (penetrate the)

Chapter 23

TRUE/FALSE Read each statement. If the statement is true, write T in the space provided. If the statement is false, write F in the space provided.

1.______ The life cycle of nonvascular plants is called homosporus alternation of generations.

2.______ The sporophyte of a nonvascular plant is larger and lives longer than the gametophyte.

3.______ In gymnosperms, each pollen grain consists of a sperm surrounded by a unicellular capsule.

4.______ Vascular green plants were the first organisms to convert oxygen in the atmosphere into the ozone.

5.______ Gymnosperm seeds are enclosed in a fruit.

6.______ Pines and all the other conifers have needle-like leaves.

7.______ Ferns need water to reproduce because their sperm must swim to eggs.

8.______ In ferns, the antheridia and archegonia are produced by different individuals.

9.______ Seeds remain dormant until moisture and temperature conditions favor seedling growth.

10.______ Plants with colorful flowers are pollinated most commonly by wind.

11.______ Pollen tubes grow through the style toward the ovule.

12.______ The fusion of a sperm with two nuclei in the ovule produces the endosperm of a seed.

MULTIPLE CHOICE Circle the letter of the term or phrase that best completes each statement or best answers each question.

13. The waxy protective covering of a land plant is called a
   a. cuticle.   c. rhizome.
   b. capsule.   d. stoma.

14. In plants, growing regions where cells divide are called
   a. sporophytes.   c. rhizomes.
   b. meristems.   d. androecia.

15. The xylem in a plant
   a. transports food from the leaves.
   b. transports water and minerals.
   c. exchanges carbon dioxide with the atmosphere.
   d. All of the above

16. Stomata are responsible for
   a. meiosis.
   b. growth at the meristems.
   c. allowing carbon dioxide to enter a leaf.
   d. transport of minerals.

17. In which of the following structures do liverworts and mosses produce eggs?
   a. antheridia
   b. capsules
   c. archegonia
   d. cones

18. The guard cells that make up a stoma
   a. have rigid walls.
   b. expand and contract.
   c. do not contain chlorophyll.
   d. develop into phloem.

19. The structures on a moss gametophyte in which spores are produced are known as
   a. meristems.
   b. pollen grains.
   c. cones.
   d. sporangia.

20. Fern sporophytes consist of rhizomes and
   a. flowers.
   b. stems.
   c. microgametophytes.
   d. fronds.

21. The mature microspore of a seed plant is called a
   a. microgametophyte.
   b. sporophyte.
   c. pollen grain.
   d. sporangium.

22. The tallest trees in the world are species of
   a. conifers.
   b. dicots.
   c. liverworts.
   d. angiosperms.

23. Immediately following pollination,
   a. the seed develops.
   b. an egg cell is formed.
   c. fertilization occurs.
   d. the pollen tube begins to form.

24. The bottom whorl of modified leaves that protects the developing flower is made up of
   a. receptacles.
   b. petals.
   c. sepals.
   d. carpels.

25. Double fertilization has great survival value because
   a. two embryos are formed.
   b. if one egg dies, the other can live.
   c. it produces very large plants.
   d. the embryo has its own temporary source of food.

26. The primary purpose of the fruit is
   a. to provide nutrition for the seed.
   b. photosynthesis.
   c. seed dispersal.
   d. to permit cross-fertilization.

27. Pollen is produced in a structure called the
   a. anther.
   b. ovary.
   c. stigma.
   d. style.

28. If a plant's flower is very colorful and produces nectar, it is probably pollinated by
   a. water.
   b. wind.
   c. insects.
   d. self-pollination.

29. Removing a flower's stigma would initially affect
   a. fertilization.
   b. seed production.
   c. pollination.
   d. seed dispersal.

30. Monocots have
   a. leaves with branching veins.
   b. flower parts in multiples of four or five.
   c. leaves with parallel veins.
   d. two cotyledons.

31. After an egg has been fertilized, a fruit develops from the
   a. style.
   b. ovary.
   c. ovule.
   d. sepal.

32. The growth and development of a seed is referred to as
   a. elongation.
   b. germination.
   c. fertilization.
   d. pollination.

33. Cross-pollination
   a. occurs only in gymnosperms.
   b. tends to increase genetic variation.
   c. occurs only in angiosperms.
   d. causes the production of sporangia.

The following question(s) refer(s) to the diagram of the plant life cycle.

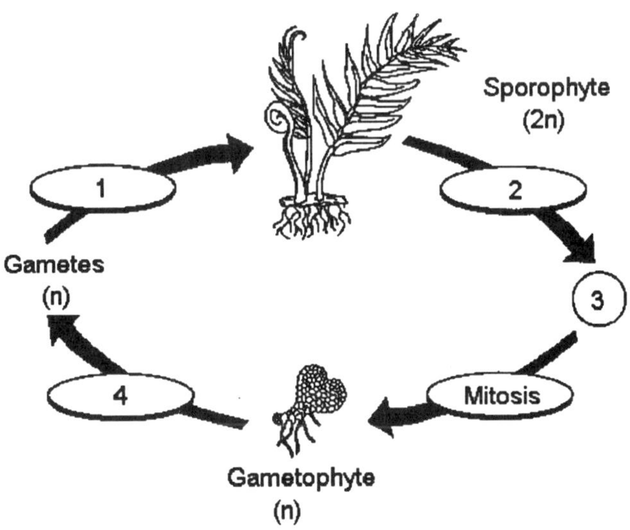

34. At which point in the life cycle does fertilization take place?
   a. 1
   b. 2
   c. 3
   d. 4

35. The structures produced at 3 are called
   a. sporangia.
   b. archegonia.
   c. spores.
   d. antheridia.

36. Haploid gametes are produced as a result of
   a. fertilization.
   b. meiosis.
   c. encapsulation.
   d. mitosis.

The following question(s) refer(s) to the diagram of the flower.

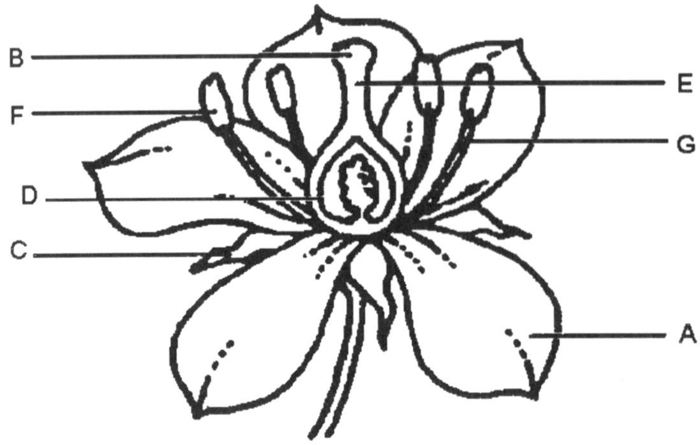

37. The structure indicated at D
    a. supports the anther.
    b. produces pollen.
    c. has tiny structures that look like leaves.
    d. develops into a fruit.

38. The structure labeled B
    a. produces pollen.
    b. contains sperm cells.
    c. is sticky to the touch.
    d. contains meristematic tissue.

39. Which of the following is part of the corolla?
    a. A
    b. B
    c. E
    d. G

SHORT ANSWER Complete each statement by writing the correct term or phrase in the space provided.

40. In leaves, the expansion and contraction of the guard cells regulate the opening and closing of the ________________________.

41. In mosses and liverworts, the ________________________ generation is the dominant generation.

42. The tissues that transport water and minerals within a plant make up the ________________________ system.

43. In alternation of generations, the ________________________ generation alternates with the diploid generation.

44. In most vascular plants, the ________________________ grows within the sporophyte.

45. Because flowering plants are rooted in the ground and cannot move from place to place, they must disperse their ________________________ so that their offspring can grow in new environments.

46. The life cycle of a conifer is characterized by a large ________________________.

47. An angiosperm that has flower parts in multiples of four or five is a(n) ________________________.

48. Seeds develop from the ______________________ after an egg has been fertilized.

49. The transfer of pollen grains from the anther to the stigma is known as ______________________ .

50. A flower that has all four whorls of appendages is called a(n) ______________________ flower.

51. Seeds are typically ______________________ before they germinate and the embryonic tissues continue their development into roots, stems, and leaves.

52. Many fruits are spread by ______________________ that are attracted to sweet, fleshy fruits, which they use for food.

The following question(s) refer(s) to the diagram of the seed.

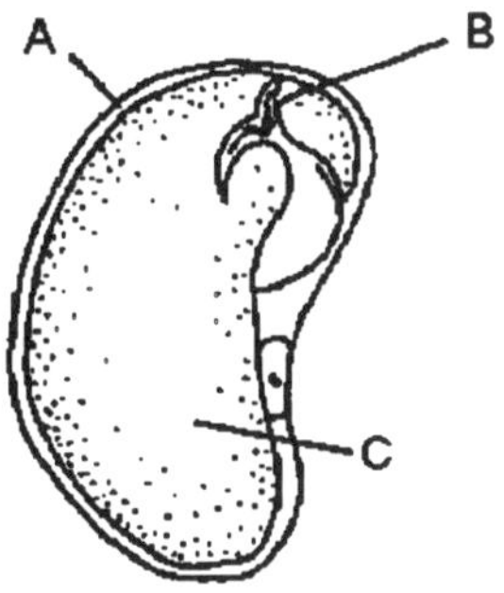

53. The structure labeled A is called the ______________________ .

54. Structure B is the ______________________ .

55. Structure C is the ______________________ .

ESSAY Read and answer each question or statement on a separate sheet of paper.

56. What adaptations had to occur in primitive plants before their descendants could live on land?

57. Describe the characteristics of flowers that attract pollinators.

58. Why is insect pollination more efficient than wind pollination?

ANALOGIES An analogy is a relationship between two pairs of words or phrases written as: a:b::c:d. The symbol : is read "is to," and the symbol :: is read "as." Write the letter of the pair of terms that best completes the analogy shown.

59. sporophytes : spores::
   a. sporophytes : gametophytes
   b. gametophytes : gametes
   c. spores : gametophytes
   d. gametes : sporophytes
   e. gametophytes : sporophytes

____

60. vascular plants : sporophytes::
   a. sporophytic plants : gametophytes
   b. gametes : sporophytes
   c. spores : gametes
   d. plants with seeds : seedless plants
   e. nonvascular plants : gametophytes

____

61. antheridia : sperm::
   a. rhizoids : gametes
   b. seeds : sporophytes
   c. archegonia : eggs
   d. sporophytes : gametophytes
   e. sperm : megaspores

____

62. gymnosperms : fruit::
   a. bakeries : meat
   b. desert islands : sand
   c. ladders : rungs
   d. airplanes : wings
   e. televisions : screens

____

63. gymnosperm : naked seed::
   a. pollen : gamete
   b. liverwort : come onto land
   c. liverwort : vascular tissue
   d. horsetail : many seeds
   e. angiosperm : case seed

____

64. calyx : sepals::
   a. androecium : fourth whorl
   b. gynoecium : third whorl
   c. whorl : sepals
   d. petal : calyx
   e. corolla : petals

____

65. imperfect flower : gynoecium or androecium::
   a. perfect flower : gynoecium
   b. perfect flower : androecium
   c. complete flower : petals
   d. incomplete flower : petals
   e. incomplete flower : color

____

66. fruit : mature ovary::
   a. gametophyte : sporophyte
   b. gymnosperm : angiosperm
   c. cotyledon : food reserve
   d. vascular plant : nonvascular plant
   e. sperm : egg

---

67. mycorrhizae : phosphorus and other minerals::
   a. fungi : fats and proteins
   b. plant : carbohydrates and other organic compounds
   c. arthropods : shelter and food
   d. mutualism : symbiosis
   e. arthropods : vertebrates

---

68. liverworts, hornworts, mosses : nonvascular plants::
   a. gymnosperms, angiosperms : bryophytes
   b. ferns : mosses
   c. gymnosperms and angiosperms : vascular plants
   d. bryophytes, liverworts : vascular plants
   e. gymnosperms : angiosperms

TEST ANSWERS

TRUE/FALSE

1.F

2.F

3.F

4.F

5.F

6.F

7.T

8.F

9.T

10.F

11.T

12.T

MULTIPLE CHOICE

13.a

14.b

15.b

16.c

17.c

18.b

19.d

20.d

21.c

22.a

23.d

24.c

25. d

26. c

27. a

28. c

29. c

30. c

31. b

32. b

33. b

34. a

35. c

36. d

37. d

38. c

39. a

SHORT ANSWER

40. stomata

41. gametophyte

42. vascular

43. haploid

44. gametophyte

45. seeds

46. sporophyte

47. dicot

48. ovules

49. pollination

50. complete

51. dormant

52. animals

53. seed coat

54. embryo

55. food supply

ESSAY

56. Before their descendants could survive in terrestrial habitats, primitive plants had to overcome three obstacles. They had to be able to absorb minerals from rocky surfaces, conserve water, and to reproduce on land.

57. Some flowers have attractive petals or scents. Others secrete a sugary nectar that pollinators use as food. For example, bees locate sources of nectar largely by odor at first, then by homing in on a flower's color and shape. Hummingbirds are attracted to red flowers, and bats are attracted to heavily scented and pale colored flowers that open at night.

58. Insects and plants have coevolved so that certain insects are attracted by particular flowers. A particular insect carries pollen from the flowers of one individual to the flowers of another individual of the same species. Thus, the flowers of insect-pollinated plant species do not need to produce as much pollen as the flowers of wind-pollinated species.

ANALOGIES

59. b (produce)

60. e (are dominated by)

61. c (produce)

62. a (do not have)

63. e (means)

64. e (consists of)

65. d (lacks)

66. c (is a)

67. b (provide)

68. c (are)

Chapter 24

TRUE/FALSE Read each statement. If the statement is true, write T in the space provided. If the statement is false, write F in the space provided.

1.______ In plants, xylem tissue is alive at maturity.

2.______ The main function of ground tissue is to conduct water and minerals.

3.______ The outer protective layer of tissue on a vascular plant is known as the meristem.

4.______ Tracheids and sieve tubes make up a xylem vessel.

5.______ The blade of a compound leaf is made up of leaflets.

6.______ Maple flowers develop several weeks after the leaves appear.

7.______ Sugars are converted to starches in roots.

8.______ The rate of water absorption in roots is influenced by the amount of water lost through transpiration.

9.______ The loss of water vapor from a plant can be explained by the pressure-flow model.

10.______ In a plant, sugar moves from where it is made to where it is needed through the lenticels.

11.______ The movement of sugar in a plant is called transpiration.

MULTIPLE CHOICE Circle the letter of the term or phrase that best completes each statement or best answers each question.

12. The ground tissue in plants that is made up of chloroplast-rich cells is called the
    a. vascular bundle.   c. pith.
    b. petiole.   d. mesophyll.

13. The phloem in a plant
    a. transports sugars.
    b. transports water and minerals.
    c. exchanges carbon dioxide and oxygen with the atmosphere.
    d. None of the above

14. In xylem tissue, water moves from tracheid to tracheid through thin, porous
    a. pits.   c. sieve tubes.
    b. vessel elements.   d. companion cells.

15. Bark contains
    a. xylem and phloem.   c. phloem and cork cells.
    b. sapwood.   d. mesophyll.

16. The conducting cells of phloem are called
    a. tracheids.   c. sieve plates.
    b. sieve tube members.   d. vessel elements.

17. The tissue of the leaf mesophyll that is located directly below the upper epidermis and consists of tightly packed column-shaped cells is the
   a. palisade layer. c. adventitious layer.
   b. cortex. d. pith.

18. Leaves connect to the stems of plants at the
   a. lateral buds. c. nodes.
   b. pith. d. internodes.

19. Lenticels
   a. permit air to pass through the bark.
   b. conduct water to the leaves.
   c. conduct water to the roots.
   d. are scars from stems that have fallen off.

20. The outer layers of ground tissue in a stem are known as the
   a. sapwood. c. pith.
   b. nodes. d. cortex.

21. The root cap
   a. protects the growing root cap. c. absorbs water.
   b. produces the bark. d. contains meristematic tissue.

22. The primary function of root hairs is
   a. support. c. absorption of water and minerals.
   b. transport of food substances. d. water storage.

23. The ground tissue in the center of roots and stems
   a. turns into meristem. c. provides support.
   b. transports food. d. germinates at least once a year.

24. The guard cells that surround a stoma
   a. have rigid walls.
   b. swell with water, causing the stoma to open.
   c. shrivel up to open the stoma.
   d. are responsible for translocation.

25. Turgor pressure refers to
   a. internal water pressure. c. internal air pressure.
   b. external water pressure. d. external water and air pressure.

26. The transport of sugar from the leaf to the rest of the plant is called
   a. translocation. c. active transport.
   b. osmosis. d. transpiration.

27. Transpiration is the process by which
   a. water is transported. c. seeds are produced.
   b. food is transported. d. water vapor is lost by a plant.

28. The movement of water through a plant is explained by
   a. the attraction of water molecules for each other.
   b. capillary action.
   c. osmosis.
   d. All of the above

29. One model that explains the movement of sugar in a plant is known as the
   a. transpiration model. c. pressure-flow model.
   b. translocation model. d. source-sink model.

30. The most commercially valuable tree in North America is the
a. white ash.
b. American basswood.
c. tulip poplar.
d. sugar maple.

31. The seeds of the sugar maple mature inside fruits that are carried away by
a. squirrels.
b. birds.
c. the wind.
d. pollinators.

The following question(s) refer(s) to the diagram showing a portion of a plant's vascular system.

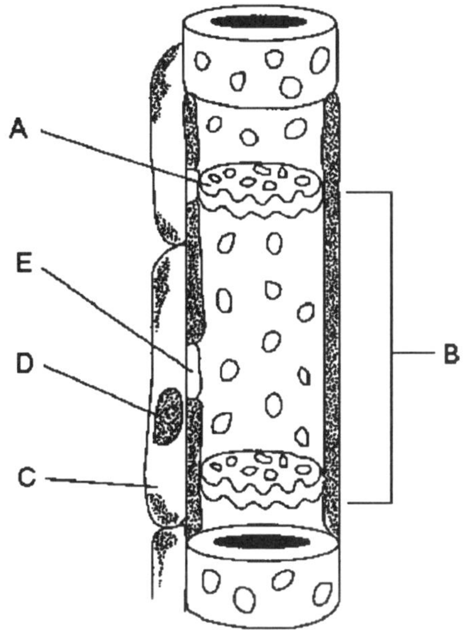

32. Structure B is known as a
a. tracheid.
b. companion cell.
c. vessel element.
d. sieve tube member.

33. Structure C is a
a. tracheid.
b. companion cell.
c. vessel element.
d. sieve tube member.

34. Which structures allow the cytoplasm of neighboring cells to connect with each other?
a. A
b. C
c. D
d. E

The following question(s) refer(s) to the diagram of the leaf cross section.

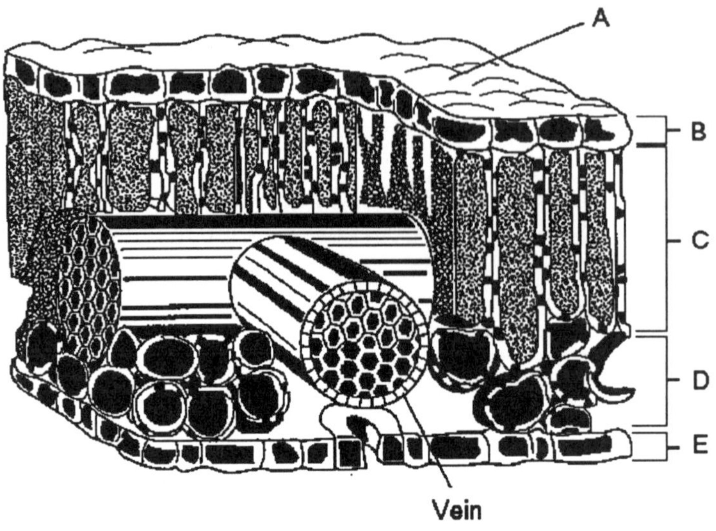

35. The spongy layer is indicated at
   a. B.   c. D.
   b. C.   d. E.

36. The vein illustrated is made up of
   a. only xylem vessels.   c. both xylem and phloem vessels.
   b. only phloem vessels.   d. neither xylem nor phloem vessels.

37. Structure A
   a. is the cuticle.   c. covers the epidermis.
   b. protects the leaf.   d. All of the above

The following question(s) refer(s) to the diagram of four different kinds of leaves.

38. Which of the following leaves is a compound leaf?
   a. Leaf A   c. Leaf B
   b. Leaf C   d. Leaf D

SHORT ANSWER Complete each statement by writing the correct term or phrase in the space provided.

39. The broad, flat portion of a typical leaf is called the ______________________.

40. The two main types of root systems are fibrous root systems and ______________________ systems.

41. Wood consists primarily of ______________________ cells.

42. ______________________ are narrow, elongated, thick-walled cells that taper at each end.

43. Plants that fix carbon only in the Calvin cycle are known as ______________________ plants.

44. The ______________________ on a plant stem grow into the branches of the stem.

45. Flexible, soft, and usually green stems are known as ______________ stems.

46. The underground stem of a potato that functions in food storage is called a(n) ______________________ .

47. The darker wood in the center of a tree trunk is called ______________________.

48. A layer of cells called ______________________ tissue covers and protects the plant body.

49. Because of adhesion, water is able to move up the side of a narrow tube by means of ______________________.

50. When the guard cells of a stoma fill with water, the stoma ______________________.

51. ______________________ ions play an important role in opening and closing stomata.

ESSAY Read and answer each question or statement on a separate sheet of paper.

52. Describe the functions of ground tissue in a vascular plant.

53. Describe how stomata open and close.

54. Define the terms *source* and *sink* in relation to the transportation of food in the phloem of plants.

55. Compare the movement of sugar and water in a plant.

ANALOGIES An analogy is a relationship between two pairs of words or phrases written as: a:b::c:d. The symbol : is read "is to," and the symbol :: is read "as." Write the letter of the pair of terms that best completes the analogy shown.

56. vascular tissue: transport of fluids::
   a. epidermis : support
   b. dermal tissue : storage
   c. dermal tissue : transport of fluids
   d. ground tissue : photosynthesis
   e. ground tissue : reducing water loss

____

57. cuticle : aboveground parts::
   a. vascular system : plant
   b. sperm : pollen
   c. guard cell : stoma
   d. wax : root system
   e. gametophytes : sporophytes

____

58. C_4 plant : photosynthesis at high temperatures and light ::

a. epidermis : storage
b. spongy layer : movement of gases and water vapor
c. palisade layer : air spaces
d. C_3 plants : Calvin cycle to produce 4C compound
e. C_4 plant : C_3 plant

---

59. cohesion : adhesion::

a. hydrogen : polar
b. book : pages
c. hand : people
d. H_2O : water
e. night : day

---

60. adventitious roots : support::

a. fibrous roots : photosynthesis
b. root hairs : surface area and absorption
c. cork cells : transport of fluids
d. lenticels : storage
e. tap roots : root caps

---

61. watery solution in sugar maple : sap::

a. flower with petals : incomplete
b. wavy grain pattern : "sapping season"
c. sugar maple winged fruit : nut
d. sugar maple wooden design : "bird's eye"
e. leaf : lenticel

---

62. water : sugar in a plant::

a. printing press : poison ivy
b. playing card : a table
c. farm crop : plow
d. snow plow : snow
e. snow : aspirin

---

63. adhesion : water molecules stick to other substances::

a. transpiration : water leaves through root hairs
b. transpiration : stoma absorb sugar
c. photosynthesis : transpiration of water vapor
d. capillary action : root caps remove water
e. cohesion : water molecules stick to each other

---

64. guard cell : stoma::
   a. can opener : can
   b. ticket : nail
   c. spice : rocking chair
   d. potholder : stream
   e. telegram : fish

---

65. source : sugar production::
   a. stored sugar : narrow tube
   b. narrow tube : stored sugar
   c. source : root tips and fruits
   d. sink : a leaf
   e. sink : delivered sugar

TEST ANSWERS

TRUE/FALSE

1. F

2. F

3. F

4. F

5. T

6. F

7. T

8. T

9. F

10. F

11. F

MULTIPLE CHOICE

12. d

13. a

14. a

15. c

16. b

17. a

18. c

19. a

20. d

21. a

22. c

23. c

24. b

25. a

26. a

27. d

28. d

29. c

30. d

31. c

32. d

33. b

34. a

35. c

36. c

37. d

38. c

SHORT ANSWER

39. blade

40. taproot

41. xylem

42. Tracheids

43. C_3

44. lateral buds

45. herbaceous

46. tuber

47. heartwood

48. dermal

49. capillary action

50. opens

51.Potassium

ESSAY

52.Much of the body of a vascular plant is made up of ground tissue. Ground tissue performs photosynthesis, stores water and carbohydrates, assists in transport, and surrounds and supports the conducting tissues.

53.Plants control water loss by opening and closing their stomata. Each stoma is formed by two guard cells that are shaped like two cupped hands. The walls of guard cells are flexible and unevenly thickened. When guard cells take in water by osmosis, they become turgid and bowed in shape. When the guard cells lose water, they collapse, and the stoma closes.

54.Sugar is the food that moves in the phloem of stems. Movement of sugar is from the source, where it is made, to the sink, where it is used or stored.

55.Water only moves upward within the xylem, while sugar moves up and down in the same sieve tube, but at different times. Water diffuses freely through a plasma membrane, but sugar cannot.

ANALOGIES

56.d (carries out)

57.c(is on the outside of)

58.b (enables)

59.e (is the opposite of)

60.b (are used for)

61.c (is called)

62.d (moves)

63.e (means that)

64.a (can open a)

65.e (is the location of)

Chapter 25

TRUE/FALSE Read each statement. If the statement is true, write T in the space provided. If the statement is false, write F in the space provided.

1.______ A plant embryo's shoot develops above the cotyledons, and its roots develop below the cotyledons.

2.______ Germination will be inhibited if water is allowed to penetrate the seed coat.

3.______ Primary plant growth occurs in apical meristems located at the tips of stems and roots.

4.______ Secondary tissues result from the extension of apical meristems.

5.______ Plant development is considered reversible because cells of certain tissues can produce unspecialized cells that can then differentiate into mature plant tissues.

6.______ Perennials can return from dormancy each year only because they grow anew from seeds left from the previous year.

7.______ In order to avoid extinction, all successful plant species must produce flowers and seeds during each and every growing season.

8.______ Auxin moves toward the lighted side of a stem, causing the cells on that side to elongate.

9.______ Gibberellins are hormones that can stimulate seed germination.

10.______ All parts of a plant exhibit both positive phototropism and positive gravitropism.

11.______ A long period of darkness induces flowering in short-day plants.

MULTIPLE CHOICE Circle the letter of the term or phrase that best completes each statement or best answers each question.

12. After double fertilization occurs within the ovary of a flower,
   a. two zygotes form.
   b. the zygote becomes endosperm.
   c. the endosperm becomes a zygote.
   d. both the zygote and endosperm undergo mitosis.

13. A mature seed forms
   a. when tissues surrounding the zygote and endosperm toughen.
   b. when tissues surrounding the zygote and endosperm become impermeable to water and oxygen.
   c. after the embryo stops growing.
   d. All of the above

14. Once the food stored in a bean's cotyledons is used up,
   a. the cotyledons begin photosynthesizing.
   b. the cotyledons shrivel up and fall off.
   c. the cotyledons become leaves.
   d. new endosperm is produced.

15. Which of the following may be needed in order for a seed to germinate?
   a. exposure to temperature changes
   b. seed coat damage
   c. penetration of water and oxygen through the seed coat
   d. All of the above

16. During periods of primary growth at apical meristems, stems and roots
   a. become wider.
   b. become longer.
   c. maintain a constant number of cells.
   d. undergo differentiation.

17. In a woody stem, cork cambium
   a. forms phloem.
   b. forms xylem.
   c. produces the cells of the outer bark.
   d. becomes vascular cambium.

18. Secondary xylem and phloem form from
   a. cork cambium.
   b. vascular cambium.
   c. apical meristems.
   d. bark.

19. A plant that completes its life cycle in a single growing season is called a(n) ___________ plant.
   a. annual
   b. perennial
   c. biennial
   d. perpetual

20. Which of the following is *not* true of a biennial plant?
   a. They take two years to complete their life cycle.
   b. Most biennial plants are weeds.
   c. Energy is stored in roots and shoots during the winter.
   d. Flowering usually occurs only during the second growing season.

21. Photosynthesis enables plants to produce most of the organic molecules that they need. This process requires the use of all of the following, *except*
   a. carbon dioxide.
   b. water.
   c. light.
   d. glucose.

22. A plant's roots
   a. generally protrude into the air to absorb oxygen.
   b. carry out photosynthesis.
   c. absorb oxygen from spaces between soil particles.
   d. produce oxygen during photosynthesis.

23. Charles and Francis Darwin observed that
   a. the weight of a clear gelatin cap on the tip of a seedling caused the shoot to bend in response to gravity.
   b. light influenced the tip of a seedling to bend toward a light source.
   c. exposure to light caused a seedling's shoots to shorten, resulting in a bending toward the light source.
   d. seedlings bend away from a light source in order to avoid excessive heat.

24. A plant hormone that is produced primarily in root tips is
   a. auxin.
   b. cytokinin.
   c. ethylene.
   d. gibberellin.

25. Apical dominance
   a. is caused by gibberellin in terminal buds.
   b. stimulates terminal buds to elongate.
   c. is the inhibition of lateral bud growth by auxin.
   d. results from gardeners cutting terminal buds in growing plants.

26. Which of the following hormones normally exists in a gaseous state?
a. auxin
b. cytokinin
c. ethylene
d. gibberellin

27. All of the following generally cause tropisms in plants, *except*
a. light.
b. gravity.
c. touch.
d. food.

28. When vines grow, they often wrap tendrils around objects for support. These plants most likely "know" the appropriate place to wrap the tendrils because of
a. phototropism.
b. gravitropism.
c. thigmotropism.
d. chance.

29. A seed or plant may remain inactive for a period of time when it is
a. dormant.
b. day-neutral.
c. photoperiodic.
d. gravitropic.

30. Plant development is very different from animal development because
a. a plant continues to develop throughout its life.
b. plants are more strongly influenced by the environment.
c. plant development can be reversed.
d. All of the above

31. Kalanchoës are succulents, meaning that they
a. have fleshy leaves and stems that store water.
b. have a large taproot that stores water.
c. require a large amount of water from the soil.
d. have stomata that open only in the daytime.

The following question(s) refer(s) to the diagram of a woody stem.

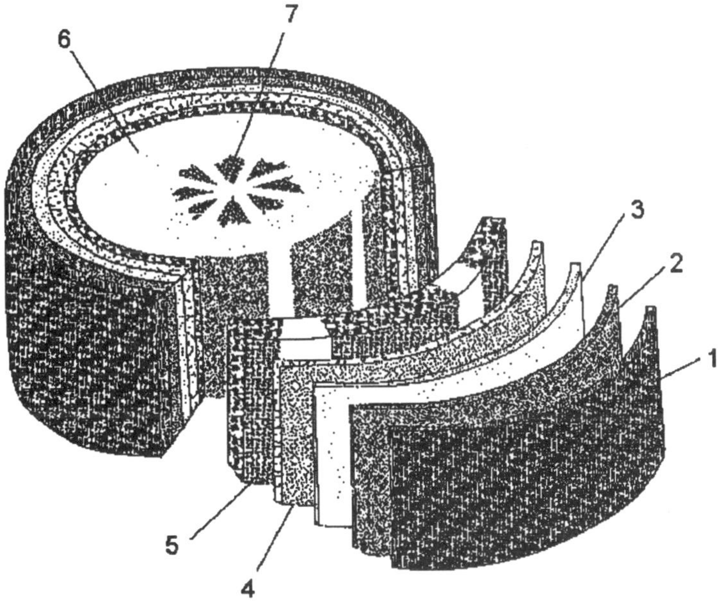

32. Secondary phloem is indicated by the cells labeled as
a. layer 1.
b. layer 3.
c. layer 5.
d. layer 7.

33.The cells that produce the secondary xylem are labeled as

a. layer 2.
b. layer 3.
c. layer 4.
d. layer 5.

34.Which tissue is responsible for transporting sugar from leaves to roots?

a. layer 1
b. layer 3
c. layer 5
d. layer 7

35.Outer bark is produced by

a. layer 2.
b. layer 4.
c. layer 6.
d. None of the above; the outer bark is inorganic material that sticks to the surface.

The following question(s) refer(s) to the illustration of a newly germinated seedling.

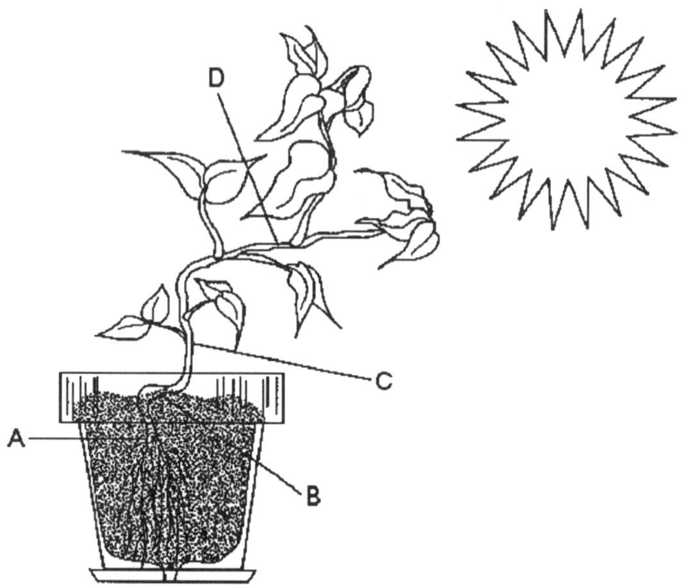

36.Which part of the plant indicates a positive phototropism?

a. A
b. B
c. C
d. D

37.Which part of the plant indicates negative geotropism?

a. A
b. B
c. C
d. D

SHORT ANSWER Complete each statement by writing the correct term or phrase in the space provided.

38.Resumption of growth by a plant embryo is called ________________.

39.Areas of a plant in which growth occurs by cell division are called ________________.

40.Growth that occurs from the formation of new cells at the tip of a plant is called ________________.

41. Plant tissues that result from primary growth are called ________________________.

42. The process by which cells become specialized in form and function is called ________________________.

43. Growth that causes a plant to increase in width is called ________________________.

44. Cell division in the part of the plant called the __________________ adds layers of new cells around the outside of a plant's body.

45. A lateral meristem that produces the cork cells of the outer bark is called ________________________.

46. A lateral meristem that produces secondary vascular tissue is called the ________________________.

47. Plants that lose their leaves at the end of each growing season, only to grow new ones the following year, are called ____________________.

48. Plants that may drop some leaves throughout the year but that retain most of their leaves year-round are ________________________.

49. A plant hormone that causes elongation of plant cells by enabling them to stretch during cell growth is named ________________________.

50. A(n) ________________________ is a chemical produced in one part of an organism and transported to another part of the organism, where it causes a response.

51. Inhibition of lateral bud growth by auxin is called ________________.

52. A hormone that can be used commercially to promote fruit-ripening is ________________________.

53. A growth response in plants in which the direction of growth is determined by the direction from which a stimulus comes is called a(n) ________________________.

54. The response of plants to changes in the lengths of days and nights is called ________________________.

55. Kalanchoës and other succulents that fix carbon by a water-conserving pathway are called ________________________ plants.

ESSAY Read and answer each question or statement on a separate sheet of paper.

56. Secondary growth adds width to a woody stem. Briefly describe the tissues involved and explain how they increase the stem's diameter.

57. Describe why plants may be more influenced by environmental factors than are animals.

58. Describe how auxin causes a plant to bend toward a light source.

ANALOGIES An analogy is a relationship between two pairs of words or phrases written as: a:b::c:d. The symbol : is read "is to," and the symbol :: is read "as." Write the letter of the pair of terms that best completes the analogy shown.

59. secondary growth : width ::
   a. secondary growth : primarily
   b. lateral meristem tissue : primarily
   c. apical meristem tissue : secondarily
   d. primary growth : length
   e. lateral growth : up

____

60. plants : no movement ::
   a. plant developments : little influence from the environment
   b. plant movements : no stimulation from the environment
   c. animals : continual development
   d. animals : movement
   e. animal developments : strong influences from the environment

____

61. biennials : two years ::
   a. perennials : more than two years
   b. perennials : less than two years
   c. annuals : two years
   d. annuals : more than two years
   e. sunflowers : more than two years

____

62. leaves : carbon dioxide from air ::
   a. leaves : water from the air
   b. roots : light from the air
   c. roots : carbon dioxide from the air
   d. leaves : nutrients from the air
   e. roots : nutrients from the soil

____

63. light, CO_2, and H_2O : carbohydrates ::
   a. photosynthesis : light
   b. light : a nutrient
   c. nutrient : light
   d. nutrients : macromolecules
   e. photosynthesis : nutrients

____

64. shoots : negative gravitropism ::
   a. light stimulations : gravitropisms
   b. gravity movements : phototropism
   c. shoots : negative phototropism
   d. roots : positive phototropism
   e. roots : positive gravitropism

____

65. xylem : inner side of vascular cambium ::
   a. vascular cambium : cork cambium
   b. cork : vascular cambium
   c. phloem : outer side of vascular cambium
   d. phloem : cork cambium
   e. vascular cambium : outer side of cork cambium

---

66. some plant genes : be reactivated ::
   a. animal development : be reversed
   b. animal genes : be reactivated
   c. animal genes : not be reactivated
   d. "turned off" animal genes : be "turned on"
   e. unspecialized plant cells : not resume differentiation

---

67. deciduous trees : at the end of the growing season ::
   a. evergreens : every two years
   b. evergreens : a few at a time throughout the year
   c. annuals : every two years
   d. annuals : throughout the year
   e. biennials : every year

TEST ANSWERS

TRUE/FALSE

1. T
2. F
3. T
4. F
5. T
6. F
7. F
8. F
9. T
10. F
11. T

MULTIPLE CHOICE

12. d
13. d
14. b
15. d
16. b
17. c
18. b
19. a
20. b
21. d
22. c
23. b
24. b

25. c

26. c

27. d

28. c

29. a

30. d

31. a

32. b

33. c

34. b

35. a

36. d

37. c

SHORT ANSWER

38. germination

39. meristems

40. primary growth

41. primary tissues

42. differentiation

43. secondary growth

44. lateral meristem

45. cork cambium

46. vascular cambium

47. deciduous

48. evergreens

49. auxin

50. hormone

51. apical dominance

52. ethylene

53. tropism

54. photoperiodism

55. CAM, or crassulacean acid metabolism

ESSAY

56. Lateral meristems called cork and vascular cambia produce tissues that contribute to the diameter of a woody stem. Cork cambium produces cells that become part of the outer bark of the plant. Secondary vascular tissues form on opposite sides of the vascular cambium. Secondary xylem forms on the inner side of the vascular cambium, while secondary phloem forms on the outer surface of the vascular cambium, becoming part of the inner bark.

57. Since plants cannot move around the way animals can, plants are likely to be intimately tied to their immediate surroundings. As the local environment changes in such aspects as light, humidity, temperature and nutrient availability, plants may respond physiologically, such as by entering or leaving dormancy, dropping or growing leaves, etc. Animals, on the other hand, are able to move to new environments when conditions change, thus reducing the influence of the environment.

58. The plant hormone auxin causes cells to stretch during growth, resulting in elongation of the cells. Under the influence of light, auxin in the plant's terminal bud moves from the lighted side of a shoot to the unlighted side. Cells in the area of the shoot beneath the auxin begin to become longer. This uneven growth of the cells in the shoot causes bending toward the light; cells on the unlighted side grow slightly longer than those on the lighted side.

ANALOGIES

59. d (is an increase in)

60. d (have)

61. a (live for)

62. e (take in)

63. d (are needed to produce)

64. e (show)

65. c (forms on the)

66. c (can)

67. b (drop their leaves)

Chapter 26

TRUE/FALSE Read each statement. If the statement is true, write T in the space provided. If the statement is false, write F in the space provided.

1.______ If a plant produces more sugar during photosynthesis than it can immediately use, it may serve as a source of food for animals.

2.______ Grains are dry, edible fruits produced by cereal grasses.

3.______ Processing rice to make white rice removes many vitamins.

4.______ Whole wheat is prepared by finely grinding the wheat endosperm from wheat grains.

5.______ Meat is called a "complete protein" because it contains all of the amino acids required in the human diet.

6.______ Genetic engineers are attempting to transfer genes for the production of essential amino acids from one species of cereal grain to another.

7.______ Rosy periwinkle, a common garden plant, is the source of two major cancer-treatment drugs.

8.______ Cortisone, a drug used to treat inflammation and allergies, is derived from yew trees.

9.______ Most of the rubber being made today is manufactured from naturally obtained latex.

10.______ Cotton cloth is made from thread prepared by spinning strong, fine fibers obtained from cotton seeds.

MULTIPLE CHOICE Circle the letter of the term or phrase that best completes each statement or best answers each question.

11. A product of photosynthesis on which humans and other animals depend for energy is
   a. carbon dioxide. c. nucleic acid.
   b. protein. d. sugar.

12. Which of the following plant parts may serve as food?
   a. leaves c. roots
   b. stems d. All of the above

13. The edible, dry fruits produced by cereal grasses are called
   a. grains. c. brans.
   b. pollens. d. flakes.

14. In the fruit of a cereal grass, the bran
   a. is dry and papery. c. includes the ovary wall and seed coat.
   b. covers the grain. d. All of the above

15. Which of the following is *not* an important cereal crop?
   a. corn c. soybeans
   b. rice d. wheat

16. Brown rice is considered to be more nutritious than white rice because
   a. brown rice has protein-rich soy sauce on it.
   b. white rice has no starch.
   c. white rice has had the bran layers removed.
   d. None of the above; both types of rice are essentially the same.

17. Wheat germ consists of
   a. embryos of wheat grains.
   b. endosperm of wheat grains.
   c. the bran of wheat grains.
   d. kernels of whole wheat.

18. The only major grain crop that originated in the Americas is
   a. wheat.
   b. rice.
   c. corn.
   d. None of the above; all grains were originally imported to the Americas.

19. Plant proteins are said to be "incomplete proteins" because
   a. the amino acids are lacking amino groups.
   b. they lack one or more essential amino acids.
   c. there are no amino acids in them.
   d. they contain only the nine essential amino acids, rather than all 20 amino acids.

20. Plants that have nitrogen-fixing bacteria living in their roots
   a. are called legumes.
   b. capture nitrogen from the air for use in making proteins.
   c. are important sources of dietary protein.
   d. All of the above

21. In order to make certain foods more nutritious, genetic engineers are attempting to transfer genes coding for essential amino acids
   a. into legumes.
   b. from rice to corn.
   c. among different species of wheat.
   d. from legumes to cereal grains.

22. A person choosing to follow a vegetarian diet can obtain a complete array of proteins by eating
   a. a variety of legumes.
   b. an assortment of cereal grains.
   c. a combination of legumes and cereal grains.
   d. beef raised on a diet of corn and soybeans.

23. Legumes are able to survive well in nitrogen-poor soil because
   a. they draw nitrogen from competing cereal crops.
   b. nitrogen is absorbed directly into cells from the air.
   c. nitrogen is not needed for the production of protein by legumes.
   d. root nodules contain symbiotic bacteria that fix nitrogen.

24. Important root crops include all of the following, *except*
   a. carrots.
   b. cassava.
   c. peanuts.
   d. beets.

25. A good reason why a farmer might alternately grow a grain crop and a legume crop in a single field is to
   a. create a varied and more attractive landscape.
   b. allow the legumes to restore nitrogen to the soil that is depleted by the grain crop.
   c. satisfy the changing needs of the consuming public.
   d. get twice as much productivity from a single piece of land.

26.Wood pulp is used to make all of the following, *except*
a. lumber. c. rayon.
b. paper. d. cardboard.

27.Recycling paper can be very expensive because
a. the manufacture of recycled paper uses much more wood.
b. recycled paper does not contain cellulose fibers.
c. the removal of inks and dyes is a costly process.
d. recycling requires that twice as many trees be chopped down as when the original paper was produced.

28.Cancer-treatment drugs originally derived from plants include all of the following, *except*
a. vincristine. c. aspirin.
b. vinblastine. d. taxol.

29.Taxol, a drug extracted from yew trees,
a. is used to treat asthma.
b. eliminates high blood pressure.
c. is used to treat certain forms of cancer.
d. provides relief from pain.

30.Which of the following is *improperly* paired?
a. aspirin--willow tree c. digitalis--foxglove
b. vincristine--rosy periwinkle d. taxol--sweet potato

31.Rubber that is used in a variety of manufacturing processes
a. can be produced from latex.
b. is often manufactured synthetically from petroleum.
c. is an important raw material that is extracted from a plant.
d. All of the above

32.Bread wheat is not a diploid (2*n*) plant, but a
a. haploid (*n*) plant. c. hexaploid (6*n*) plant.
b. triploid (3*n*) plant. d. octaploid (8*n*) plant.

The following question(s) refer(s) to the drawings of three plant structures.

A

B

C

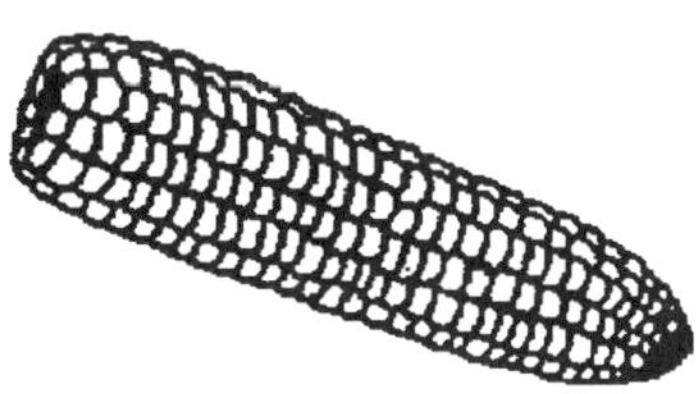

33. Which of the plants shown would contain nitrogen-fixing bacteria?
   a. plant A
   b. plant B
   c. plant C
   d. None of them would have nitrogen-fixing bacteria.

34. The plant that would be highest in protein content is
   a. plant A.
   b. plant B.
   c. plant C.
   d. None of them; all plants are high in carbohydrates, but very low in protein.

35. The plant that is a grain crop is
   a. plant A.
   b. plant B.
   c. plant C.
   d. None of them is a grain crop.

36. Which plant is a root crop?
   a. plant A
   b. plant B
   c. plant C
   d. None of them is a root crop.

SHORT ANSWER Complete each statement by writing the correct term or phrase in the space provided.

37. Botanically speaking, a(n) ______________________ is the ripened ovary of a flower.

38. Any nonreproductive part of a plant is referred to as a(n) ______________________ part of the plant.

39. Grasses grown as food for humans and livestock are called __________.

40. The ______________________ is a dry, papery husk that covers a grain.

41. The embryos of wheat grains are referred to as the ________________.

42. Many Middle Eastern recipes use cracked wheat grains called ______________________.

43. The human body can only make 11 of the 20 amino acids that are needed to make proteins. The remaining 9 are called ______________________ and must be eaten as part of the diet.

44. Plants, such as peas, that produce protein-rich seeds in long pods are called ______________________.

45. Legumes may thrive in nitrogen-poor soil because they contain root nodules in which there are ______________________.

46. Certain plants, such as potatoes, have modified underground stems called ______________________, which store starch.

47. ______________________ is stored in root crops, making them an important source of calories for human populations.

48. Salicin is an extract of willow bark and is used in the production of ______________________.

49. The milky white sap of tropical rubber trees is called ______________________.

50. Bread wheat is a hybridized species produced naturally from ______________________ related species.

51. An important evolutionary tool that enables fertile hydbrids to result from crosses among different species is called ____________________.

ESSAY Read and answer each question or statement on a separate sheet of paper.

52. How can a person following a vegetarian diet be sure of obtaining all of the amino acids necessary for protein production?

53. Explain why crop rotation is used as a method to increase crop yields.

54. In spite of the widely recognized importance of recycling paper, landfills still contain tremendous amounts of wastepaper. Why is it not commercially feasible to recycle paper more extensively?

ANALOGIES An analogy is a relationship between two pairs of words or phrases written as: a:b::c:d. The symbol : is read "is to," and the symbol :: is read "as." Write the letter of the pair of terms that best completes the analogy shown.

55. rice and corn : cereal::
  a. beets and wheat : leaves
  b. sweet potatoes and cassava : roots
  c. turnips and radishes : stems
  d. wheat and corn : roots
  e. corn and rice : leaves

____

56. meat : cereals::
  a. rice : beans
  b. corn : peas
  c. wheat : peanuts
  d. a legume : rice
  e. beet : soybeans

____

57. wood : wood pulp::
  a. wrinkle : door
  b. milk : cheese
  c. weapon : herb
  d. canoe : violin
  e. horse : paragraph

____

58. cloth : cotton::
  a. magnet : a shelf
  b. staple : an estate
  c. notecard : a lagoon
  d. mask : a termite
  e. pottery : clay

____

59. rice from 10,000 years ago : modern-day rice::
   a. wild wheat : whole wheat
   b. teosinte : modern-day corn
   c. bulgar : modern agricultural wheat
   d. modern-day rice : grass
   e. corn pone : rice

---

60. building construction : lumber::
   a. development of medicines : cellulose fibers
   b. furniture : wood pulp
   c. lumber : cotton
   d. fuel : wood
   e. linen : bamboo

---

61. vincristine : rosy periwinkle::
   a. cortisone : foxglove
   b. digitalis : willow trees
   c. reserpine : sweet potatoes
   d. taxol : yew trees
   e. birth control pill : wormwood shrub

---

62. cotton : fibers ::
   a. cotton : rubber
   b. latex : cotton
   c. fiber : sap
   d. cotton : latex
   e. rubber : sap

---

63. white flour : wheat endosperm::
   a. whole wheat : endosperm, germ, and bran
   b. white flour : bran
   c. wheat germ : endosperm
   d. whole wheat : wheat bran and germ
   e. wheat germ : pasta

TEST ANSWERS

TRUE/FALSE

1.T

2.T

3.T

4.F

5.F

6.F

7.T

8.F

9.F

10.T

MULTIPLE CHOICE

11.d

12.d

13.a

14.d

15.c

16.c

17.a

18.c

19.b

20.d

21.d

22.c

23.d

24.c

25. b

26. a

27. c

28. c

29. c

30. d

31. d

32. c

33. b

34. b

35. c

36. a

SHORT ANSWER

37. fruit

38. vegetative

39. cereals

40. bran

41. wheat germ

42. bulgur

43. essential amino acids

44. legumes

45. nitrogen-fixing bacteria

46. tubers

47. Starch

48. aspirin

49. latex

50. closely

51. polyploidy

ESSAY

52. Humans can produce only 11 of the 20 amino acids needed for protein production; the remaining nine essential amino acids are a necessary part of the diet. While meat provides complete proteins, with all the required amino acids, a vegetarian must be sure to eat a variety of plant types. Since plants provide incomplete proteins, lacking one or more essential amino acids, vegetarians must eat a combination of grains and legumes to obtain complete proteins.

53. Alternating grain and legume crops in a single field is known as crop rotation. The principle behind this farming method is that legumes, which contain nitrogen-fixing bacteria, convert atmospheric nitrogen into nutrient molecules that plants can use. These nitrogen compounds are used by the legume crop and also remain in the soil as nutrients that can be used by the successive grain crop.

54. Paper is recycled by dissolving it in water, removing inks and dyes, and re-pressing the fibers into new paper. Ink and dye removal is a very expensive process, and the cost can often exceed the expense of making new paper from wood pulp.

ANALOGIES

55. b (are examples of)

56. d (has more protein than)

57. b (can be used to make)

58. e (is made from)

59. b (developed into)

60. d (is a major use of)

61. d (is made from)

62. e (comes from)

63. a (is made from)

Chapter 27

TRUE/FALSE Read each statement. If the statement is true, write T in the space provided. If the statement is false, write F in the space provided.

1.______ The development of a particular animal's body plan is a function of the animal's environment, rather than of its genetic heritage.

2.______ Without exception, all animals are heterotrophs.

3.______ The majority of animal species are classified as invertebrates.

4.______ Although animal life is believed to have evolved originally in the sea, the vast majority of animals today are terrestrial.

5.______ In most animal species, the gametes are the only haploid cells in the life cycle.

6.______ The evolutionary significance of flatworms is that they were the first animals to have both a mouth for ingesting food and an anus for eliminating waste.

7.______ The fluid-filled pseudocoelom of nematodes serves as a simple circulatory system.

8.______ The development of a true coelom between the ectoderm and the mesoderm represents a true evolutionary advance over the pseudocoelomate body plan.

9.______ Unlike arthropods, mollusks are unsegmented; instead, they are divided into three regions.

10.______ Segmentation occurs in the bodies of higher vertebrates (annelids, arthropods, and echinoderms), but is not present in chordates.

11.______ Jellyfish and echinoderms both exhibit radial symmetry, demonstrating that they are clearly close relatives.

12.______ All chordates are classified as vertebrates.

MULTIPLE CHOICE Circle the letter of the term or phrase that best completes each statement or best answers each question.

13. Which of the following is *not* a characteristic of animals?
   a. All are heterotrophic.
   b. External digestion is more common than internal digestion.
   c. Their bodies are multicellular.
   d. They are able to move in order to catch food.

14. The cells of animals are
   a. similar to plants in that they have a cell wall.
   b. always organized into tissues.
   c. usually quite flexible.
   d. usually multi-nucleated.

15. A most striking characteristic of animals that distinguishes them from the members of the other kingdoms is that members of the animal kingdom
   a. are capable of more complex and rapid movements than the other kingdoms.
   b. possess cells with rigid cell walls.
   c. contain choanocytes within their tissues.
   d. include both aquatic and terrestrial species.

16. The presence of flight is
   a. observed in at least some species of all kingdoms.
   b. unique to certain animals.
   c. restricted to certain invertebrate species.
   d. blamed for the extinction of pterosaurs.

17. Sponges
   a. are nonsymmetrical.
   b. lack organization into tissues and organs.
   c. possess cells that are capable of recognizing other sponge cells.
   d. All of the above are true.

18. Choanocytes of a sponge
   a. create a current that is necessary for the sponge to obtain food.
   b. form primitive tissues.
   c. are the equivalent of sperm cells of higher animals.
   d. None of the above is true.

19. The three embryonic tissue layers of all eumetazoans include all of the following *except*
   a. ectoderm. c. mesoderm.
   b. endoderm. d. pyroderm.

20. Which of the following is an *incorrect* match?
   a. ectoderm—nervous system c. endoderm—muscular system
   b. mesoderm—skeletal system d. endoderm—digestive system

21. Radially symmetrical phyla include all of the following *except*
   a. jellyfish. c. hydra.
   b. comb jellies. d. sponges.

22. Nematocysts
   a. contain harpoonlike structures called cnidocytes.
   b. create water currents in sponges.
   c. can spear a sea anemone's prey.
   d. are found in most predatory eumetazoans.

23. A characteristic shared between all fungi and cnidarians is
   a. the possession of nematocysts. c. tentacles for catching food.
   b. extracellular digestion. d. radial symmetry.

24. Which of the following has radial symmetry?
   a. flatworm c. echinoderm larva
   b. annelid d. hydra

25. The acoelomate body type is exemplified by
   a. flatworms. c. mollusks.
   b. roundworms. d. annelids.

26. The first organisms to develop an internal body cavity were the
   a. flatworms. c. mollusks.
   b. nematodes. d. arthropods.

27. The evolution of body cavities was important because
   a. fluids within the body cavity aid in circulation of materials from one part of the body to another.
   b. fluids in the cavity make the body rigid and offer resistance to muscles, aiding in movement.
   c. organs are better able to function if they can move freely within the body cavity.
   d. All of the above are true.

28. Which of the following is *correctly* paired?
   a. phylum Platyhelminthes—hydra
   b. phylum Nematoda—planaria
   c. phylum Mollusca—octopus
   d. phylum Annelida—roundworm

29. As animals evolved, the coelom
   a. became the cavity within the digestive tract.
   b. developed between the mesoderm and the endoderm.
   c. became located between the ectoderm and the mesoderm.
   d. developed completely within the mesoderm.

30. All of the animal phyla that evolved after the mollusks
   a. are vertebrates.
   b. are prokaryotes.
   c. have a coelom.
   d. lack mesoderm.

31. Each of the following groups of invertebrates is a coelomate, *except*
   a. annelids.
   b. echinoderms.
   c. mollusks.
   d. nematodes.

32. All of the following are classes of the phylum Mollusca, *except*
   a. bivalves.
   b. cephalopods.
   c. gastropods.
   d. pseudopods.

33. A characteristic structure found in many mollusks is the radula, which is involved in
   a. jet propulsion.
   b. opening and closing of the shells.
   c. eating.
   d. reproduction.

34. Segmented worms are known as
   a. nematodes.
   b. annelids.
   c. planarians.
   d. arthropods.

35. A similarity between annelids and arthropods is that they both have
   a. exoskeletons made of chitin.
   b. the ability to fly.
   c. segmented body patterns.
   d. well-developed lungs for respiration.

36. Jointed appendages of arthropods may
   a. become specialized for particular functions.
   b. function in locomotion.
   c. function in feeding.
   d. All of the above are true.

37. Characteristics of the arthropods include
   a. segmentation.
   b. a chitinous exoskeleton.
   c. jointed appendages.
   d. All of the above

38. The first organisms to develop a hardened endoskeleton were the
   a. echinoderms.
   b. annelids.
   c. arthropods.
   d. chordates.

39. A characteristic shared by all chordates is
a. a dorsal hollow nerve cord.
b. a notochord.
c. pharyngeal gill slits.
d. All of the above

40. All vertebrates
a. have a backbone.
b. are craniate.
c. Both a and b are correct.
d. None of the above are correct.

The following question(s) refer(s) to the diagrams of a fish and a jellyfish.

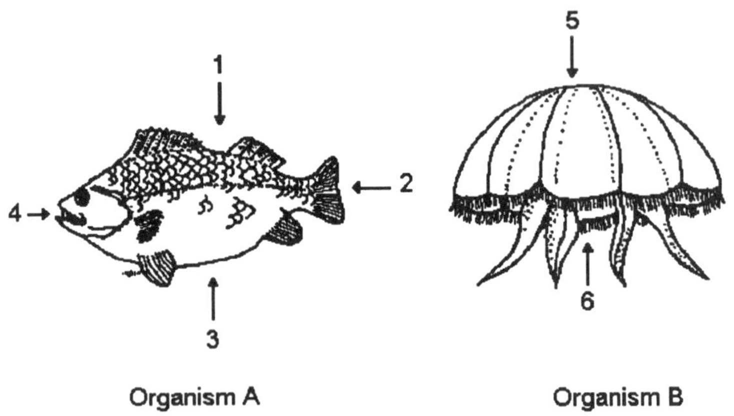

41. Organism B is
a. radially symmetrical.
b. bilaterally symmetrical.
c. unilaterally symmetrical.
d. nonsymmetrical.

42. The position of the arrow labeled ____________ can be referred to as anterior.
a. 1
b. 4
c. 5
d. 6

SHORT ANSWER Complete each statement by writing the correct term or phrase in the space provided.

43. Since animals cannot make their own food before digesting it, they are said to be ____________ .

44. Since sponges do not have body parts that grow around a central point as do all other animals, the sponges are said to lack ____________.

45. Water is moved through the body of a sponge by the action of flagellated cells called ____________.

46. "True animals," which have tissues and symmetry, are called ____________.

47. Animals that have their body parts arranged around a central point are said to exhibit ____________.

48. A free-floating, gelatinous body form found in some cnidarians is called a(n) ____________, while an attached body form in these animals is called a(n) ____________.

49. Organisms that have left and right halves that mirror each other when divided by an imaginary longitudinal plane are said to have ____________ symmetry.

50. The evolution of a definite head end of an organism is called ______________________.

51. A(n) ______________________ is a collection of different tissues that work together as a unit to perform a particular function.

52. The ______________________ is a fluid-filled cavity that develops within the mesoderm of higher invertebrates and vertebrates.

53. The development of specialized body tissues as a result of contact between the primary embryonic tissues is called ____________________.

54. An organ known as the ______________________ is a rasping, tonguelike structure found in some mollusks.

55. In one anterior segment of annelids there is a(n) __________________, which is the brain of these organisms.

56. A fluid-filled system of interconnected internal canals and tube feet in echinoderms is called the ______________________.

ESSAY Read and answer each question or statement on a separate sheet of paper.

57. Compare the symmetry of body forms observed in the phylum Porifera and the phylum Cnidaria.

58. From a developmental point of view, what is the advantage of the coelomate body plan over the pseudocoelomate body plan?

59. Using examples, distinguish between protostomes and deuterostomes.

ANALOGIES An analogy is a relationship between two pairs of words or phrases written as: a:b::c:d. The symbol : is read "is to," and the symbol :: is read "as." Write the letter of the pair of terms that best completes the analogy shown.

60. body of radially symmetrical animal : central point ::
   a. greeting card : dump truck
   b. toaster : fireplace
   c. shoestring : street sign
   d. fringe : light socket
   e. wheel : central axis

____

61. sponges : fewer than three body layers ::
   a. eumetazoans : no body symmetry
   b. sponges : bilateral symmetry
   c. cnidarians : extracellular digestion
   d. cnidarians : choanocytes
   e. sponges : nematocysts

____

62. bilaterally symmetrical animals : sides ::
a. bilaterally symmetrical animals : ends
b. bilaterally symmetrical animals : body layers
c. radially symmetrical animals : surfaces
d. radially symmetrical animals : pores
e. radially symmetrical animals : body cavities

---

63. first airplane : jumbo jet ::
a. three dermal tissues : two dermal tissues
b. a two-way gut : a gut with one opening
c. coelomate : acoelomate
d. evolution of pseudocoelom : evolution of coelom
e. extracellular digestion : intracellular digestion

---

64. roundworm : a pseudocoelom ::
a. radially symmetrical animal : a coelom
b. sponge : a pseudocoelom
c. sponge : a coelom
d. flatworm : absence of a coelom
e. flatworm : a pseudocoelom

---

65. mollusks : a mantle ::
a. pseudocoelomate animals : true body cavities
b. solid worms : a gut with one opening
c. solid worms : coelom
d. mollusks : a pseudocoelom
e. mollusks : a coelom

---

66. annelids : pseudocoeloms ::
a. mollusks : repeated segments
b. mollusks : a visceral mass
c. mollusks : a muscular region called a foot
d. annelids : connections between segments
e. annelids : cerebral ganglia

---

67. arthropod exoskeletons : muscle attachment ::
a. antennae : walking or flying
b. mouthparts : flying or walking
c. arthropod appendages : walking or chewing
d. arthropod muscles : impeding water loss
e. chitinous materials : muscles

---

68. mouth of deuterostome : second ::
   a. anus of protostome : first
   b. mouth of protostome : first
   c. blastopore of echinoderms and chordates : into mouth
   d. blastopore of mollusks and arthropods : into anus
   e. protostome : a mouth, second

---

69. echinoderm endoskeleton : a hard shell that encases the body ::
   a. chordate : a water vascular system
   b. chordate : a hard exoskeleton
   c. chordate endoskeleton : a stiff rod along the back
   d. echinoderm : a notochord
   e. echinoderm : pharyngeal gill slits

TEST ANSWERS

TRUE/FALSE

1.F

2.T

3.T

4.F

5.T

6.F

7.T

8.F

9.T

10.F

11.F

12.F

MULTIPLE CHOICE

13.b

14.c

15.a

16.b

17.d

18.a

19.d

20.c

21.d

22.c

23.b

24.d

25. d

26. b

27. d

28. c

29. d

30. c

31. d

32. d

33. c

34. b

35. c

36. d

37. d

38. a

39. d

40. c

41. a

42. b

SHORT ANSWER

43. heterotrophs

44. symmetry

45. choanocytes

46. eumetazoans

47. radial symmetry

48. medusa; polyp

49. bilateral

50. cephalization

51. organ

52. coelom

53. primary induction

54. radula

55. cerebral ganglion

56. water vascular system

ESSAY

57. The members of the phylum Porifera, the sponges, are asymmetrical. In other words, their bodies are not arranged in any particular way around a central point or axis. On the other hand, the animals belonging to the phylum Cnidaria, including the jellyfish, hydra, and sea anemones, are radially symmetrical. Their bodies are arranged around a central point.

58. The development of the coelom totally within the mesoderm allows contact between the mesoderm and the endoderm. Such contact permits primary induction, physical and chemical contact between the cells of the two germ layers, to occur. This results in the development of organ systems more complex than those in pseudocoelomate organisms.

59. Protostomes are animals (including acoelomates, pseudocoelomates, annelids, mollusks, and arthropods) in which the blastopore of the embryo becomes the mouth of the organism. Another pattern of embryonic development characterizes coelomates that evolved after the arthropods. In these animals, the anus develops from or near the blastopore. This pattern of development is seen in echinoderms and chordates.

ANALOGIES

60. e (has parts arranged around a)

61. c (have)

62. a (have two)

63. d (came before)

64. d (is characterized by)

65. e (have)

66. a (lack)

67. c (are used for)

68. b (forms)

69. c (is characterized by)

Chapter 28

TRUE/FALSE Read each statement. If the statement is true, write T in the space provided. If the statement is false, write F in the space provided.

1._______ None of the stages in a sponge life cycle are capable of moving through the environment.

2._______ Choanocytes are the only cells of a sponge that receive nutrients.

3._______ Sponges are filter feeders.

4._______ Sponges are capable of total regeneration, even from the smallest pieces of their bodies.

5._______ Polyps are a body form of cnidarians that are specialized for swimming.

6._______ Cnidocytes function in capturing prey.

7._______ The life cycle of anthozoans includes medusae, planulae, and polyps.

8._______ Tapeworms absorb food from their host's intestine directly through their skin.

9._______ Humans can avoid trichinosis by wearing shoes when they walk through fields.

10._______ Planarians have a branched digestive tract with both a mouth and an anus.

MULTIPLE CHOICE Circle the letter of the term or phrase that best completes each statement or best answers each question.

11. Adult sponges
   a. have body walls with many pores.
   b. possess true tissues.
   c. are active swimmers.
   d. use stinging cells to capture prey.

12. Which of the following is *not* sessile as an adult?
   a. sponge
   b. hydra
   c. sea anemone
   d. Portugese man-of-war

13. In sponges, currents that draw water through the organism are created by
   a. amoebocytes.
   b. choanocytes.
   c. gemmules.
   d. spicules.

14. Cells that move throughout the sponge mesenchyme to deliver food to the organism's cells are
   a. amoebocytes.
   b. choanocytes.
   c. gemmules.
   d. spicules.

15. Water leaves the internal cavity of a sponge through the
   a. food vacuoles.
   b. spicules.
   c. mesenchyme.
   d. osculum.

16. Skeletal support in sponges may be provided by
   a. spicules of calcium carbonate.
   b. spicules of silica.
   c. spongin fibers.
   d. All of the above

17. The gemmules of sponges
   a. create water currents for feeding.
   b. are equivalent to the sperm cells of higher animals.
   c. are equivalent to the egg cells of higher animals.
   d. are necessary for one form of asexual reproduction.

18. Hermaphroditic organisms
   a. reproduce only by asexual means.
   b. produce both eggs and sperm.
   c. have gemmules that are fertilized by amoebocytes.
   d. possess only male amoebocytes.

19. Self-fertilization is avoided in sponges because
   a. there is no way for sperm to enter the sponge.
   b. the sperm do not possess the enzyme needed to penetrate the egg.
   c. sperm and eggs are not produced at the same time.
   d. None of the above is true; self-fertilization is common in sponges.

20. Which of the following is a characteristic associated *only* with cnidarians?
   a. a digestive tract with a single opening
   b. tentacles with stinging cells
   c. choanocytes containing nematocysts
   d. a parasitic life cycle

21. The class of cnidarians that typically live only as polyps is the
   a. Anthozoa.   c. Scyphozoa.
   b. Hydrozoa.   d. None of the above

22. Scyphozoans, such as jellyfish, spend most of their lives as
   a. polyps.   c. corals.
   b. medusae.   d. parasites.

23. The hydra is unique among the hydrozoans because it
   a. is an active swimmer.   c. is strictly a marine species.
   b. feeds without tentacles.   d. has no medusa stage.

24. Planula larvae of hydrozoans
   a. result from fertilization of eggs by sperm.
   b. swim freely through the water.
   c. settle to the ocean bottom and grow into polyps.
   d. All of the above

25. Anthozoans include organisms known as
   a. jellyfish.   c. Portuguese men-of-war.
   b. hydras.   d. sea anemones and corals.

26. Which of the following is *not* a structure formed by corals?
   a. planulo reefs   c. barrier reefs
   b. fringing reefs   d. atolls

27. Flatworms have no need for circulatory and respiratory systems because
   a. the digestive system performs these functions.
   b. their cells are close to the animal's exterior surface.
   c. the spherical body shape allows diffusion of materials into tissues.
   d. the coelom is bathed in blood and oxygen.

28. Schistosomiasis is a disease caused by a
   a. roundworm.   c. cestode.
   b. trematode.   d. planarian.

29. A type of roundworm that lives a parasitic life is
a. *Ascaris*.
b. *Necator*.
c. *Trichinella*.
d. All of the above are parasites.

30. Which of the following statements about tapeworms is *false*?
a. They can infect a person who eats improperly cooked beef.
b. They belong to the genus *Schistosoma*.
c. Tapeworms can grow to be large in human intestines.
d. These flatworms do not have a digestive system.

The following question(s) refer(s) to the diagram of the jellyfish life cycle.

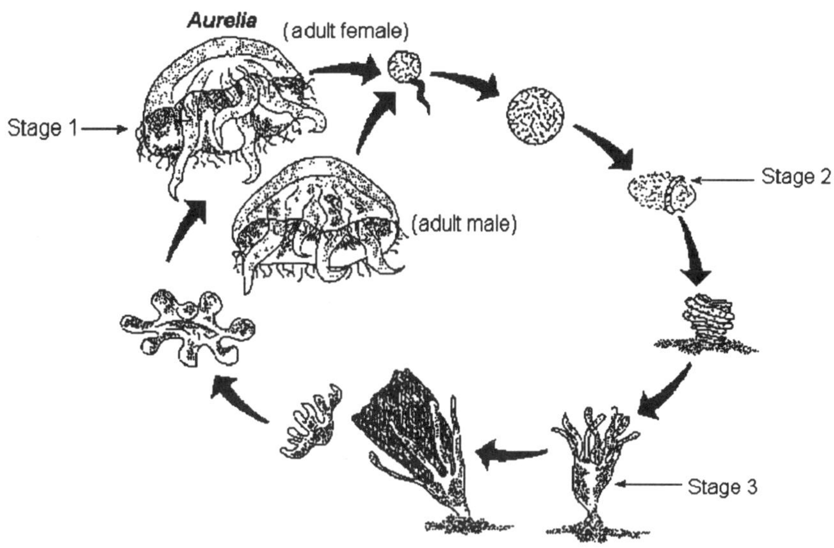

31. Which of the indicated stages reproduces asexually?
a. stage 1
b. stage 2
c. stage 3
d. None of the above; only sexual reproduction occurs.

32. Which stage is called a planula?
a. stage 1
b. stage 2
c. stage 3
d. None of the above is correct.

SHORT ANSWER Complete each statement by writing the correct term or phrase in the space provided.

33. Facing into the internal cavity of a sponge are flagellated collar cells called ______________________.

34. The jellylike middle layer of a sponge is the ______________________.

35. Food molecules are carried throughout a sponge's body by ______________________.

36. ______________________ is drawn into a sponge through pores and leaves through the osculum.

37. A network of tough protein fibers that provide support in some sponges is called ______________.

38. Spicules may be composed of either ______________ or ______________.

39. Food-filled balls of sponge amoebocytes that are involved in asexual reproduction are called ______________.

40. An organism that produces both eggs and sperm is called a(n) ______________.

41. Hydras attach themselves to their substrate by means of a sticky secretion produced by the area called the ______________.

42. The Portuguese man-of-war is a complex colony containing both ______________ and ______________.

43. In *Obelia*, medusae release gametes into the water. After fertilization a(n) ______________ larva is formed.

44. The class of cnidarians that typically have a thick, stalk-like body crowned by tentacles that occur in groups of six is known as the ______________.

45. The ______________ is a thick protective cellular covering of the bodies of endoparasitic flukes that prevents them from being digested by their hosts.

46. ______________ are rectangular body sections of tapeworms.

47. The organisms that are the simplest animals with a complete digestive system consisting of a tube that is open at both ends are ______________.

48. Organisms that feed by the use of sweeping cilia, resembling a rotating wheel, near their mouths are called ______________.

ESSAY Read and answer each question or statement on a separate sheet of paper.

49. Name and describe the three types of sponges.

50. Sea anemones and corals belong to the same group of cnidarians, the class Anthozoa. Distinguish between these two groups of organisms.

51. Describe a major similarity and a major difference between ribbon worms and roundworms.

ANALOGIES An analogy is a relationship between two pairs of words or phrases written as: a:b::c:d. The symbol : is read "is to," and the symbol :: is read "as." Write the letter of the pair of terms that best completes the analogy shown.

52. collar cell : water ::
   a. mesenchyme : wastes
   b. spongin : food
   c. spicule : water
   d. osculum : mesenchyme
   e. amoebocyte : nutrients and wastes

____

53. freshwater sponges : gemmules ::
a. hermaphrodites : eggs and sperm
b. gemmules : eggs
c. gemmules : sperms
d. amoebocytes : eggs
e. amoebocytes : sperms

---

54. *Hydra* : one polyp ::
a. class Hydrozoa : jellyfish
b. Portuguese man-of-war : a colony
c. Cnidaria : 4 classes
d. *Hydra* : medusa stage
e. class Scyphozoa : *Hydra*

---

55. *Obelia* : polyp ::
a. *Hydra* : jellyfish
b. jellyfish : polyp
c. *Aurelia* : colony
d. *Aurelia* : medusa
e. *Aurelia* : polyp

---

56. turbellarians : free living ::
a. planaria : parasitic
b. tapeworms : free living
c. cestodes : free living
d. flukes : parasitic
e. jellyfish : parasitic

---

57. cow infestation by tapeworm : grass contaminated by tapeworm eggs::
a. cow infestation by tapeworm : beef muscle with larval worm
b. cow infestation by tapeworm : infected beef
c. human infestation : cow intestine with tapeworm eggs
d. human infestation : grass with tapeworm eggs
e. human infestation : undercooked beef muscle with larval worms

---

58. ribbon worm : body openings ::
a. roundworm : body openings
b. ribbon worm : pseudocoeloms
c. roundworm : proboscises
d. roundworm : intestines
e. roundworm : poisonous stingers

---

59. *Necator* : hookworm infestation ::
   a. ribbon worm : human waste
   b. *Trichinella* : trichinosis
   c. *Ascaris* : free swimming
   d. *Ascaris* : proglottids
   e. tapeworm : rotifer infestation

____

60. a ribbon worm feeding : proboscis ::
   a. a roundworm feeding : pseudocoelom
   b. a rotifer feeding : tail
   c. a fluke feeding : mouth
   d. a *Dugesia* feeding : pharynx
   e. a *Dugesia* feeding : flame cells

____

61. corals : stonelike outer covering ::
   a. ectoparasites : tegument
   b. endoparasites : tegument
   c. flatworms : circulatory systems
   d. flatworms : respiratory systems
   e. trematodes : large number of free-living species

____

TEST ANSWERS

TRUE/FALSE

1. F
2. F
3. T
4. T
5. F
6. T
7. F
8. T
9. F
10. F

MULTIPLE CHOICE

11. a
12. d
13. b
14. a
15. d
16. d
17. d
18. b
19. c
20. b
21. a
22. b
23. d
24. d

25. d

26. a

27. b

28. b

29. d

30. b

31. c

32. b

SHORT ANSWER

33. choanocytes

34. mesenchyme

35. amoebocytes

36. Water

37. spongin

38. calcium carbonate; silica

39. gemmules

40. hermaphrodite

41. basal disk

42. medusae; polyps

43. planula

44. anthozoans

45. tegument

46. Proglottids

47. ribbon worms

48. rotifers

ESSAY

49. The three types of sponges may be distinguished from each other based on the nature of their skeletons. Calcareous sponges are those that have skeletons composed of calcium carbonate spicules. Sponges with spicules made of silica are called glass sponges. Demosponges, the most common type of sponges, may have skeletons composed of either silica spicules, spongin fibers, or both.

50. While both types of organisms exist only in polyp forms, their adult structures are different. Sea anemones are generally solitary, soft-bodied polyps. Corals, on the other hand, usually live in colonies that are surrounded by a calcium-rich skeleton.

51. These two types of worms are similar in that both have a complete digestive system consisting of digestive tubes that are open at both ends, the mouth and the anus. A major difference between these phyla of worms is that the ribbon worms are acoelomates, while the roundworms are pseudocoelomates.

ANALOGIES

52. e (circulates)

53. a (produce)

54. b (is composed of)

55. d (spends most of its life as a)

56. d (are)

57. e (occurs when they eat)

58. a (has two)

59. b (causes)

60. d (uses its)

61. b (have)

Chapter 29

TRUE/FALSE Read each statement. If the statement is true, write T in the space provided. If the statement is false, write F in the space provided.

1.______ Mollusks and annelids are believed to be the first animals to develop a true coelom.

2.______ Mollusks have four distinct body parts: a foot, a head, a visceral mass, and a tail.

3.______ In terrestrial snails, the mantle cavity functions as a simple lung.

4.______ Mollusks must eat constantly because the functioning of their nephridia causes the disposal of useful molecules as well as of waste products.

5.______ Most bivalves have separate sexes, but some are hermaphroditic.

6.______ Due to the presence of a shell, all gastropods are capable of living on land for at least a portion of their lives.

7.______ Since cephalopods have tentacles, they have no need for a radula.

8.______ The eyes of cephalopods are very similar to the compound eyes of insects.

9.______ While all but one group of mollusks have open circulatory systems, all annelids possess a closed circulatory system.

10.______ Some leeches have become specialized to a parasitic way of life by developing the ability to suck blood from the bodies of other organisms.

MULTIPLE CHOICE Circle the letter of the term or phrase that best completes each statement or best answers each question.

11. Which of the following has a true coelom?
   a. flatworm
   b. roundworm
   c. rotifer
   d. mollusk

12. Trochophore larvae
   a. are found in molluscan life cycles.
   b. possess a belt of cilia around their bodies.
   c. occur in annelid life cycles.
   d. All of the above

13. All of the following are characteristics of mollusks, *except* that they have ______
   a. a pseudocoelomate body plan.
   b. bilateral symmetry.
   c. a complete digestive tract.
   d. an open circulatory system.

14. The cilia of a trochophore larva
   a. allow attachment to the ocean bottom.
   b. create currents for drawing in food.
   c. cover the entire larval body.
   d. are necessary for reproduction.

15. The evolution of a coelom was significant because
   a. more food could be stored within it.
   b. more wastes could be stored before excretion.
   c. it enabled development of more complex organ systems.
   d. it eliminated the need for a circulatory system.

16. In an open circulatory system
   a. water is drawn into the mantle cavity to provide oxygen to body tissues.
   b. lungs branch into small tubules to provide oxygen to tissues.
   c. wastes are eliminated directly to the environment from the tissues.
   d. body tissues are bathed directly in blood.

17. The only group of mollusks that have a closed circulatory system is the
   a. bivalves. c. chitons.
   b. cephalopods. d. gastropods.

18. The most primitive of the following classes of mollusks is the
   a. bivalves. c. chitons.
   b. cephalopods. d. gastropods.

19. Mollusks are believed to have evolved from
   a. primitive segmented worms. c. unsegmented wormlike animals.
   b. sponge-like ancestors. d. shelled sea anemones.

20. Adductor muscles are responsible for
   a. moving the valves of bivalves.
   b. extending the feet of mollusks.
   c. pumping the hearts of mollusks.
   d. fanning the gills of aquatic mollusks.

21. Bivalve gills are covered with a sticky mucus
   a. because they live in polluted waters.
   b. in order to trap small organisms consumed as food.
   c. to protect them from irritation.
   d. in order to cause water to pass over them more rapidly.

22. During reproduction in bivalves, fertilization
   a. occurs internally.
   b. occurs externally.
   c. takes place in the mantle cavity.
   d. does not occur, since bivalves are hermaphroditic.

23. Twisting of the visceral mass of gastropods is called
   a. extension. c. torsion.
   b. inversion. d. conversion.

24. Among the various species of gastropods, respiration may take place
   a. with gills. c. with lungs.
   b. through the skin. d. All of the above

25. All of the cephalopods
   a. have eight tentacles. c. possess protective shells.
   b. are active predators. d. are filter feeders.

26. Jet propulsion in an octopus or squid is the result of
   a. rapid closing of the organism's shell.
   b. strong contractions of the tentacles.
   c. high-pressure "spitting" of fluid from the organism's mouth.
   d. forcefully squeezing water from the siphon.

27. The only cephalopod that has retained its external shell is the
a. cuttlefish.
b. nautilus.
c. octopus.
d. squid.

28. The most significant evolutionary advancement of annelids over mollusks is believed to be
a. the ability to burrow.
b. the existence of a true coelom.
c. segmentation.
d. cephalization.

29. Coordinated movements of an annelid's body segments are possible because of the
a. development of an advanced brain.
b. ventral nerve cord and paired segmental ganglia.
c. presence of a true coelom in the annelid body.
d. existence of a complete digestive system.

30. The digestive tube of annelids is divided into three regions. Which of the following is *not* among these regions?
a. crop
b. intestine
c. radula
d. stomach

31. The advantage of a closed circulatory system over an open circulatory system is that
a. blood moves more efficiently through the tubes of a closed circulatory system.
b. a closed circulatory system prevents blood from leaking out of the body.
c. blood is able to be pumped by a muscular heart in a closed circulatory system.
d. lungs are able to function in animals with a closed circulatory system.

32. Of the following organs, the one that does not exist in annelids is the
a. heart.
b. gills.
c. nephridia.
d. stomach.

33. Parapodia are involved in all of the following *except*
a. swimming.
b. gas exchange.
c. burrowing.
d. reproduction.

34. The clitellum of an earthworm
a. contains the heart.
b. is associated with reproduction.
c. acts as a primitive respiratory system.
d. is necessary for movement.

35. Leeches
a. use suckers to aid in movement.
b. may be parasitic.
c. are segmented.
d. All of the above

The following question(s) refer(s) to the diagram of the clam.

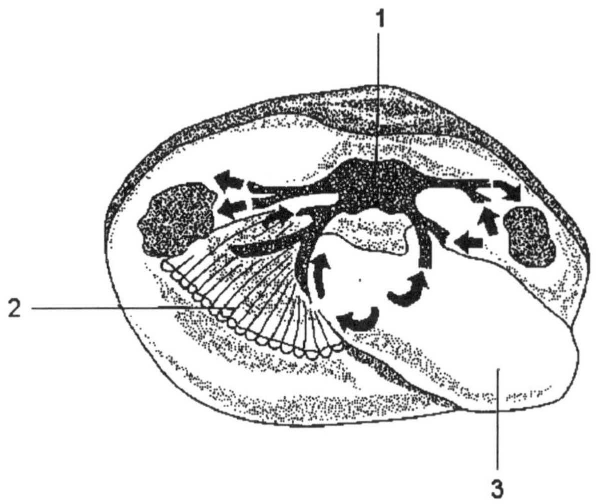

36. The organism shown in this diagram is a
   a. bivalve.   c. chiton.
   b. cephalopod.   d. gastropod.

37. This organism possesses a(n)
   a. simple lung.   c. internal beak.
   b. closed circulatory system.   d. open circulatory system.

38. Movement of this organism is dependent upon the structure labeled
   a. 1.   c. 3.
   b. 2.   d. None of the above

SHORT ANSWER Complete each statement by writing the correct term or phrase in the space provided.

39. The ______________________ larva is a distinguishing characteristic of mollusks and annelids.

40. Constant beating of ______________________ in the mantle cavity of mollusks causes a continuous stream of water to pass over the gills.

41. Structures in mollusks that function as simple kidneys are called ______________________.

42. Of the major classes of mollusks, the one that contains sessile filter-feeders is the ______________________.

43. A clam's shells close when the pair of ______________________ contract.

44. Water is drawn into the bodies of bivalves through tubes called ______________________.

45. The ______________________ larva is a second larval stage that develops from the trochophore larva of bivalves and gastropods.

46. Snails and slugs belong to the molluscan class ______________________.

47. A snail can pull its head into its mantle cavity because the cavity has been moved to the anterior of the animal during a twisting process called ________________________ that occurs during development.

48. A tonguelike scraping organ used by gastropods in feeding is called a(n) ________________________.

49. The foot of cephalopods has been divided into numerous ________________________.

50. The only living cephalopod that has retained its external shell is the ________________________.

51. Bristles that exist along the sides of an annelid are called ________________________.

52. The ________________________ are annelids that are marine and that have many setae and parapodia.

53. A muscular portion of the annelid digestive system that grinds up organic material is called the ________________________.

54. A mucus cocoon that will contain the eggs and sperm and allow development of an earthworm's offspring is produced by the ________________________.

ESSAY Read and answer each question or statement on a separate sheet of paper.

55. Describe the function of molluscan nephridia.

56. How is it possible for a snail to pull its head into its mantle cavity in times of danger?

57. Distinguish between the three classes of annelids.

ANALOGIES An analogy is a relationship between two pairs of words or phrases written as: a:b::c:d. The symbol : is read "is to," and the symbol :: is read "as." Write the letter of the pair of terms that best completes the analogy shown.

58. beating of molluscan ciliated gills : water ::
   a. gliding of molluscan foot : waste fluids
   b. gliding of molluscan foot : blood
   c. pumping of molluscan heart : blood
   d. pumping of molluscan heart : water
   e. filtering of nephridia : mucus

____

59. true coelom : mollusks and annelids ::
   a. tentacle : annelids
   b. siphon : annelids
   c. valve : annelids
   d. segmentation : annelids
   e. muscular foot : annelids

____

60. cephalopod motion : siphon
   a. annelid motion : siphon
   b. annelid breathing : nephridia
   c. annelid circulation : closed circulatory system
   d. annelid motion : tentacles
   e. annelid feeding : filter feeding

____

61. annelids : mantle ::
   a. annelids : organ systems
   b. annelids : gills or lungs
   c. annelids : body cavity
   d. annelids : segmentation
   e. annelids : setae

____

62. leeches : blood ::
   a. earthworms : small animals
   b. earthworms : blood
   c. marine worms : small animals
   d. marine worms : blood
   e. marine worms : soil

____

63. marine worm parapodia : many ::
   a. earthworm parapodia : many
   b. earthworm setae : missing
   c. leech parapodia : missing
   d. leech parapodia : many
   e. leech segments : few

____

64. earthworm movement : circular muscles and setae ::
   a. earthworm digestion : circular muscles and setae
   b. earthworm digestion : sperm receptacles
   c. earthworm reproduction : pharynx
   d. earthworm respiration : gizzard
   e. earthworm respiration : skin

____

65. mollusks : a mantle ::
   a. pseudocoelomate animals : true body cavities
   b. solid worms : a gut with one opening
   c. solid worms : coelom
   d. mollusks : a pseudocoelom
   e. mollusks : a coelom

____

TEST ANSWERS

TRUE/FALSE

1. T

2. F

3. T

4. F

5. T

6. F

7. F

8. F

9. T

10. T

MULTIPLE CHOICE

11. d

12. d

13. a

14. b

15. c

16. d

17. b

18. c

19. c

20. a

21. b

22. b

23. c

24. d

25. b

26. d

27. b

28. c

29. b

30. c

31. a

32. b

33. d

34. b

35. d

36. a

37. d

38. c

SHORT ANSWER

39. trochophore

40. cilia

41. nephridia

42. bivalves

43. adductor muscles

44. siphons

45. veliger

46. Gastropoda

47. torsion

48. radula

49. tentacles

50. nautilus

51. setae

52. polychaetes

53. gizzard

54. clitellum

ESSAY

55. Ciliary beating draws coelomic fluid into the nephridia. Useful molecules in this fluid are reabsorbed into the mollusk's body tissues. Wastes are then passed out of the nephridia through a pore that leads into the mantle cavity. The wastes then leave the body.

56. Snails are able to protect themselves when threatened by pulling their head into their mantle cavity. This is possible because the mantle cavity of snails moved to the anterior end of the body during development. Such movement is brought about by a process called torsion, in which the visceral mass of the organism twists 180°, bringing the mantle cavity to the front in a position just over and behind the head.

57. The phylum Annelida is divided into three classes: polychaetes, oligochaetes, and hirudineans. The polychaetes, believed to be the most ancient of the three classes, possess many setae and parapodia on each body segment. Oligochaetes have fewer setae per segment and completely lack parapodia. Hirudineans, the leeches, have neither setae nor parapodia.

ANALOGIES

58. c (moves)

59. d (is found in)

60. c (is performed by)

61. b (do not have)

62. c (feed on)

63. c (are)

64. e (is performed by the)

65. e (have)

Chapter 30

TRUE/FALSE Read each statement. If the statement is true, write T in the space provided. If the statement is false, write F in the space provided.

1.______ Insects were the first animals to evolve wings.

2.______ Arthropods may have either an open or closed circulatory system.

3.______ The ancestor of arthropods was probably some sort of echinoderm that had a coelom.

4.______ Species of animals belonging to several phyla possess a cephalothorax.

5.______ The bodies of arthropods are encased in a shell-like exoskeleton.

6.______ The barnacle is an example of a crustacean that is sessile as an adult.

7.______ The swimmerets of lobsters and crayfish are attached to their thorax.

8.______ The adult form of all echinoderms exhibits radial symmetry.

9.______ The hard, spiny skin of an echinoderm is called an exoskeleton.

10.______ In some echinoderms, respiration and waste removal are performed by skin gills.

11.______ Chordates and echinoderms have a common ancestor.

12.______ None of the echinoderms living in the oceans today are sessile.

13.______ Sea stars are carnivores and are among the most important predators in many marine ecosystems.

MULTIPLE CHOICE Circle the letter of the term or phrase that best completes each statement or best answers each question.

14. Spiders, scorpions, and mites belong to the subphylum
   a. Arthropoda. c. Chordata.
   b. Crustacea. d. Chelicerata.

15. Chelicerae
   a. are biting mouthparts used for chewing.
   b. contain bundles of nerves.
   c. are extremely light-sensitive and give spiders excellent night vision.
   d. are clawlike fangs.

16. The appendages that scorpions and spiders use to capture and handle their prey are called
   a. diptera. c. palps.
   b. walking legs. d. uropods.

17. Malpighian tubules in insects
   a. remove wastes. c. carry Malpighian fluid.
   b. carry blood. d. are important in respiration.

18. The individual units of the compound eye of an arthropod are called
   a. ocelli. c. tunicates.
   b. ossicles. d. ommatidia.

19. An insect's caste is determined by
a. its grade.
b. how much it can eat in its lifetime.
c. its genes.
d. its size.

20. The ovipositors of a grasshopper
a. are used to deposit ovaries.
b. serve as a place to incubate eggs.
c. are the organs in which ovaries develop.
d. are used to dig holes.

21. Spiders typically have
a. three body segments and six legs.
b. two body segments and four legs.
c. two body segments and eight legs.
d. None of the above

22. The word *millipede* means
a. 1,000 feet.
b. 100 feet.
c. metric feet.
d. 1,000 peas.

23. In incomplete metamorphosis, the young insect hatches from the egg as
a. an adolescent.
b. a chrysalis.
c. a caterpillar.
d. a nymph.

24. In a termite colony, the members that defend the colony against attack are called
a. snipers.
b. workers.
c. soldiers.
d. kings of the colony.

25. What kind of crustacean might a gardener encounter?
a. crayfish
b. millipede
c. pill bug
d. There are no land-dwelling crustaceans.

26. Copepods are important
a. as a direct source of food for people.
b. sources of photosynthetic oxygen.
c. members of the aquatic food chains.
d. None of the above

27. Shrimps, lobsters, and crabs are examples of
a. uropods.
b. copepods.
c. hemipods.
d. decapods.

28. The structures on a spider's abdomen that direct the flow of silk from silk-producing glands are called
a. spiracles.
b. pedipalps.
c. chelicerates.
d. spinnerets.

29. The horseshoe crab
a. is a common garden pest in the northeastern United States.
b. is found in large freshwater lakes.
c. is not really a crab.
d. is the direct ancestor of echinoderms.

30. The skeleton of an echinoderm is composed of individual plates called
a. ocelli.
b. ossicles.
c. Odonata.
d. isopods.

31. Which of the following are sessile echinoderms?
a. feather stars
b. sand dollars
c. sea urchins
d. sea lilies

32. Which of the following is *not* a subphylum of arthropods?
a. Chelicerata
b. Uniramia
c. Crustacea
d. They are all arthropod subphyla.

33. Which of the following is *not* a food source for sea stars?
a. clams
b. cnidarians
c. oysters
d. sponges

The following question(s) refer(s) to the diagram of an insect.

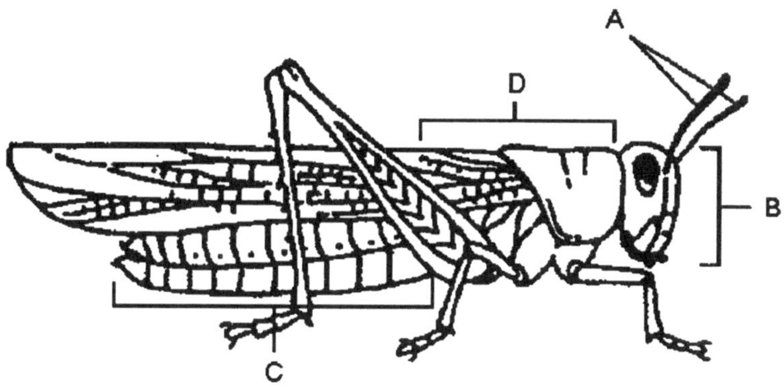

34. The structures labeled A are
a. similar to structures found on spiders.
b. used to take in air and water.
c. specialized for sensing the environment.
d. reproductive organs.

35. Mandibles are attached to the structure labeled
a. A.
b. B.
c. C.
d. D.

36. The structure labeled C is called the
a. diptera.
b. thorax.
c. boll.
d. abdomen.

37. The structure labeled D is called the
a. abdomen.
b. orthopterus.
c. thorax.
d. cuticle.

The following question(s) refer(s) to the diagram of the life cycle of a butterfly.

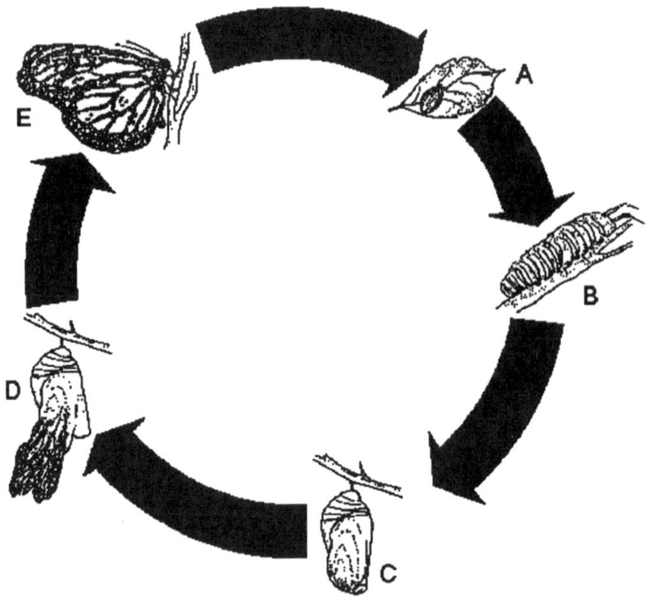

38. The life cycle illustrated is an example of
   a. direct development.
   b. complete metamorphosis.
   c. seasonal development.
   d. incomplete metamorphosis.

39. The developmental stage shown in diagram C is known as the
   a. larva.
   b. caterpillar.
   c. nymph.
   d. chrysalis.

40. A larva is shown in diagram
   a. A.
   b. B.
   c. C.
   d. E.

The following question(s) refer(s) to the drawing of the crayfish.

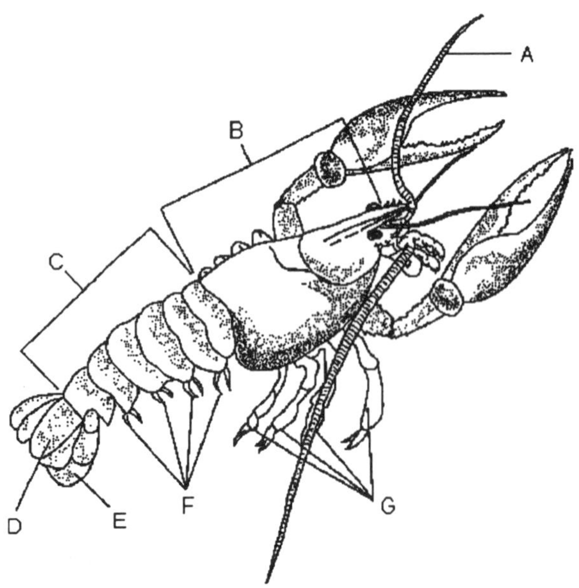

41. The organism shown is

a. an insect.
b. a chelicerate.
c. a crustacean.
d. a nymph.

42. The structure labeled C is the

a. cephalothorax.
b. tail.
c. abdomen.
d. gill case.

43. "A flattened, paddle-like appendage" best describes the structure labeled

a. D.
b. E.
c. F.
d. G.

44. Structure B is covered on top by a shield called the

a. nauplius.
b. carapace.
c. thorax.
d. ossicle.

SHORT ANSWER Complete each statement by writing the correct term or phrase in the space provided.

45. Arthropods such as centipedes that have jaws are called ________________________.

46. Simple, single-lens eyes found on some arthropods are called ________________________.

47. The process whereby arthropods shed and discard their exoskeletons periodically is called molting or ________________________.

48. Insects, millipedes, and centipedes belong to the subphylum ________________________.

49. Some decapods have a tail spine called a(n) ________________________.

50. The system of interconnected canals and tube feet in echinoderms is called a(n) ____________ system.

51. Sea cucumbers belong to the phylum ________________________.

52. ________________________ are the direct ancestors of the fishes.

53. Adult tunicates have a tough sac made out of ________________________ that surrounds their bodies.

The following question(s) refer(s) to the diagram of the respiratory system of an insect.

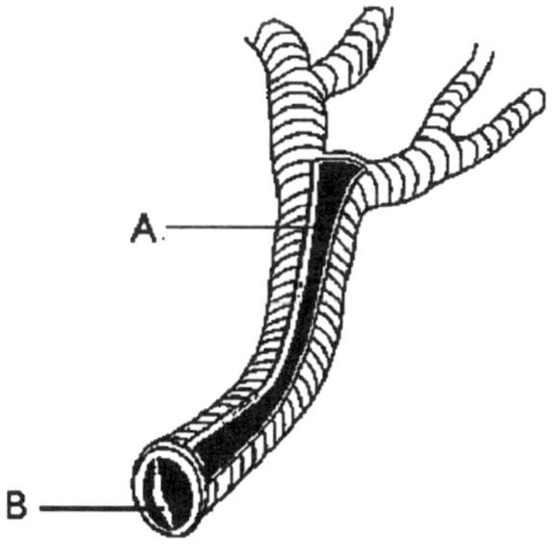

54. Air from the outside passes into the insect through structure A, which is known as a(n) ______________________.

55. Structure B is called a(n) ______________________.

The following question(s) refer(s) to the diagram of the organism.

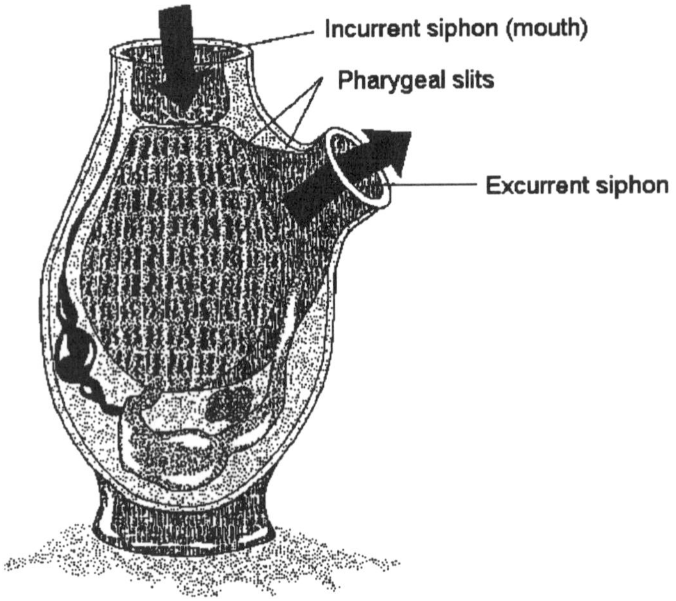

56. The organism illustrated is a member of the subphylum Urochordata known as a(n) ______________________.

ESSAY Read and answer each question or statement on a separate sheet of paper.

57. Describe the functions of the different members of a termite colony.

58. If a mutation occurred during a crustacean's embryonic development that prevented it from molting, how might the crustacean be affected?

59. Why are tunicates classified in the phylum Chordata?

ANALOGIES An analogy is a relationship between two pairs of words or phrases written as: a:b::c:d. The symbol : is read "is to," and the symbol :: is read "as." Write the letter of the pair of terms that best completes the analogy shown.

60. evolutionary origin of arthropod circulatory system : mollusks::
   a. annelid evolution : arthropods
   b. annelid evolution : trilobite
   c. arthropod segmentation : annelids
   d. arthropod segmentation : mollusks
   e. evolution of annelid wormlike body : arthropod

____

61. evolution of jointed appendages : arthropods to come onto land::
   a. evolution of spiracles with valves : prevent drying out
   b. evolution of compound eyes : respiration
   c. evolution of Malpighian tubules : vision
   d. evolution of exoskeleton : circulation
   e. evolution of tracheae : flight

---

62. exoskeleton : protection::
   a. calcium carbonate : fluids
   b. ecdysis : growth
   c. wing : excretion
   d. spiracle : vision
   e. Malpighian tubule : protection

---

63. spiders : subphylum Chelicerata::
   a. scorpions : class Crustacea
   b. millipedes : class Crustacea
   c. centipedes : class Crustacea
   d. insects : class Crustacea
   e. lobsters : class Crustacea

---

64. arachnid 2nd pair of appendages : palps::
   a. arachnid legs : pincers
   b. arachnid 1st pair of appendages : walking legs
   c. spider 1st pair of appendages : chelicerae
   d. spider fangs : palps
   e. scorpion pincers : fangs

---

65. crayfish : swimmerets for swimming and reproduction::
   a. lobsters : uropods for reproduction
   b. shrimp : telson for chewing
   c. crabs : telson for eating
   d. crustaceans : gills for breathing
   e. crustaceans : nauplius for breathing

---

66. grasshopper egg : ovipositor to hole in the ground::
   a. grasshopper chewed food : crop to gizzard
   b. grasshopper chewed food : gizzard to crop
   c. grasshopper chewed food : midgut to gizzard
   d. air : tracheae to spiracles
   e. grasshopper abdomen : head to the thorax

---

67. complete metamorphosis : pupa stage::
a. grasshopper : dorsal nerve cord
b. grasshopper : closed circulatory system
c. incomplete metamorphosis : pupa stage
d. incomplete metamorphosis : pupa and nymph stage
e. incomplete metamorphosis : nymph stage

____

68. sea cucumbers : a fused skeleton::
a. sea urchins : a 5-part body plan
b. sea urchins : distinct arms
c. sand dollars : endoskeletons
d. sea urchins : endoskeletons
e. sand dollars : a 5-part body plan

____

69. tunicates : in shallow and deep water environments::
a. lancelets : buried in mud or sand
b. lancelets : swimming near the water's surface
c. tentacles : in sea urchin mouths
d. ossicles : in lancelets
e. feather stars and sea lilies : in the chordate classification

____

TEST ANSWERS

TRUE/FALSE

1. T

2. F

3. F

4. F

5. T

6. T

7. F

8. T

9. F

10. T

11. T

12. F

13. T

MULTIPLE CHOICE

14. d

15. d

16. c

17. a

18. d

19. c

20. d

21. c

22. a

23. d

24. c

25. c

26. c

27. d

28. d

29. c

30. b

31. d

32. d

33. d

34. c

35. b

36. d

37. c

38. b

39. d

40. b

41. c

42. c

43. b

44. b

SHORT ANSWER

45. mandibulates

46. ocelli

47. ecdysis

48. Uniramia

49. telson

50. water-vascular

51. Echinodermata

52. Lancelets

53. cellulose

54. spiracle

55. trachea

56. tunicate

ESSAY

57. Different individuals serve the colony by performing different functions. Workers gather food, tend offspring, and excavate the tunnels and chambers of the colony. Soldiers defend the colony against attack. The king and queen of the colony reproduce.

58. Because crustaceans must molt in order to grow, the crustacean would be unable to grow larger than the size it attained before emerging from the egg.

59. Tunicate larvae have a body cavity, nerve cord, and notochord. These are all characteristics possessed by chordates.

ANALOGIES

60. c (is from the)

61. a (helped)

62. b (offers)

63. e (belong to the)

64. c (are called)

65. d (use)

66. a (travels from the)

67. e (has a)

68. b (lack)

69. a (are found)

Chapter 31

TRUE/FALSE Read each statement. If the statement is true, write T in the space provided. If the statement is false, write F in the space provided.

1.______ Small, jawed fishes are the first vertebrates for which there is fossil evidence.

2.______ Lobe-finned fishes were the ancestors of amphibians.

3.______ A land animal needs stronger bones and muscles than an aquatic animal does because of the increased pull of gravity on terrestrial body structures.

4.______ In most amphibians, fertilization takes place externally.

5.______ Animals in which body temperature is largely determined by the environment are called endothermic.

6.______ In a fish, countercurrent flow prevents a steady supply of oxygen from reaching the gill filaments.

7.______ Waterproof skin is a reptilian adaptation to life on land.

8.______ Since they live in salt water, marine fishes do not have a problem maintaining the proper balance of water and salt in their body.

9.______ Ovoviviparous reptiles carry their eggs in their bodies throughout development, and their young are born alive.

10.______ Marsupials are egg-laying mammals.

11.______ Reptiles must return to the water in order to reproduce.

12.______ Like reptiles, monotremes are oviparous.

13.______ Egg-laying mammals nourish their young after birth with milk.

MULTIPLE CHOICE Circle the letter of the term or phrase that best completes each statement or best answers each question.

14. Endothermic animals
    a. are animals that have the ability to maintain a constant internal temperature.
    b. are animals whose body temperature is not stable.
    c. are mostly found among living reptiles.
    d. are coldblooded animals.

15. Mammals evolved
    a. from reptiles earlier in history than did birds.
    b. from reptiles well after birds appeared on Earth.
    c. at the same time as birds.
    d. directly from birds.

16. Mammals arose from early reptiles called
    a. mesosaurs.
    b. ichthyosaurs.
    c. therapsids.
    d. pterosaurs.

17. Which of the following is a reptilian adaptation to living on land?
   a. external fertilization
   b. endothermic temperature regulation
   c. respiration through gills
   d. the amniotic egg

18. In flying birds, large flight muscles are directly attached to
   a. leg muscles.
   b. feathers.
   c. the keel of the breastbone.
   d. air sacs.

19. Bird respiration is very efficient because
   a. bird lungs are small and hollow.
   b. only one lung functions at a time.
   c. of a special set of blood vessels.
   d. their system of air sacs permits air to flow in only one direction through the lungs.

20. The countercurrent flow of water and blood found in the gills of fishes
   a. allows blood and water to flow in the same direction.
   b. ensures that oxygen diffuses into the blood over the whole length of the blood vessels in the gills.
   c. results in an uneven supply of oxygen reaching the blood vessels in the gills.
   d. hampers the diffusion of oxygen and carbon dioxide between the blood and the water.

21. In amphibians, gases are exchanged through lung breathing and through the
   a. heart.
   b. air bladder.
   c. tracheids.
   d. skin.

22. The amount of oxygen a lung can absorb depends primarily on
   a. its thickness.
   b. its position in the body of an animal.
   c. its internal surface area.
   d. the diameter of the bronchioles in the lung.

23. A collection chamber that reduces the resistance of blood flow into the heart of a fish is called the
   a. sinus venosus.
   b. ventricle.
   c. conus arteriosus.
   d. atrium.

24. Pulmonary veins carry blood from the
   a. body to the heart.
   b. heart to the lungs.
   c. lungs to the heart.
   d. heart to the body.

25. The amphibian heart
   a. pumps only deoxygenated blood.
   b. has four chambers.
   c. pumps only oxygenated blood.
   d. pumps both deoxygenated and oxygenated blood.

26. Reptiles, except for alligators and crocodiles, have a heart that has
   a. two atria and two ventricles.
   b. two atria and one ventricle.
   c. one atrium and two ventricles.
   d. two atria and two ventricles.

27. A four-chambered heart is a characteristic of
   a. ectotherms.
   b. reptiles.
   c. amphibians.
   d. endotherms.

28. Mammals share with birds
   a. teeth.
   b. a four-chambered heart.
   c. air sacs.
   d. All of the above

29. The concentration of the urine an animal produces depends primarily on
   a. the time of the year.
   b. the size of the organism.
   c. the environment in which the animal lives.
   d. the diet of the organism.

30. The ion concentration in the body of a shark
   a. changes constantly as the shark swims at different depths.
   b. is higher than that in the surrounding sea water.
   c. is lower than that in the surrounding sea water.
   d. is the same as that in the surrounding sea water.

31. Some animals are oviparous. This means that their young
   a. are nourished by a placenta.
   b. are born live from eggs that hatch within the mother's body.
   c. hatch from eggs laid outside their mother's body.
   d. continue to develop in their mother's pouch.

32. Although adapted to land, toads must have access to a watery environment in order to
   a. obtain food and oxygen.
   b. excrete wastes.
   c. reproduce.
   d. All of the above

33. The major difference between marsupials and placentals is
   a. their ability to maintain a steady body temperature.
   b. their teeth.
   c. their pattern of embryonic development.
   d. the size of their eggs.

34. Offspring remain inside the mother until development is essentially complete in
   a. placental mammals.
   b. monotremes.
   c. marsupials.
   d. All of the above

The following question(s) refer(s) to the diagrams of the vertebrate organs.

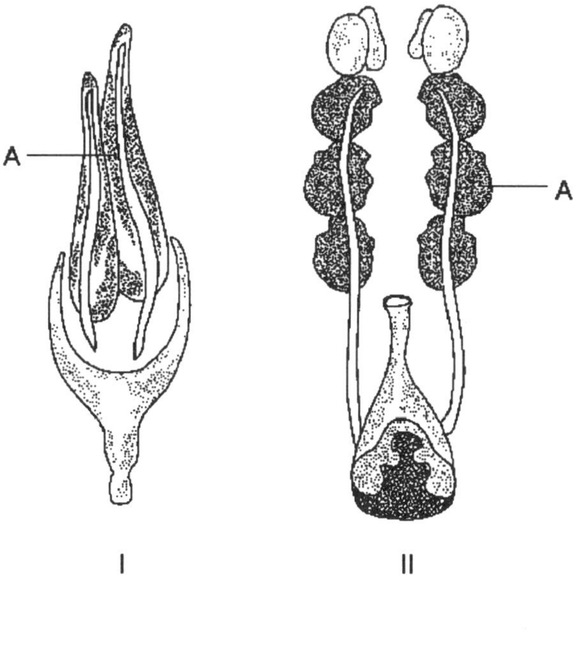

35. The diagrams represent vertebrate
a. respiratory systems.
b. digestive systems.
c. excretory systems.
d. circulatory systems.

36. The structure labeled A is a(n)
a. lung.
b. heart.
c. amnion.
d. kidney.

37. Diagram II represents a system that would be found in a(n)
a. fish.
b. amphibian.
c. reptile.
d. bird.

The following question(s) refer(s) to the diagram showing the stages of frog development.

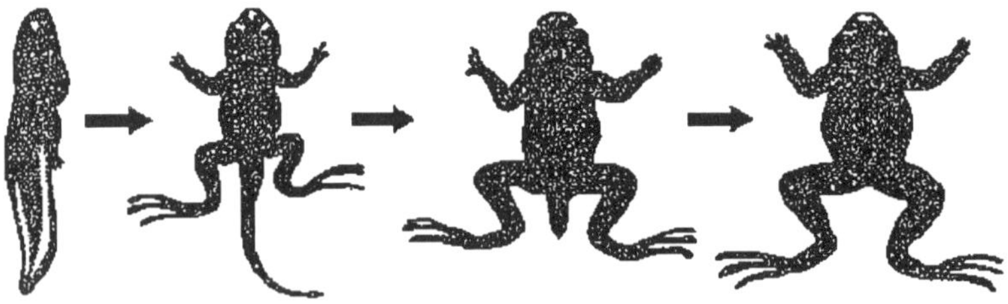

38. The process illustrated in the diagram is
a. amniocentesis.
b. metamorphosis.
c. evolution.
d. synapsis.

SHORT ANSWER Complete each statement by writing the correct term or phrase in the space provided.

39. Lobe-finned fishes evolved ______________________, which let them extract oxygen from the air.

40. The inner surface of reptile lungs consists of small chambers called ______________________.

41. Each gill in a fish is made up of two rows of gill ________________ that are stacked on top of one another.

42. In some species of amphibians, a ______________________ partly or completely divides the atrium into right and left halves.

43. Over the course of vertebrate evolution, the ______________________ has become reduced in size. It has become the pacemaker in the hearts of birds and mammals.

44. A kidney is made up of repeating units called ____________________ that filter the blood.

45. Modern reptiles differ from amphibians in that they have a ______________________ skin.

46. The term used to describe mammals that give birth to live offspring that develop within the mother's body is ______________________.

47. In oviparous and ovoviviparous animals, the ______________________ provides a rich supply of food for the developing embryo.

The following question(s) refer(s) to the diagrams of three different hearts.

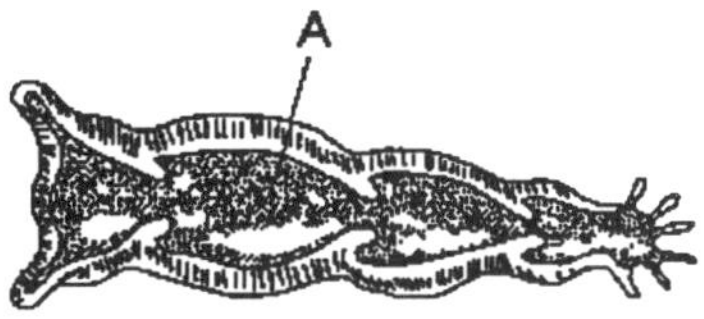

I

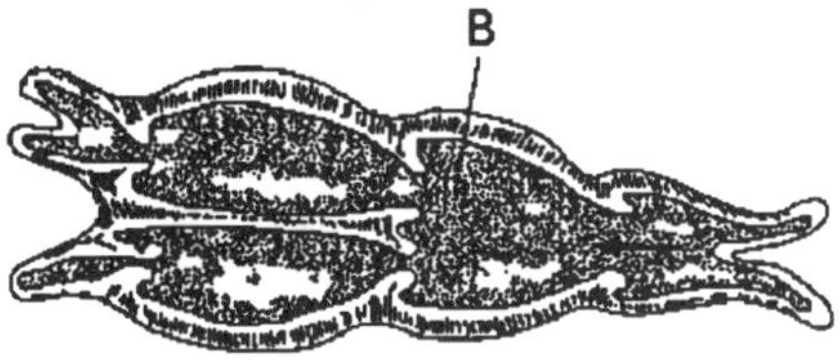

II

III

48.The structure labeled A is a(n) ______________________.

49.Structure B is a(n) ______________________.

50.Diagram I represents the heart of a(n) ______________________.

51.Diagram III represents the heart of a(n) ______________________.

ESSAY Read and answer each question or statement on a separate sheet of paper.

52.Explain the importance of the swim bladder to bony fishes.

53.Birds live in the Arctic, but reptiles do not. Offer an explanation for this fact.

54.Flying requires a great amount of energy. What adaptations in the respiratory systems of birds allow them to meet their high energy needs?

55.The right and left sides of a bird's heart are completely separated. This means that oxygen-rich blood is never mixed with oxygen-poor blood. Why is this complete separation necessary for birds?

56.What features of an amniotic egg are adaptations to life on land?

ANALOGIES An analogy is a relationship between two pairs of words or phrases written as: a:b::c:d. The symbol : is read "is to," and the symbol :: is read "as." Write the letter of the pair of terms that best completes the analogy shown.

57. amphibian : reptile::
   a. land : jaw
   b. feathers : scales
   c. reptile : bird
   d. mammal : bird
   e. hammer : nail

____

58. jaws : lampreys::
   a. cartilage : shark
   b. fins : jawed fishes
   c. teeth : jaws
   d. paired fins : jawless fishes
   e. bone : bony fishe

____

59. amphibian : moist::
   a. ball : square
   b. oxygen : water
   c. reptile : scaly
   d. lung : primitive
   e. moist : dry

____

60. metabolism : walking on land::
   a. jawless fishes : jaw
   b. mammal : size
   c. frog : amphibian
   d. lung surface area : activity
   e. breathing : filter-feeding

____

61. gills : fish::
   a. skin : reptile
   b. blood vessels : circulation
   c. tree : plant
   d. lungs : mammal
   e. egg : vertebrate

____

62. folds : lung::
   a. fire : wood
   b. fish : gill filaments
   c. oxygen : atmosphere
   d. lung : fish
   e. gill filaments : gills

____

63. oxygen delivery : pulmonary blood vessels::
   a. lungs : gills
   b. blood separation : septum
   c. oxygen-rich : oxygen-poor
   d. heart : amphibian
   e. blood mixing : septum

___

64. freshwater fish kidney : dilute urine::
   a. nephron : kidney
   b. car : exhaust
   c. milk : kitten
   d. salt water intake : absorption
   e. blood : gills

___

65. reptile : dry::
   a. amphibian : watertight
   b. terrestrial : moist
   c. swim bladder : active
   d. amphibian : slippery
   e. both : ventricle

___

66. water : mammalian kidney::
   a. spilled liquid : sponge
   b. sponge : water
   c. water : sprinkler
   d. rain : pavement
   e. snow : rain

___

TEST ANSWERS

TRUE/FALSE

1. F
2. T
3. T
4. T
5. F
6. F
7. T
8. F
9. T
10. F
11. F
12. T
13. T

MULTIPLE CHOICE

14. a
15. a
16. c
17. d
18. c
19. d
20. b
21. d
22. c
23. a
24. c

25. d

26. b

27. d

28. b

29. c

30. d

31. c

32. c

33. c

34. a

35. c

36. d

37. d

38. b

SHORT ANSWER

39. lungs

40. alveoli

41. filaments

42. septum

43. sinus venosus

44. nephrons

45. watertight

46. viviparous

47. yolk

48. atrium

49. ventricle

50. fish

51. reptile

ESSAY

52. The swim bladder is a gas-filled sac that gives the fishes active control of their buoyancy. By regulating the amount of gas in the swim bladder, bony fishes can remain at different depths without expending energy through swimming.

53. Reptiles are ectothermic and cannot survive extreme temperatures. Birds are endothermic and can adapt to some extremes.

54. In order to release large amounts of energy, birds must have a constant supply of oxygen. Birds have a system of air sacs that maintain a constant one-way flow (and supply of oxygen) to the lungs. In addition, the network of lung capillaries is arranged across the air flow at a 90-degree angle. Blood leaving a bird's lung can contain more oxygen than exhaled air does.

55. Birds need blood with a high oxygen content because flight requires large amounts of energy and oxygen.

56. Answers may include any of the following features: series of protective membranes—amnion and chorion—and a shell that enclose the embryo in a watery environment; membranes and shell are resistant to loss of water and allow for exchange of gases; yolk sac for food storage.

ANALOGIES

57. c (evolved before the)

58. d (are not found in)

59. c (has skin that is)

60. d (must be greater for)

61. d (obtain oxygen in a)

62. e (increase surface area of)

63. b (is accomplished by)

64. b (releases)

65. d (has skin that is)

66. a (is absorbed)

Chapter 32

TRUE/FALSE Read each statement. If the statement is true, write T in the space provided. If the statement is false, write F in the space provided.

1.______ The first fishes that appeared in the ancient seas were jawless and toothless.

2.______ Fishes and amphibians first appeared on the Earth during the Cambrian period, about 550 million years ago.

3.______ The first fishes to develop jaws were called spiny fishes, members of the class Acanthodia.

4.______ Sharks underwent a great burst of evolution during the age of dinosaurs.

5.______ Both sharks and bony fishes evolved in the ancient seas.

6.______ In fishes, oxygenated blood passes from the gills to the heart and then to the rest of the body.

7.______ In order to fill their swim bladders, bony fishes have to come to the surface to gulp air.

8.______ Some fossil amphibians had bony plates and armor covering their bodies.

9.______ The earliest known amphibian fossil is that of an extinct amphibian called *Ichthyostega*.

10.______ All adult anurans are strictly herbivores.

11.______ In amphibians, the blood that is pumped from the heart to the body is completely oxygenated.

12.______ In amphibians, reproductive cells and wastes exit the body through the cloacal opening.

MULTIPLE CHOICE Circle the letter of the term or phrase that best completes each statement or best answers each question.

13. Fishes are unable to synthesize
    a. ATP. c. proteins.
    b. aromatic amino acids. d. nucleic acids.

14. The first fishes to appear in the seas were members of the class
    a. Vertebrata. c. Placodermii.
    b. Chondrichthyes. d. Agnatha.

15. Hagfishes are scavengers that are found
    a. only in rivers. c. in rivers and the ocean.
    b. in lakes and streams. d. only in marine environments.

16. Chondrichthyes means
    a. bony fish. c. big fish.
    b. gilled fish. d. cartilage fish.

17. Members of the class Osteichthyes
   a. have skeletons made of bone.
   b. do not have jaws.
   c. include the rays and skates.
   d. All of the above

18. The living agnathans are the
   a. lampreys and sharks.
   b. sharks and rays.
   c. hagfishes and coelacanths.
   d. lampreys and hagfishes.

19. Coelacanths and lungfishes are
   a. ray-finned fishes.
   b. cartilaginous fishes.
   c. lobe-finned fishes.
   d. jawless fishes.

20. Cartilaginous fishes have all of the following except
   a. spiracles.
   b. internal fertilization.
   c. a swim bladder.
   d. gill slits.

21. Jaws probably evolved from the
   a. pectoral fins of jawless fishes.
   b. gills slits of sharks and rays.
   c. paired pelvic fins of jawless fishes.
   d. gill arches of jawless fishes.

22. The two major groups of bony fishes are the lobe-finned fishes and the
   a. lungfishes.
   b. ray-finned fishes.
   c. coelacanths.
   d. placoderms.

23. Shark eggs
   a. all develop externally.
   b. are fertilized internally.
   c. are fertilized externally.
   d. have leathery shells.

24. In bony fishes
   a. internal fertilization takes place.
   b. external fertilization takes place.
   c. internal and external development take place.
   d. All of the above

25. The swim bladder
   a. preserves air for breathing.
   b. contains antibodies.
   c. is found in all amphibians.
   d. allows fishes to become more buoyant.

26. The word *amphibian* means
   a. fishlike.
   b. frog.
   c. double life.
   d. first animal with lungs.

27. Amphibians without tails are classified in the order
   a. Apoda.
   b. Anura.
   c. Urodela.
   d. Hydrodela.

28. The ancestors of the amphibians are the
   a. lobe-finned fishes.
   b. ray-finned fishes.
   c. caecilians.
   d. Urodela.

29. Adult frogs, like other amphibians, are
   a. herbivores.
   b. omnivores.
   c. parasites.
   d. carnivores.

30. During metamorphosis in frogs,
   a. lungs replace gills.
   b. limbs develop.
   c. the tail disappears.
   d. All of the above

31. In amphibians, gases are exchanged through lung breathing and through the
   a. heart.
   b. air bladder.
   c. tracheids.
   d. skin.

32. Amphibians that have slender bodies and no limbs are classified as
   a. anurans.
   b. caecilians.
   c. salamanders.
   d. newts.

33. The amphibian heart
   a. pumps only deoxygenated blood.
   b. has four chambers.
   c. pumps only oxygenated blood.
   d. pumps both deoxygenated and oxygenated blood.

The following question(s) refer(s) to the diagram of the external structure of a bony fish.

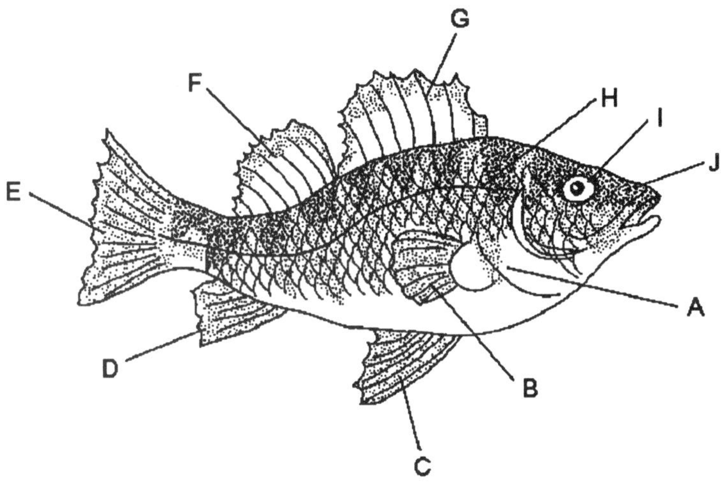

34. In order to move forward, the fish uses the fin(s) labeled
   a. B.
   b. C.
   c. E.
   d. F.

35. The structure labeled A, which draws water into the mouth of the fish, is the
   a. pharynx.
   b. esophagus.
   c. gills.
   d. operculum.

The following question(s) refer(s) to the diagram of part of the lateral line system of a fish.

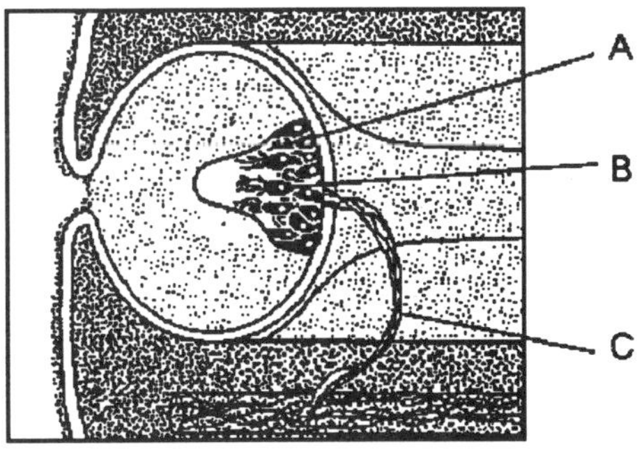

36. The lateral line system enables a fish to
a. smell.
b. see objects at a distance.
c. detect vibrations in the water.
d. swim rapidly.

37. The structure labeled C is a
a. cilium.
b. scale.
c. supporting cell.
d. nerve.

SHORT ANSWER Complete each statement by writing the correct term or phrase in the space provided.

38. Lampreys and hagfishes that are found on the Earth today are descendants of the armored ________________________.

39. A support known as the ________________________ enables the mouths of sharks and bony fishes to open very wide.

40. The ________________________ are the most advanced of the ray-finned bony fishes.

41. Bony fishes sense pressure changes in the water with their ________________________ system.

42. During reproduction a male bony fish deposits ______________________, a fluid containing sperm, on eggs laid by the female.

43. Studies of DNA suggest that ________________________ are the closest relative of modern amphibians.

44. Salamanders and ________________________ are amphibians with a distinct head, tail, and limbs.

45. The changes that transform a tadpole into an adult frog are called ________________________.

46. Amphibians supplement the use of their lungs by respiring directly through their skin. This "skin breathing" is known as ___________________ respiration.

47. In amphibians, the blood vessels that go from the lungs to the heart are called the ________ veins.

48. In frogs, the eardrum is called the ________________________.

The following question(s) refer(s) to the diagram showing the jawless fish.

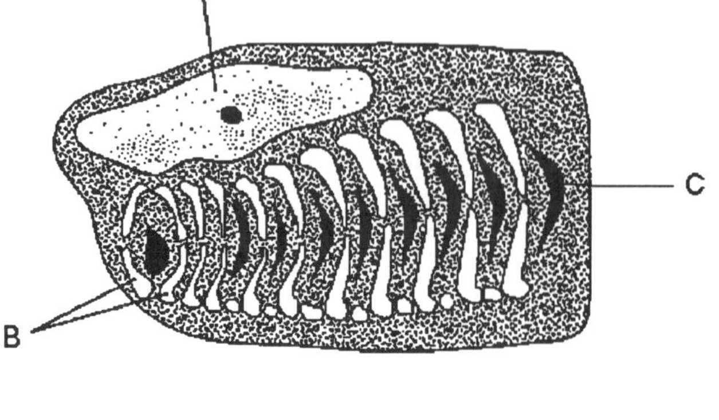

49. Structure A is the ______________________.

50. The letter B is pointing to the ______________________.

51. The letter C is pointing to the ______________________.

ESSAY Read and answer each question or statement on a separate sheet of paper.

52. How is the body design of bony fishes and sharks better adapted for swimming than the body design of primitive fishes?

53. Why must a shark constantly swim through the water?

54. Explain the importance of the swim bladder to bony fishes.

55. Describe how jaws are thought to have evolved.

ANALOGIES An analogy is a relationship between two pairs of words or phrases written as: a:b::c:d. The symbol : is read "is to," and the symbol :: is read "as." Write the letter of the pair of terms that best completes the analogy shown.

56. placoderms : armor::
 a. bony fishes : cartilaginous skeleton
 b. sharks : no teeth
 c. lampreys : jaws
 d. sharks and bony fishes : streamlined bodies
 e. hagfishes and lampreys : teeth

____

57. bony fishes : stronger muscles::
 a. bony fishes : lungs
 b. sharks : bony skeleton
 c. sharks : ray fins
 d. sharks : rows of teeth
 e. teleosts : cartilaginous skeletons

____

58. "amphi" : double::
 a. external : inside
 b. internal : outside
 c. "gnathos" : skin
 d. "bios" : water
 e. "anura" : no tail

____

59. lateral line : motion detector::
 a. swim bladder : roof on a house
 b. operculum : wheel
 c. operculum : water pump
 d. fin : microwave oven
 e. scale : telephone

____

60. amphibian skin : surface area for respiration::
   a. flowers : ugliness
   b. happiness : sadness
   c. speech : communication
   d. water : dryness
   e. sunglasses : glare

---

61. tympanic membrane : hearing::
   a. cloaca : breathing
   b. frog teeth : chewing food
   c. skin : balance
   d. frog tongue : vocalization
   e. frog tongue : catching prey

TEST ANSWERS

TRUE/FALSE

1. T
2. F
3. T
4. T
5. F
6. F
7. F
8. T
9. T
10. F
11. F
12. T

MULTIPLE CHOICE

13. b
14. d
15. d
16. d
17. a
18. d
19. c
20. c
21. d
22. b
23. b
24. b

25. d

26. c

27. b

28. a

29. d

30. d

31. d

32. b

33. d

34. c

35. d

36. c

37. d

SHORT ANSWER

38. ostracoderms

39. hyomandibular

40. teleosts

41. lateral line

42. milt

43. lungfishes

44. newts

45. metamorphosis

46. cutaneous

47. pulmonary

48. tympanic membrane

49. braincase

50. gill arches

51. gill slits

ESSAY

52. Most bony fishes and sharks have streamlined bodies that are well adapted for movement through the water. The head acts as a wedge that cleaves the water, and the body tapers back to the tail, allowing the fish to slip through the water with minimal resistance. In addition, sharks and bony fishes have an assortment of movable fins that aid their swimming.

53. A shark must constantly swim through the water because it has no swim bladder and its body is denser than water.

54. The swim bladder is a gas-filled sac that gives the fish buoyancy. By regulating the amount of gas in the swim bladder, a fish can remain at different depths without expending energy through swimming.

55. Scientists think that jaws evolved from one or more of the gill arches that support the pharynx in agnathans.

ANALOGIES

56. d (have)

57. d (developed)

58. e (means)

59. c (acts as)

60. c (is/are used to increase)

61. e (is for)

Chapter 33

TRUE/FALSE Read each statement. If the statement is true, write T in the space provided. If the statement is false, write F in the space provided.

1.______ Lizards and snakes are found on every continent of the Earth.

2.______ Crocodiles have undergone enormous changes since they first appeared 200 million years ago.

3.______ Lizards have modified diapsid skulls.

4.______ Modern reptiles are classified by the number of membranes found in their eggs.

5.______ The majority of reptiles are oviparous.

6.______ Many reptiles regulate their temperature by their behavior.

7.______ Turtles generally live on land, while tortoises generally live in the water.

8.______ Lizards appeared in the late Permian period, about 250 million years ago.

9.______ The age of a rattlesnake can be accurately determined by counting the number of rings on its rattle.

10.______ The Jurassic period is often referred to as "the golden age of dinosaurs."

11.______ Because of the position of their legs, dinosaurs were faster and more agile runners than the thecodonts.

12.______ The large plates on the back of a sauropod may have played a part in heating and cooling the animal.

13.______ Some skeletons of fossil birds have teeth.

14.______ In eagles and other birds, the optic lobe processes information from the eyes.

MULTIPLE CHOICE Circle the letter of the term or phrase that best completes each statement or best answers each question.

15. The most ancient surviving group of reptiles are the
   a. rhynchocephalians.   c. lizards.
   b. snakes.   d. turtles.

16. Early reptiles evolved into
   a. lizards.   c. mammals.
   b. birds.   d. All of the above

17. The outermost membrane of a reptilian egg is the
   a. allantois.   c. mitotic membrane.
   b. yolk sac.   d. chorion.

18. Unlike other reptiles, turtles and tortoises
   a. live only in water.   c. are endangered.
   b. are prehistoric.   d. do not have teeth.

19. Which of the following is true of snakes?
   a. They lack limbs.
   b. They lack movable eyelids.
   c. They lack external ears.
   d. All of the above

20. The tuatara
   a. represents the last species of reptiles to have evolved.
   b. has a four-chambered heart.
   c. represents the last surviving species of a group of reptiles that appeared more than 225 million years ago.
   d. is found throughout the world except for the Antarctic.

21. One disadvantage of endothermy is that it
   a. limits the size of an organism.
   b. requires a relatively greater amount of food.
   c. impedes circulation.
   d. can only occur at night.

22. The heat-sensing organs between the eyes and nostril on each side of the head of a rattlesnake are the
   a. tracheal organs.
   b. Jacobson's organs.
   c. thermal organs.
   d. pit organs.

23. Dinosaurs first appeared during the
   a. Permian period.
   b. Triassic period.
   c. Jurassic period.
   d. Tertiary period.

24. The most widely accepted hypothesis to explain the mass extinction of dinosaurs proposes
   a. a massive volcanic explosion.
   b. sunspots that disrupted the Earth's weather.
   c. giant carnivorous mammals that also became extinct.
   d. that the Earth was struck by a meteorite that caused thick dust clouds.

25. Scientific evidence suggests that birds arose from
   a. mammals.
   b. amphibians.
   c. reptiles.
   d. protozoa.

26. The link between birds and reptiles is the extinct animal called
   a. *Archaeopteryx*.
   b. *Ichthyosaurus*.
   c. lancelet.
   d. hagfish.

27. Relationships among the families of modern birds are mostly inferred from studies of
   a. fossils.
   b. bones.
   c. feathers.
   d. DNA.

28. The bones of birds
   a. are composed primarily of keratin.
   b. are solid.
   c. are found sparingly throughout the body.
   d. are hollow and contain air pockets.

29. Birds excrete most of their nitrogenous wastes as
   a. urea.
   b. ammonia.
   c. uric acid.
   d. urine.

30. A bird's skeleton
   a. is composed of thin, hollow bones.
   b. is more rigid than a reptile's.
   c. is composed of many fused bones.
   d. All of the above

31. Talons would most likely be found among birds that
   a. eat seeds.
   b. capture their prey.
   c. live in water.
   d. drink the nectar of flowers.

32. A bird's crop
   a. temporarily stores food.
   b. is the first chamber of its stomach.
   c. is critical for flight.
   d. often contains small stones that the bird swallowed.

33. In flying birds, large flight muscles are directly attached to
   a. leg muscles.
   b. feathers.
   c. the keeled breastbone.
   d. air sacs.

34. Unlike other living reptiles, crocodilians
   a. are viviparous.
   b. have a three-chambered heart.
   c. care for their young after hatching.
   d. have a synapsid skull.

35. Many herbivorous dinosaurs became extinct due to the rise of
   a. mammals.
   b. flowering plants.
   c. climatic temperatures.
   d. meteor impacts.

The following question(s) refer(s) to the diagrams of reptilian skulls.

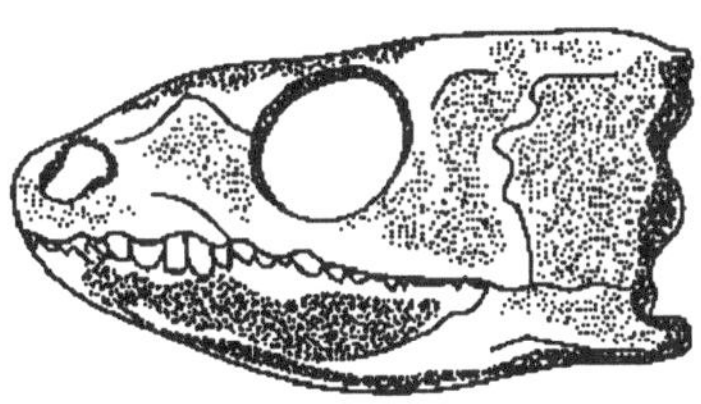

A

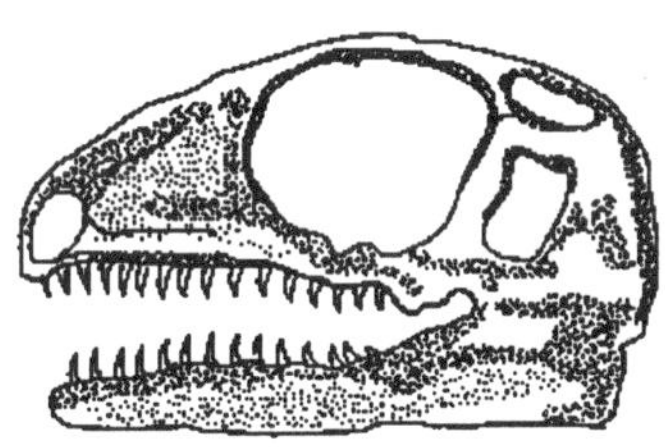

B

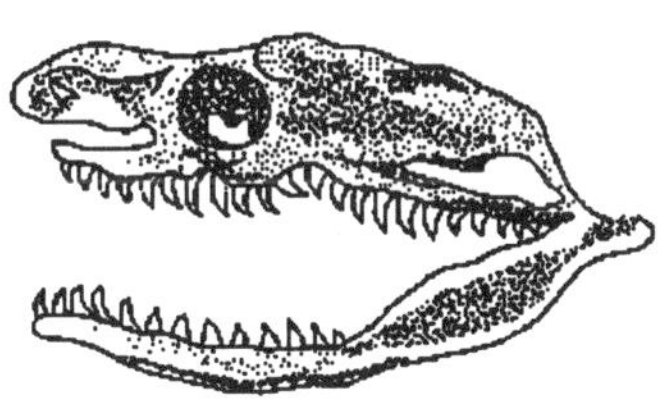

C

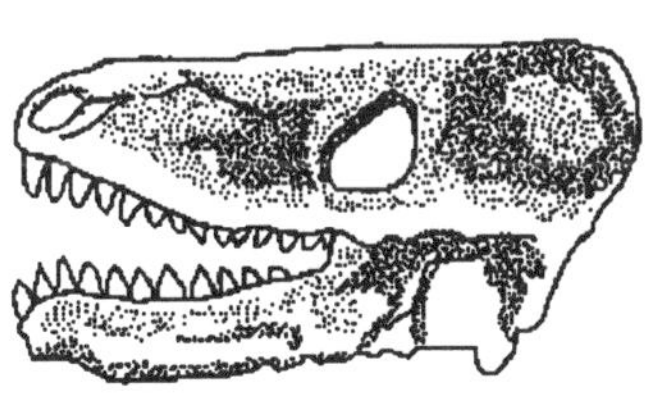

D

36. Which type of skull would most probably have been found on an early reptile?
   a. A
   b. B
   c. C
   d. D

37. The skull illustrated in diagram B is known as a(n)
   a. synapsid skull.
   b. anapsid skull.
   c. therapsid skull.
   d. diapsid skull.

38. The skull that most closely resembles the skull of a mammal is illustrated in diagram
   a. A.
   b. B.
   c. C.
   d. D.

39. Skull C is most likely the skull of a
   a. turtle.
   b. rhynchocephalian.
   c. snake.
   d. crocodile.

The following question(s) refer(s) to the diagram of a turtle's shell.

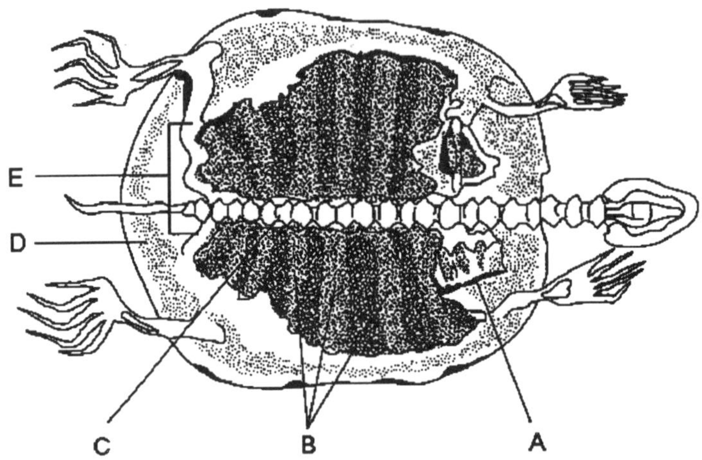

40. Structure D is called the
   a. carapace.
   b. lumbar cover.
   c. cervical cape.
   d. therapsid.

41. The pectoral girdle is labeled
   a. A.
   b. B.
   c. C.
   d. D.

42. In a turtle,
   a. the ribs are reduced or missing.
   b. the shells are made of calcified cartilage.
   c. the vertebrae are fused to the ventral shell.
   d. None of the above

SHORT ANSWER Complete each statement by writing the correct term or phrase in the space provided.

43. Early reptiles, like their amphibian ancestors, had ________________ skulls.

44. Reptiles whose body temperature changes with the temperature of their surroundings are known as ______________________.

45. The ______________________ is the food source in a reptilian egg.

46. The membrane in a reptilian egg that encloses the developing embryo within a fluid-filled cavity is called the ______________________.

47. A reptile's body temperature is largely determined by the temperature of its ______________________.

48. The lower (ventral) portion of a tortoise's shell is called the ______________________.

49. All of the continents were once joined together in a single supercontinent called ______________________.

50.Feathers are modified ______________________ scales.

51.Birds, like mammals, are ______________________.

The following question(s) refer(s) to the diagram of the mouth of a rattlesnake.

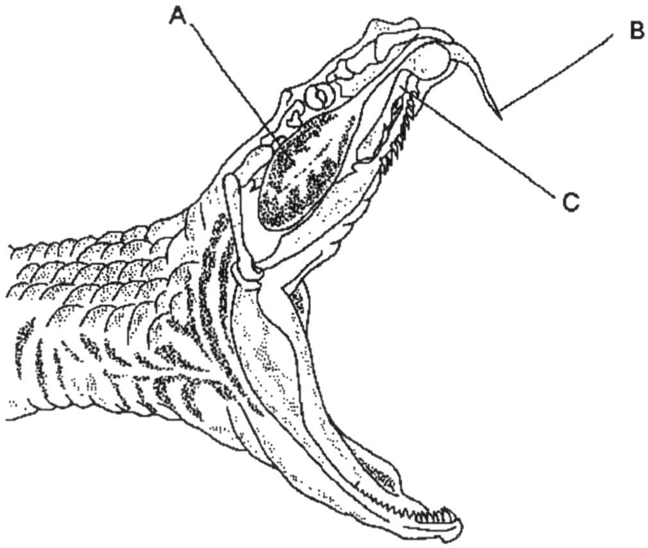

52.Structure A is the ______________________ gland.

53.Structure B is a ______________________.

54.Structure C, which is analogous to the taste buds found in other vertebrates, is called ______________________.

The following question(s) refer(s) to the diagram of the skeleton of a bird.

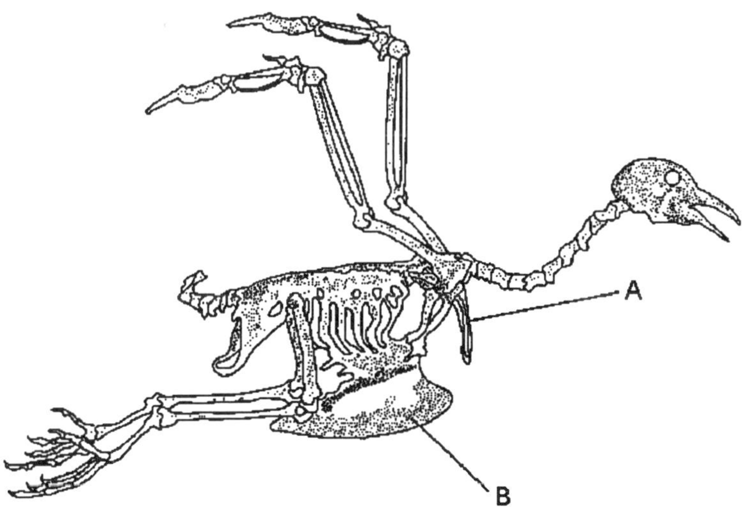

55.Structure A is known as the wishbone or ______________________.

56.Structure B is called the ______________________.

ESSAY Read and answer each question or statement on a separate sheet of paper.

57. Support the statement that "crocodiles resemble birds far more than they resemble living reptiles."

58. What improvements in jaw design have contributed to the success of snakes and lizards as predators?

59. What may have caused the sudden extinction of the dinosaurs about 65 million years ago?

ANALOGIES An analogy is a relationship between two pairs of words or phrases written as: a:b::c:d. The symbol : is read "is to," and the symbol :: is read "as." Write the letter of the pair of terms that best completes the analogy shown.

60. crocodile : a 4-chambered heart::
   a. member of the order Squamata : shell
   b. bird : a 4-chambered heart
   c. bird : a 3-chambered heart
   d. lizard and snake : a 4-chambered heart
   e. crocodile : no legs

____

61. rattlesnake pit organ : heat::
   a. camera light meter : vibrations
   b. cat's whisker : light
   c. crocodile eye : heat
   d. earthquake seismograph : chemicals
   e. human ear : sound

____

62. rattlesnake : rattles when growth is rapid::
   a. tortoise : teeth than lizards and snakes
   b. lizard : shells than turtles
   c. lizard : jaw arches than crocodiles
   d. crocodile : care of the young compared with other reptiles
   e. tortoise : jaw muscles than snakes

____

63. water supply increase : continents separating::
   a. dinosaur leg structure change : dry climate
   b. angiosperm evolution : iguanodonts
   c. dry climate occurrence : shallow inland seas
   d. angiosperm-eating iguanodont evolution : dominance of angiosperms
   e. shallow inland sea : angiosperm

____

64. dust cloud from meteorite : sun::
   a. hole in a wall : a wall
   b. sewer pipe : sewer water
   c. road : forest
   d. window : view
   e. football player : another football player

____

65. bird : hollow bones::

a. *Compsognathus* : feathers
b. reptiles : hollow bones
c. *Archaeopteryx* : feathers
d. bird : dinosaur forelimbs
e. dinosaur : wishbone

____

66. bird feathers : body heat::

a. ectotherms : body heat
b. bird tails : flight
c. ectotherms : energy by low activity
d. endotherms : cold temperatures
e. muscles : energy by high activity

____

TEST ANSWERS

TRUE/FALSE

1. F
2. F
3. T
4. F
5. T
6. T
7. F
8. T
9. F
10. T
11. T
12. F
13. T
14. T

MULTIPLE CHOICE

15. d
16. d
17. d
18. d
19. d
20. c
21. b
22. d
23. b
24. d

25. c

26. a

27. d

28. d

29. c

30. d

31. b

32. a

33. c

34. c

35. b

36. a

37. d

38. d

39. c

40. a

41. a

42. d

SHORT ANSWER

43. anapsid

44. ectotherms

45. yolk

46. amnion

47. environment

48. plastron

49. Pangaea

50. reptilian

51. endothermic

52. venom

53. fang

54. Jacobson's organs

55. collarbone

56. breastbone

ESSAY

57. Crocodiles are the only living reptiles that, like birds, care for their young. They are also the only living reptiles that have a four-chambered heart like that of birds. In many other anatomical features, crocodiles differ from all other living reptiles and resemble birds.

58. The lower jaw of snakes and lizards is loosely connected to the skull. This loose connection allows the mouth to open to accommodate large prey. In addition, the loss of the lower arch of bone below the lower opening in the skull of lizards makes room for large muscles to move the jaws.

59. Most biologists now agree that the most likely cause was the impact of a gigantic meteorite. The impact threw massive amounts of material into the atmosphere, creating a worldwide period of low temperatures. With no way to retain body heat and with no insulation, the cold may have killed the dinosaurs.

ANALOGIES

60. b (has/have)

61. e (senses)

62. d (has more)

63. d (was due to)

64. e (blocks out)

65. c (has/had)

66. c (conserve)

Chapter 34

TRUE/FALSE Read each statement. If the statement is true, write T in the space provided. If the statement is false, write F in the space provided.

1.______ Mammals have been the dominant land animals on Earth for over 500 million years.

2.______ One of the first steps leading to the evolution of mammals was the evolution of reptiles known as pelycosaurs.

3.______ Pelycosaurs had small, uniformly shaped teeth that were primarily used for grinding and chewing.

4.______ The dominant land animals from the middle of the Permian until the middle of the Triassic were the therapsids.

5.______ The first mammals to appear on Earth were about the size of a large dog.

6.______ During the past few million years, the number of mammalian species on Earth has continued to increase.

7.______ Whales are examples of mammals that do not have any hair.

8.______ Hair evolved from reptilian scales.

9.______ Each hair filament is produced by a hair follicle beneath the surface of the skin.

10.______ The gestation period in marsupials is far shorter than the gestation period in placental mammals.

11.______ Most species of mammals are placental.

12.______ Animals of the order Cetacea are the only mammals that live entirely in the water.

13.______ The incisors of a rodent continue to grow throughout its lifetime.

MULTIPLE CHOICE Circle the letter of the term or phrase that best completes each statement or best answers each question.

14. Early pelycosaurs
   a. were small.
   b. had synapsid skulls.
   c. were lizardlike.
   d. All of the above

15. Pelycosaurs evolved into, and were replaced by, the
   a. dinosaurs.
   b. Pinnipedia.
   c. lizards.
   d. therapsids.

16. The sheet of bone that separates the mouth cavity from the nasal cavity in mammals is called the
   a. nasal septum.
   b. secondary palate.
   c. osseous plate.
   d. oral palate.

17. What adaptations did therapsids and mammals share?
   a. complex teeth
   b. a secondary palate
   c. large synapsid openings
   d. All of the above

18. The large eye sockets on the early mammals suggest that the mammals
   a. could see in the back of their heads.
   b. did not have to turn their heads to see to the side.
   c. were active at night.
   d. lived in caves.

19. One of the main ways to distinguish the skull of a mammal from the skull of a therapsid is that mammals have
   a. a variety of teeth. c. a single jawbone.
   b. eye sockets. d. None of the above

20. The duckbill platypus and spiny anteater are examples of
   a. marsupials. c. macroscelidea.
   b. monotremes. d. placentals.

21. Kangaroos and opossums are
   a. marsupials. c. macroscelidea.
   b. monotremes. d. placentals.

22. Mammals that lay eggs are
   a. placental mammals. c. monotremes.
   b. pouched mammals. d. semipouched mammals.

23. Mammals reached their maximal diversity during the
   a. Jurassic period. c. Triassic period.
   b. Permian period. d. Tertiary period.

24. Half of the energy in milk comes from
   a. keratin. c. protein.
   b. carbohydrates. d. fat.

25. The sheet of muscle at the bottom of the rib cage of mammals is called the
   a. secondary palate. c. diaphragm.
   b. metabolic sheet. d. placenta.

26. Egg-laying mammals are
   a. oviparous. c. ovoviviparous.
   b. viviparous. d. None of the above

27. Marsupials are found in
   a. Australia. c. North America.
   b. Asia. d. All of the above

28. The major difference between marsupials and placentals is
   a. the length of their hairs. c. their pattern of development.
   b. their teeth. d. their hooves.

29. Bats
   a. are all carnivores.
   b. only fly by gliding.
   c. have wings that are modified hind limbs.
   d. use sound to help them navigate.

30. Offspring remain inside the mother until development is essentially complete in
   a. placental mammals. c. marsupials.
   b. monotremes. d. All of the above

31.Seals and walruses belong to the order

a. Cetacea.
b. Pinnipedia.
c. Proboscidea.
d. Insectivora.

32.Dugongs and sea cows are

a. toothless mammals.
b. hoofed mammals.
c. sirenians.
d. marine hunters.

33.The largest animal that ever lived is

a. a cetacean.
b. the blue whale.
c. a filter feeder.
d. All of the above

34.Unlike reptiles, mammalian young are dependent on parental care for

a. food.
b. protection.
c. learning.
d. All of the above

35.Today all monotremes and most marsupials live in Australia and New Guinea. This limited distribution is due to

a. predation by other mammals.
b. continental drift.
c. inability of these animals to survive in cold climates.
d. All of the above

36.A dog is a member of the order

a. Rodentia.
b. Insectivora.
c. Carnivora.
d. Cetacea.

37.Carnivores usually

a. have a keen sense of smell.
b. have excellent eyesight.
c. can run quickly.
d. All of the above

The following question(s) refer(s) to the diagram of a coyote's skull.

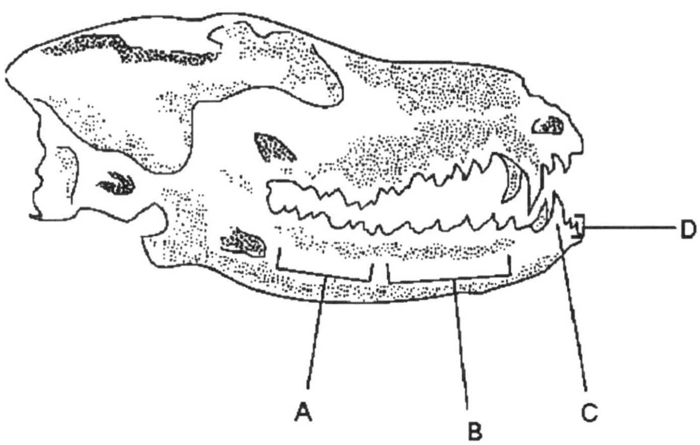

38.The fact that the coyote is a predator can be inferred from

a. the shape of the skull.
b. the position of the eye sockets.
c. the length of the jawbone.
d. the shape of the teeth.

39.The teeth labeled A are called

a. premolars.
b. incisors.
c. molars.
d. canines.

40. The teeth labeled C are called
   a. premolars.   c. molars.
   b. incisors.   d. canines.

41. The teeth primarily used for biting and cutting are
   a. A.   c. C.
   b. B.   d. D.

SHORT ANSWER Complete each statement by writing the correct term or phrase in the space provided.

42. A synapsid skull has a single ____________________ behind each eye socket.

43. The ____________________ are the direct ancestors of the mammals.

44. An animal with a secondary palate can ____________________ and chew at the same time.

45. Scientists think that therapsids may have declined in numbers because of competition from ____________________.

46. The only prototherians surviving today are the ____________________.

47. Hair is a filament composed mainly of dead cells filled with the protein ____________________.

48. Milk contains water to prevent ____________________.

49. An animal that maintains a high, nearly constant body temperature through metabolism is said to be ____________________.

50. The period of time between fertilization and birth in placentals and marsupials is known as the ____________________.

51. Marsupial animals complete their development in the mother's ____________________.

52. The only mammals with binocular vision are the __________.

53. ____________________ are the only mammals capable of powered flight.

ESSAY Read and answer each question or statement on a separate sheet of paper.

54. While no living therapsids have ever been found, scientists are fairly certain they were endothermic. Why?

55. What are the functions of hair on a mammal?

56. Why do scientists think monotremes are more closely related to the early mammals than are any other living mammals?

ANALOGIES **An analogy is a relationship between two pairs of words or phrases written as: a:b::c:d. The symbol : is read "is to," and the symbol :: is read "as." Write the letter of the pair of terms that best completes the analogy shown.**

57. hair : insulation::
   a. endothermy : keeping mammals warm in hot weather
   b. synapsid opening : swallowing
   c. diaphragm : helping to draw air into lungs
   d. complex teeth : breathing
   e. secondary palate : stabbing

____

58. breathing and eating : secondary palate::
   a. higher body temperature : large synapsid opening
   b. better sensory functions : complex teeth
   c. bigger jaw muscles : mammalian hair
   d. bigger jaw muscles : large synapsid opening
   e. night activity decreasing : mammalian hair

____

59. young mammals : nourishment from mother::
   a. monotreme embryos : oxygen from mother
   b. young reptiles : no nourishment from mother
   c. mammalian tissues : deoxygenated blood
   d. monotremes : no nourishment from mother
   e. marsupials : no nourishment from mother

____

60. monotremes : Australia and New Guinea::
   a. placentals : Australia
   b. Virginia opossums : North America
   c. placentals : New Guinea
   d. marsupials : Central America
   e. monotremes : North America

____

61. cetaceans : oceans::
   a. lagomorphs : fresh water
   b. native carnivores : Australia
   c. placental mammals : Antarctica
   d. manatees : terrestrial areas
   e. edentates : South and Central America

____

TEST ANSWERS

TRUE/FALSE

1.F

2.T

3.F

4.T

5.F

6.F

7.F

8.F

9.T

10.T

11.T

12.T

13.T

MULTIPLE CHOICE

14.d

15.d

16.b

17.d

18.c

19.c

20.b

21.a

22.c

23.d

24.d

25. c

26. a

27. d

28. c

29. d

30. a

31. b

32. c

33. d

34. d

35. b

36. c

37. d

38. d

39. c

40. d

41. d

SHORT ANSWER

42. opening

43. therapsids

44. breathe

45. thecodonts

46. monotremes

47. keratin

48. dehydration

49. endothermic

50. gestation period

51. pouch

52. primates

53. Bats

ESSAY

54. Therapsid adaptations for efficient feeding suggest that they had a high demand for energy. In addition, therapsid fossils have been found in areas that had cold winters, and there is some evidence that they had fur to insulate their babies.

55. The primary function of hair is insulation. Hair may also serve as camouflage or serve as a signal to other animals. In some mammals, hair serves a sensory function.

56. Monotremes, like reptiles, lay eggs and have a cloaca. In addition, monotreme shoulders and forelimbs resemble those of reptiles.

ANALOGIES

57. c (is for)

58. d (was/were made possible with)

59. b (receive)

60. b (live only in)

61. e (live in)

Chapter 35

TRUE/FALSE Read each statement. If the statement is true, write T in the space provided. If the statement is false, write F in the space provided.

1.______ One of the most important features of vertebrates is their endoskeleton.

2.______ There are over 2,000 bones, some of which are fused together, in the human skeleton.

3.______ In humans, extracellular digestion occurs in the gut.

4.______ Glandular tissue is a type of connective tissue.

5.______ Stratified epithelium is most abundant in the skin.

6.______ Fibroblasts make antibodies or attack virus-infected cells and cancer cells.

7.______ The endocrine system includes various glands that regulate and coordinate the body's functions.

8.______ The marble-sized region of the brain that monitors temperature and ion concentrations throughout the body is called the hypothalamus.

9.______ An open-system series of vessels in the body that collects fluid and returns it to the blood is called the endocrine system.

10.______ Air travels from the mouth or nose to the lungs through a long tube called the esophagus.

11.______ The urinary system filters metabolic wastes from the bloodstream and controls the ionic composition of the blood.

12.______ Carbohydrates, proteins, and fats are broken down into sugars, amino acids, and fatty acids in the large intestine.

MULTIPLE CHOICE Circle the letter of the term or phrase that best completes each statement or best answers each question.

13. Ligaments attach
   a. bone to bone.
   b. muscle to bone.
   c. muscle to muscle.
   d. cartilage to bone.

14. The point where two or more bones meet is called a
   a. sprain.
   b. joint.
   c. point of intersection.
   d. growth region.

15. The nerves that bring information to the brain and transmit commands from it are called the
   a. lateral nervous system.
   b. conjunctive nervous system.
   c. dynamic nervous system.
   d. peripheral nervous system.

16. A group of similar cells that work together to perform a common function is called
   a. a body system.
   b. an organ.
   c. a tissue.
   d. a neuron.

17. Blood, bone, and cartilage are examples of
   a. three basic tissue types found in the body.
   b. connective tissue.
   c. epithelial tissues.
   d. organs of the body.

18. Tissue that is specialized to cover the inner and outer surfaces of your internal organs is
   a. epithelial tissue.
   b. connective tissue.
   c. muscle tissue.
   d. nerve tissue.

19. From the smallest functional units to the largest, the body is organized as follows:
   a. cell, system, organ, tissue, body.
   b. organ, cell, tissue, system, body.
   c. system, organ, tissue, cell, body.
   d. cell, tissue, organ, system, body.

20. Tightly connected cells that are arranged in flat sheets is indicative of
   a. epithelial tissue.
   b. connective tissue.
   c. muscle tissue.
   d. nerve tissue.

21. Of the four basic types of tissue, the tissue that is specialized to transmit and receive messages from throughout the body is
   a. epithelial tissue.
   b. connective tissue.
   c. muscle tissue.
   d. nerve tissue.

22. Collagen is produced by
   a. erythrocytes.
   b. glial cells.
   c. cartilage.
   d. fibroblasts.

23. Cells that insulate and provide neurons with nutrients are called
   a. nurturing cells.
   b. fibroblasts.
   c. glial cells.
   d. amniotic cells.

24. Smooth muscle can be found
   a. attached to the skeleton.
   b. in the wrist bones.
   c. at the knee joint.
   d. in some internal organs.

25. Keratin
   a. is a strong, fibrous protein.
   b. fills dead cells in the dermis.
   c. is a skin pigment.
   d. All of the above

26. The system that defends the body against pathogens and cancer is called the
   a. nervous system.
   b. endocrine system.
   c. respiratory system.
   d. immune system.

27. Muscle tissue functions to move
   a. blood.
   b. food in the digestive tract.
   c. bones.
   d. All of the above

28. Functions of the skeleton include
   a. blood cell production.
   b. calcium storage.
   c. protection of organs and movement.
   d. All of the above

29. Tendons connect
   a. bone to bone.
   b. muscle to bone.
   c. muscle to muscle.
   d. cartilage to bone.

30. Homeostasis in the body is maintained by
   a. endothermy.
   b. feedback loops.
   c. homeostatic nodes.
   d. the plasma.

31. The part of the brain that is responsible for learning and thought is the
  a. cerebrum. c. brain stem.
  b. cerebellum. d. hypothalamus.

32. Homeostasis in the body is maintained by a battery of specific molecules called
  a. erythrocytes. c. plasmas.
  b. collagens. d. hormones.

The following question(s) refer(s) to the diagram of the internal anatomy of the human body.

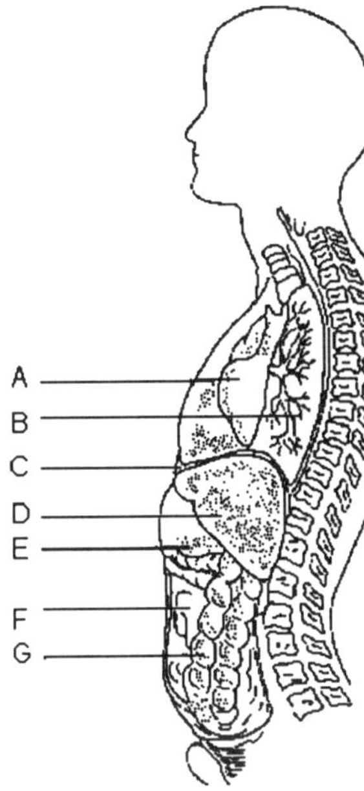

33. The large cavity that is found in all vertebrates is called the
  a. chorion. c. colon.
  b. coelom. d. cloaca.

34. Arrow C is pointing to the
  a. dysenteric membrane. c. mid-body muscle.
  b. lungs. d. diaphragm.

35. Arrow E is pointing to the
  a. stomach. c. pharyngeal cavity.
  b. pancreas. d. liver.

SHORT ANSWER Complete each statement by writing the correct term or phrase in the space provided.

36. All organs of the body contain epithelial, connective, muscle, and ________________ tissues.

37. The ________________ nervous system includes the brain and spinal cord.

38. Like insects, humans have ________________ appendages.

39. The coelom develops entirely within the ________________ tissue of the developing embryo.

40. ________________ tissue joins, supports, and protects other types of tissue.

41.______________________ tissue can be smooth, skeletal, or cardiac.

42.Tissues only one cell thick that line the respiratory tract and major cavities of the body are known as ______________________ epithelium.

43.Mobile cells of the immune system that engulf bacteria are called ______________________.

44.Red blood cells are called ______________________.

45.The ______________________ is a cytoplasmic extension from a neuron that can be over 1 m in length.

46.The proteins actin and ______________________ in muscle enable the cells to contract.

47.The fluid in your body that is not inside your cells is called ______________________ fluid.

48.The bones of the skull and backbone are part of the ______________________ skeleton.

The following question(s) refer(s) to the diagram of an artery and a vein.

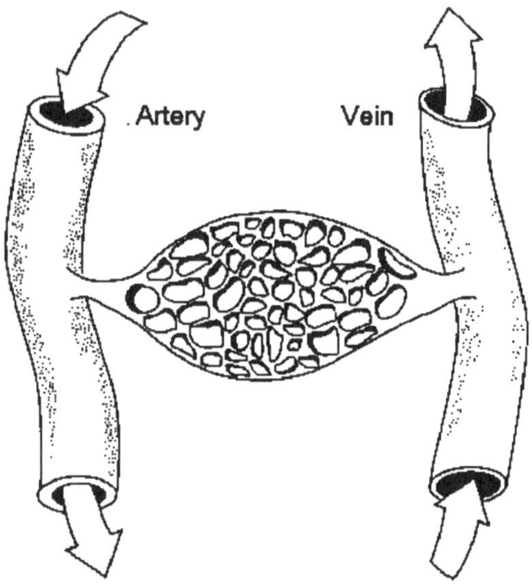

49.The vessels in the diagram that connect an artery and a vein are called ______________________.

ESSAY Read and answer each question or statement on a separate sheet of paper.

50.What are the two advantages of a closed circulatory system?

51.How does maintaining a relatively warm body temperature (endothermy) help humans?

52.Explain how a negative feedback loop functions in homeostasis.

53.Why aren't all communications between cells in the body handled by chemical signals?

54. Why is it essential for the body to maintain homeostasis?

ANALOGIES An analogy is a relationship between two pairs of words or phrases written as: a:b::c:d. The symbol : is read "is to," and the symbol :: is read "as." Write the letter of the pair of terms that best completes the analogy shown.

55. bone : vertebrate skeletal strength::
   a. skin : skeletal strength
   b. ligament : outer body covering
   c. joint : motion
   d. cartilage : holding joints together
   e. skin : holding joints together

____

56. central nervous system : a computer::
   a. peripheral nervous system : a lubricant
   b. peripheral nervous system : bands
   c. peripheral nervous system : pads
   d. peripheral nervous system : telephone lines
   e. peripheral nervous system : cushions

____

57. fibroblast : construction worker::
   a. cartilage : a scuba diver's oxygen tank
   b. epithelial tissue : cellophane wrapping
   c. erythrocyte : shoe heel pad
   d. bone : lubricating oil
   e. lymphocyte : oil well

____

58. skeletal system : major organs and body parts::
   a. integumentary system : against injury and infection
   b. nervous system : lungs and heart
   c. digestive system : kidneys
   d. muscular system : brain and sense organs
   e. circulatory system : against inhaling oxygen

____

59. negative feedback system : variables back to set point::
   a. key ion concentration level : positive feedback system
   b. set point : normal values
   c. positive feedback system : variables back to set point
   d. positive feedback system : variables farther away from set point
   e. homeostasis : set points

____

60. chemical signals : a long time::
   a. electrical signals : a time longer than hormonal signals
   b. electrical signals : a period slower than chemical signals
   c. electrical signals : brief periods
   d. electrical signals : a long time
   e. electrical signals : same period of time as chemical signals

____

61. artery : blood away from the heart::
   a. lymphatic system : erythrocytes
   b. blood : capillaries away from the heart
   c. white blood cell : carbon dioxide
   d. white blood cell : oxygen
   e. vein : blood toward the heart

____

62. grinding up food in mouth : digestion in the stomach::
   a. concentration of solids : digestion and absorption by the small intestine
   b. concentration of solids : digestion in the stomach
   c. absorption by small intestine : digestion in the stomach
   d. concentration of solids : grinding up food in the mouth
   e. absorption by small intestine : concentration of solids

____

63. brain stem : heart rate, blood pressure, and breathing::
   a. cerebrum : transmitting motor nerves to sensory nerves
   b. cerebellum : balance and coordination
   c. spinal cord : learning, memory, and thought
   d. sensory nerve : the transmitting of commands to the body
   e. motor nerve : the relay of sensory information to the brain

____

64. atrium : heart::
   a. nephron : kidney
   b. kidney : nephron
   c. heart : atrium
   d. heart : septum
   e. land : water

____

TEST ANSWERS

TRUE/FALSE

1.T

2.F

3.T

4.F

5.T

6.F

7.T

8.T

9.F

10.F

11.T

12.F

MULTIPLE CHOICE

13.a

14.b

15.d

16.c

17.b

18.a

19.d

20.a

21.d

22.d

23.c

24.d

25. a

26. d

27. d

28. d

29. b

30. b

31. a

32. d

33. b

34. d

35. a

SHORT ANSWER

36. nerve

37. central

38. jointed

39. mesoderm

40. Connective

41. Muscle

42. simple

43. macrophages

44. erythrocytes

45. axon

46. myosin

47. extracellular

48. axial

49. capillaries

ESSAY

50. With a closed circulatory system the body can maintain different circulation rates in different organs by changing the diameter of blood vessels. A second advantage of a closed circulatory system is that it allows the blood to be pumped under pressure. The heart pushes the blood out through the body's arteries, sending the blood through the body much faster than would be possible with an open system.

51. Endothermy permits the body to maintain its activity at all times and in many different places, regardless of the temperature of the surroundings. In addition, endothermy allows the body to sustain strenuous activity. Ectotherms, by contrast, are limited to short periods of exertion.

52. Negative feedback prevents a variable such as pH or temperature from deviating from its normal value, its set point. Any changes that increase the difference between the variable and its set point initiate responses that tend to oppose the changes and restore the variable to its set point.

53. Their transmission by the circulatory system is very slow.

54. A stable fluid environment is necessary for multicellular organisms. Any extreme change in extracellular fluid can have very serious effects on the activities inside cells. If factors such as pH level, ion concentration, or sugar level vary outside the range that cells require, death may result.

ANALOGIES

55. c (is for)

56. d (is like)

57. b (acts like a)

58. a (protects)

59. d (changes)

60. c (last for)

61. e (carries)

62. e (comes before)

63. b (controls)

64. a (is a part of a)

Chapter 36

TRUE/FALSE Read each statement. If the statement is true, write T in the space provided. If the statement is false, write F in the space provided.

1.______ The vertebrate skeleton is similar to the arthropod skeleton, in that both are composed of chitin.

2.______ The Haversian canals of long bones are filled with yellow marrow.

3.______ The chances of developing osteoporosis later in life are greatly reduced by a calcium-rich diet and regular exercise earlier in life.

4.______ The majority of the bones of the skeleton are part of the axial skeleton.

5.______ The hip attachment of bones is called the pectoral girdle and the shoulder attachment is called the pelvic girdle.

6.______ In order for muscles to move parts of the body, they must be attached to something they can pull against.

7.______ Extensor muscles cause bones to bend at a joint, while flexors cause them to straighten.

8.______ Resistance exercises, such as chin-ups, are designed to increase the efficiency of the body's muscles to produce ATP.

9.______ The skin is the largest organ of the human body.

10.______ The distribution of melanin in the epidermis of the skin determines the complexion of an individual.

11.______ Psoriasis is a form of acne that is common in adults, rather than in teens.

MULTIPLE CHOICE Circle the letter of the term or phrase that best completes each statement or best answers each question.

12. The endoskeleton of humans
   a. contains chitinous bones.
   b. is too brittle to be sufficiently strong.
   c. does not provide enough protection to internal organs.
   d. is composed of calcium-hardened bones.

13. The type of bone that provides the greatest strength for support is
   a. spongy bone.   c. compact bone.
   b. chitinous bone.   d. marrow bone.

14. Yellow marrow
   a. provides internal support to spongy bone.
   b. produces red blood cells.
   c. is found only in lower vertebrates.
   d. stores fat.

15. Young bone cells that are responsible for the formation of new bone are
   a. osteoblasts.   c. red marrow cells.
   b. periosteal cells.   d. yellow marrow cells.

16. In an embryo, the skeleton is originally made of
   a. red and yellow marrow.
   b. calcium phosphate.
   c. cartilage.
   d. osteopores.

17. Osteocytes trapped within spaces in the bone surrounding them
   a. are called marrow cells.
   b. are provided with food and oxygen by Haversian canals.
   c. receive nutrients directly from the cells of the surrounding cartilage.
   d. eventually become osteoblasts as the bone matures.

18. Osteoporosis is a medical condition that results from
   a. too much calcium in a person's diet.
   b. failure to convert cartilage to bone during development.
   c. an excessive rate of bone replacement.
   d. inefficient bone replacement later in life.

19. Women are more susceptible than men to the effects of osteoporosis because
   a. their bones are smaller than those of men.
   b. they lose calcium monthly during the menstrual cycle.
   c. sex hormone production declines during menopause.
   d. All of the above

20. Of the following, the structure that is *not* part of the axial skeleton is the
   a. backbone.
   b. pectoral girdle.
   c. rib cage.
   d. skull.

21. Muscle cells are able to exert force by
   a. converting ADP and inorganic phosphate into ATP.
   b. interfering with the forces of gravity and friction.
   c. rapid relaxation of muscle fibers.
   d. pulling on surrounding tissues.

22. A muscle's insertion
   a. is located on a bone that remains stationary when the muscle contracts.
   b. moves away from the origin during muscle contraction.
   c. is attached to a bone by a ligament.
   d. None of the above

23. The origin of a muscle
   a. is at the opposite end of the muscle from the insertion.
   b. is located on a bone that remains stationary when the muscle contracts.
   c. does not move when the muscle contracts.
   d. All of the above

24. Which of the following is/are the smallest?
   a. muscle
   b. muscle fibers
   c. myofibrils
   d. myosin and actin filaments

25. Repeating units of myosin and actin filaments bounded by two Z lines are
   a. muscles.
   b. myofibrils.
   c. sarcomeres.
   d. extensors.

26. The total amount of force that a muscle can exert
   a. is determined by the strength of the nerve impulse that caused the contraction.
   b. depends on the total number of individual muscle fibers that have been stimulated.
   c. is dependent upon the weight of the object being moved.
   d. correlates to the number of Z lines contained within the sarcomeres of the muscle.

27. Aerobic exercises are beneficial because they
   a. increase muscle size and strength.
   b. stimulate an increased efficiency of ATP production.
   c. use up glycogen stores quickly.
   d. prevent myosin and actin linkages in muscles.

28. The physical danger of abusing anabolic steroids is that they
   a. increase muscle size, but not endurance.
   b. exaggerate endurance of muscles during exercise.
   c. give an unfair advantage to the abusers.
   d. reduce the size of certain skeletal muscles.

29. The functions of the skin include
   a. defense against microbes.
   b. regulation of body temperature.
   c. prevention of dehydration.
   d. All of the above

30. The dermis of the skin is
   a. composed of corneal and basal layers.
   b. the innermost layer of the skin.
   c. the location of melanocytes.
   d. involved in temperature regulation.

31. Skin tanning may occur as a result of
   a. exposure to excessive heat.
   b. increases in melanin production.
   c. deposits of fat-soluble vitamins in subcutaneous tissue.
   d. a deficiency of UV radiation in the environment.

32. A skin disorder caused by blockage of oil ducts is called
   a. acne.
   b. carcinoma.
   c. osteoporosis.
   d. psoriasis.

33. Hair and nails are produced by the cells of the
   a. dermis.
   b. subcutaneous layer.
   c. subcutaneous glands.
   d. epidermis.

34. The heart and lungs are protected by the
   a. pectoral girdle.
   b. pelvic girdle.
   c. rib cage.
   d. periosteum.

The following question(s) refer(s) to the diagrams of several joints.

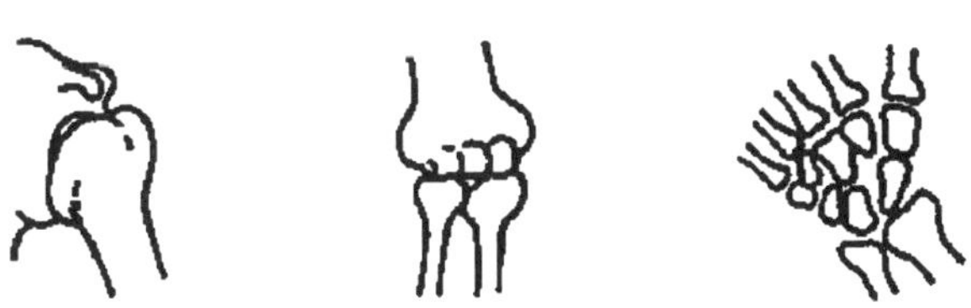

Joint "1" Joint "2" Joint "3"

35. Which of the joints shown represents a ball-and-socket joint?
   a. joint 1
   b. joint 2
   c. joint 3
   d. None of the above is a ball-and-socket joint.

36. Joint 2 is most likely an illustration of a(n)
   a. shoulder.
   b. elbow.
   c. ankle.
   d. hand.

37. Movement occurs in a sliding manner in plane joints as indicated in the illustration of
   a. joint 1.
   b. joint 2.
   c. joint 3.
   d. None of the above

SHORT ANSWER Complete each statement by writing the correct term or phrase in the space provided.

38. ______________________ bone is dense, almost solid, providing a great deal of support.

39. The ______________________ is a membrane that surrounds individual bones.

40. The cells that produce new bone by secreting collagen fibers on which calcium phosphate is deposited are called ______________________.

41. The skull, backbone, and rib cage compose the ______________________ skeleton.

42. The rib cage is formed by curved ribs extending from the backbone and joining together in the front at the bone called the ______________________.

43. The bones of the arms, legs, pectoral girdle, and pelvic girdle make up the ______________________ skeleton.

44. The heart and lungs are protected by the ______________________.

45. The junction of two bones is called a(n) ______________________.

46. ______________________ is a painful degeneration of movable joints caused by attacks of the joints by cells of the immune system.

47. Strips of dense connective tissue that attach muscles to bones are called ______________________.

48. Muscles that function to straighten a joint are called ____________.

49. ______________________ are muscles that cause limbs to bend toward each other at a joint.

50. Repeating units of actin and myosin filaments are called ______________________.

51. ______________________ are a group of drugs that can cause increased muscle size without an accompanying increase in muscle endurance.

52. The outermost layer of epidermis, consisting of a layer of flattened, mostly dead cells filled with keratin is called the ______________________ layer.

53. The most common type of skin cancers are ______________________, which originate in the non-pigment-producing cells of the epidermis.

54. Cells of the ____________________ layer of the epidermis are constantly being damaged and replaced.

ESSAY Read and answer each question or statement on a separate sheet of paper.

55. There are three basic types of joints. Using examples, describe these types of joints and their primary functions.

56. Briefly describe the operation of muscle according to the sliding filament theory.

57. The skin is composed of three layers. What are they, and what do they do?

ANALOGIES An analogy is a relationship between two pairs of words or phrases written as: a:b::c:d. The symbol : is read "is to," and the symbol :: is read "as." Write the letter of the pair of terms that best completes the analogy shown.

58. human skeleton : internal organs::
  a. spoon : fork
  b. construction worker's hard hat : a skull
  c. whiskbroom : dust
  d. turnstile : a ticket
  e. puzzle : a picture frame

____

59. compact bone : periosteum::
  a. periosteum : compact bone
  b. compact bone : spongy bone
  c. spongy bone : compact bone
  d. marrow : compact bone
  e. compact bone : marrow

____

60. freely movable joint : slightly movable joint::
  a. slightly movable joint : immovable joint
  b. immovable joint : slightly movable joint
  c. immovable joint : freely movable joint
  d. joint with sutures : cartilaginous joints
  e. joint with sutures : freely movable joints

____

61. flexors : bend::
  a. immovable joint : bend
  b. slightly movable joint : be immovable
  c. extensors : straighten
  d. sutures : move a great deal
  e. tendons : get weak

____

62. sarcomere contraction : ATP::
   a. Z lines : actin and myosin
   b. actin and myosin : sarcomeres
   c. muscle forces : number of Z lines
   d. muscle forces : number of muscle fibers contracting
   e. sarcomeres : number of muscle fibers contracting

---

63. specialized epidermal cell : nails::
   a. "half-moon area" : hair
   b. keratin : melanocytes
   c. melanocyte : keratin
   d. hair follicle : hair
   e. melanocyte : hair

---

64. heat loss : relaxation of blood vessels::
   a. decrease in body temperature : exercise
   b. increase of external body temperature : rest
   c. removal of heat from body surface : evaporation of sweat
   d. build-up of body heat : evaporation of sweat
   e. reddish skin coloration : drop in body temperature

---

65. psoriasis : skin disorder::
   a. burned skin area : carcinoma
   b. psoriasis : malignant melanoma
   c. psoriasis : skin cancer
   d. acne : psoriasis
   e. malignant melanoma : skin cancer

TEST ANSWERS

TRUE/FALSE

1.F

2.F

3.T

4.F

5.F

6.T

7.F

8.F

9.T

10.T

11.F

MULTIPLE CHOICE

12.d

13.c

14.d

15.a

16.c

17.b

18.d

19.d

20.b

21.d

22.d

23.d

24.d

25. c

26. b

27. b

28. a

29. d

30. d

31. b

32. a

33. d

34. c

35. a

36. b

37. c

SHORT ANSWER

38. Compact

39. periosteum

40. osteoblasts

41. axial

42. sternum or breastbone

43. appendicular

44. rib cage

45. joint

46. Rheumatoid arthritis

47. tendons

48. extensors

49. Flexors

50. sarcomeres

51. Anabolic steroids

52. corneal

53. carcinomas

54. corneal

ESSAY

55. Immovable joints are very tight joints that hold adjacent bones together, permitting very little or no movement. The cranial bones of the skull are held together by immovable joints called sutures. Limited mobility is permitted by slightly movable joints, the second basic type of joint. In these joints, a bridge of cartilage joins two bones together, as in the joints between the vertebrae of the spine. The third type of joint is freely movable joints. These joints allow the greatest degree of movement, between bones that are held together by ligaments.

56. The sliding filament theory explains how a muscle contracts. Myosin and actin filaments lie in parallel lines along the length of a myofibril in units called sarcomeres. The myosin heads touch the adjacent actin filaments. When a muscle contracts, the myosin heads attach to the actin filaments, and, when the heads bend inward, they pull the actin filaments along with them toward each other. This bending of the myosin head requires the use of an ATP molecule. As the actin filaments move toward each other along the myosin filament, they pull their attached Z lines with them, thus shortening the sarcomere. As sarcomeres are shortened along the entire muscle fiber, the entire muscle contracts.

57. The outermost layer of the skin is the epidermis. This layer is responsible for protection from interaction with the environment. It produces new cells from its lower layer (the basal layer) that move upward to become the corneal layer. Accessory structures, such as hair and nails, are produced by cells of the epidermis. The middle, and thickest, layer of the skin is called the dermis, which is the major framework of the skin. This layer provides structural support and is a matrix for nerve endings, blood vessels, and specialized cells of the skin. The subcutaneous tissue is the innermost layer of the skin. This is a layer of fat-rich cells that act as shock absorbers, provide insulation, and store vitamins.

ANALOGIES

58. b (protects)

59. d (is found inside)

60. a (moves more than)

61. c (cause a joint to)

62. d (depend on)

63. d (produces)

64. c (occurs with)

65. e (is a type of)

Chapter 37

TRUE/FALSE Read each statement. If the statement is true, write T in the space provided. If the statement is false, write F in the space provided.

1.______ One of the main functions of the human circulatory system is to distribute heat to all parts of the body.

2.______ The innermost layer of an artery is a layer of elastic, smooth muscle tissue.

3.______ New red blood cells are produced by root cells in the bone marrow.

4.______ When a platelet encounters a damaged blood vessel it releases fibrin which causes a clot to form.

5.______ Antibodies are produced by lymphocytes.

6.______ A person with blood type AB can donate blood to a person with any other blood type.

7.______ The superior vena cava drains blood from the head and neck.

8.______ The bridge of cardiac muscle cells that connects the upper chambers of the heart to the lower chambers is called the Bundle of His.

9.______ Blood pressure is measured with a device called a sphygmomanometer.

10.______ Asthma is a respiratory disease in which the alveoli of the lungs become narrowed.

11.______ The voice box is sometimes called the pharynx.

12.______ The effect of carbon dioxide on the unloading of oxygen by the red blood cells is called the Bohr effect.

MULTIPLE CHOICE Circle the letter of the term or phrase that best completes each statement or best answers each question.

13. Excess fluids and proteins in the body are returned to the bloodstream by
   a. the heart. c. the respiratory system.
   b. arteries. d. the lymphatic system.

14. An artery has a much thicker muscle layer than
   a. a vein. c. a venule.
   b. a capillary d. All of the above

15. The smallest and most numerous blood vessels in the body are the
   a. venules. c. arteries.
   b. veins. d. capillaries.

16. Red blood cells
   a. transport respiratory gases. c. destroy viruses.
   b. combat bacterial invasion. d. remove nutrients from the blood.

17. Defending the body against bacterial infection and invasion by other foreign substances is the function of
   a. red blood cells. c. platelets.
   b. plasma. d. leukocytes.

18. A person with antigen A on their red blood cells can give blood to someone with blood type(s)
a. A, AB.
c. AB.
b. B, AB.
d. O.

19. An abnormality involving the platelets would probably affect the process of
a. breathing.
c. digestion.
b. locomotion.
d. clotting.

20. The circulatory system is responsible for
a. the distribution of oxygen throughout the body.
b. carrying nutrients.
c. transmitting hormones.
d. All of the above

The following question(s) refer(s) to the diagram of the human heart.

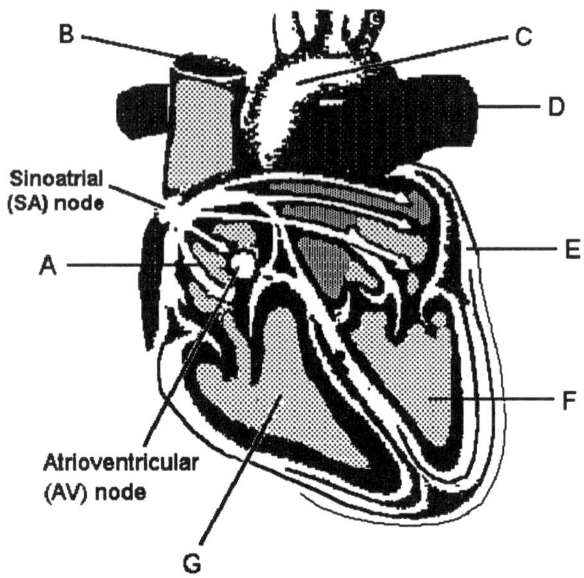

21. Blood in chamber A
a. is full of oxygen.
c. is oxygen poor.
b. is returning from the lungs.
d. has very little plasma.

22. The vessel indicated at D carries deoxygenated blood. This vessel is
a. the pulmonary artery.
c. part of the aorta.
b. the pulmonary vein.
d. part of the atria.

23. The chamber indicated at E is
a. the right atrium.
c. the right ventricle.
b. the left atrium.
d. the left ventricle.

24. The aorta is indicated by
a. C.
c. F.
b. D.
d. B.

25. The pacemaker responsible for starting a heartbeat
a. is located in the brain.
b. squeezes the ventricles shut.
c. is a small cluster of cells embedded in the right atrium.
d. is in the aorta.

26. If a blood vessel has valves, it probably
   a. is a vein.
   b. is an artery.
   c. is a capillary.
   d. has a muscular wall.

27. A condition known as atherosclerosis results in
   a. increased circulation to the heart.
   b. larger muscles.
   c. a narrowing of the inner walls of blood vessels.
   d. a widening of the inner walls of blood vessels.

28. The higher blood pressure that occurs when a pulse of blood is forced into the arteries is called the
   a. sphygmotic pressure.
   b. diastolic pressure.
   c. alveolar pressure.
   d. systolic pressure.

29. The pulmonary circulation is circulation to the
   a. kidneys.
   b. liver.
   c. intestines.
   d. lungs.

30. Vessels that carry blood away from the heart are called
   a. veins.
   b. capillaries.
   c. arteries.
   d. venules.

31. The ventricles are
   a. the upper chambers of the heart.
   b. the chambers of the heart that pump blood to the lungs and to the rest of the body.
   c. the chambers of the heart that receive blood from the lungs and the rest of the body.
   d. less muscular than the atria.

32. Carbon dioxide is transported in the blood in all of the following ways except
   a. dissolved in plasma.
   b. combined with hemoglobin.
   c. as bicarbonate ions.
   d. by white blood cells.

33. Alveoli in the lungs are connected to the bronchi by a network of tiny tubes called
   a. arterioles.
   b. venules.
   c. capillaries.
   d. bronchioles.

34. A respiratory disease in which airways in the lungs become narrow because of sensitivity to certain stimuli is called
   a. asthma.
   b. lung cancer.
   c. emphysema.
   d. alveolar reduction.

35. The respiratory control center in the brain is sensitive to the
   a. concentration of oxygen.
   b. concentration of carbon dioxide in the lungs.
   c. concentration of carbon dioxide in the blood.
   d. the amount of carbon dioxide in the cells.

36. Each alveolus
   a. contains many air sacs.
   b. attaches directly to the larynx.
   c. is surrounded by capillaries.
   d. consists of several layers of tissue.

The following question(s) refer(s) to the diagram of one of the components of the blood.

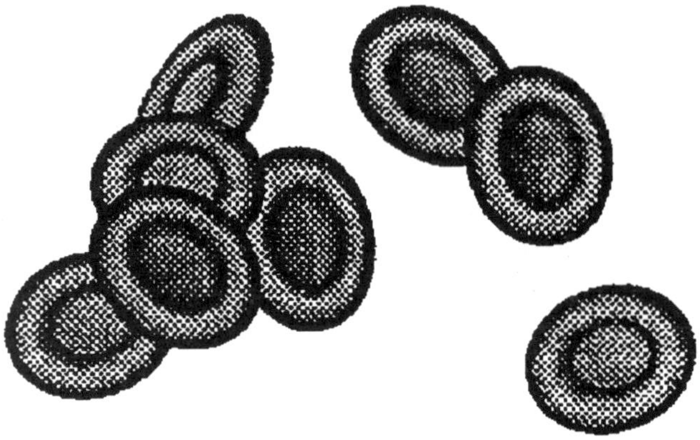

37. The cells illustrated are
   a. part of the plasma.
   b. platelets.
   c. red blood cells.
   d. white blood cells.

38. These cells
   a. can live for about a year.
   b. are the largest cells in the circulatory system.
   c. promote clotting.
   d. do not have a nucleus.

SHORT ANSWER Complete each statement by writing the correct term or phrase in the space provided.

39. As a result of metabolizing proteins, ______________________ wastes are released into the bloodstream.

40. From the arteries, the blood enters a large network of ______________,
or small arteries.

41. The largest vein in the human body is the ______________________.

42. ______________________ are blood cells that do not always stay in the cardiovascular system.

43. Large cells in bone marrow called ______________________ pinch off bits of their cytoplasm to form platelets.

44. The oxygen-carrying molecule in red blood cells is called ______________.

45. Excess fluids and proteins in the body are returned to the blood by a system of vessels called the ______________________ system.

46. The ______________________ valve prevents blood from going from the left ventricle to the left atrium.

47. The first arteries to branch from the aorta are the ______________________ arteries.

48. Continually higher-than-normal blood pressure is called ______________________.

49. The ______________________ node starts each contraction of the heart.

50. When you swallow, the ________________________ prevents food from entering the respiratory system.

51. The ________________________ is a long, straight tube that carries air from the back of the throat to the lungs.

52. An enzyme within the red blood cells, called ______________________, combines carbon dioxide molecules with water to form carbonic acid.

53. ______________________ is a disease in which the lung tissue loses its elasticity.

ESSAY Read and answer each question or statement on a separate sheet of paper.

54. How does the circulatory system help the body maintain a constant temperature?

55. Explain how the RH factor can cause problems with pregnancies.

56. How is air moved into and out of the lungs?

ANALOGIES An analogy is a relationship between two pairs of words or phrases written as: a:b::c:d. The symbol : is read "is to," and the symbol :: is read "as." Write the letter of the pair of terms that best completes the analogy shown.

57. carbon dioxide : lungs::
   a. nutrient : kidneys
   b. nutrient : lungs
   c. nitrogen waste : kidneys
   d. nitrogen waste : lungs
   e. blood : kidneys

____

58. fluid : out of circulatory system capillaries::
   a. fluid : into arteries
   b. fluid : out of veins
   c. fluid : into veins
   d. fluid : into lymphatic capillaries
   e. fluid : out of arteries

____

59. vitamins, salts, and proteins : plasma solutes::
   a. arteries and veins : lymphatic vessels
   b. erythrocytes, leukocytes, and megakeryocytes : blood cells
   c. platelets and megakeryocytes : leukocytes
   d. lymphocytes and macrophages : erythrocytes
   e. erythrocytes and leukocytes : megakeryocytes

____

60. blood pressure : sphygmomanometer::
   a. heart electrical impulse : blood pressure device
   b. heart electrical impulse : heart muscle contractor
   c. heart electrical impulse : electrocardiogram
   d. diastolic pressure : pressure in the left ventricle
   e. systolic pressure : pressure in the arteries

---

61. cholesterol buildup : atherosclerosis::
   a. calcium buildup : atherosclerosis
   b. arteriosclerosis : the heart to work more easily
   c. low intake of saturated fats : heart attacks
   d. low cholesterol : heart attacks
   e. atherosclerosis : heart attacks

---

62. bronchiole : alveoli::
   a. alveoli : bronchi
   b. bronchi : larynx
   c. larynx : pharynx
   d. esophagus : pancreas
   e. bronchi : bronchioles

---

63. contraction of the diaphragm : inhalation::
   a. expanded rib cage : inhalation
   b. lowered rib cage : inhalation
   c. relaxation of the diaphragm : inhalation
   d. reduced air pressure in lungs : exhalation
   e. increased air pressure in lungs : inhalation

---

64. acidity in blood : signals to rib muscles and diaphragm::
   a. chest cavity size : with exhalation
   b. carbon dioxide in blood : unloading of oxygen
   c. increased rib muscle activity : lower breathing rate
   d. increased breathing rate : relaxation
   e. diaphragm : rib muscle contractions

TEST ANSWERS

TRUE/FALSE

1.T

2.F

3.F

4.F

5.T

6.F

7.T

8.F

9.T

10.F

11.F

12.T

MULTIPLE CHOICE

13.d

14.d

15.d

16.a

17.d

18.a

19.d

20.d

21.c

22.a

23.b

24.a

25. c

26. a

27. c

28. d

29. d

30. c

31. b

32. d

33. d

34. a

35. c

36. c

37. c

38. d

SHORT ANSWER

39. nitrogenous

40. arterioles

41. vena cava

42. Macrophages

43. megakeryocytes

44. hemoglobin

45. lymphatic

46. mitral

47. coronary

48. hypertension

49. sinoatrial

50. epiglottis

51. trachea

52. carbonic anhydrase

53. Emphysema

ESSAY

54. The body maintains a relatively constant temperature by using metabolic energy to make heat. The circulatory system distributes the heat to all parts of the body. Passing just below the skin, the circulatory system absorbs the heat from a hot environment and gives off heat to a cold one. When it is cold outside, blood vessels constrict. In the opposite manner, the blood vessels under the skin dilate when it is hot outside, thereby delivering more blood to the surface and allowing more heat to escape.

55. The greatest danger in Rh incompatibility arises in pregnancies in which an Rh^- mother is carrying an Rh^+ child. No Rh problems usually occur during the pregnancy. However, at delivery, some Rh^+ blood passes into the mother's bloodstream. After delivery, if the mother does not receive a special drug, her body will produce anti-Rh^+ proteins. If the mother becomes pregnant again, her blood will attack an Rh^+ child's blood cells, causing severe complications for the child.

56. When the diaphragm and rib muscles contract, the diaphragm moves downward and the rib cage is moved up and outward. This expands the chest cavity, lowering the air pressure in the lungs and causing air to flow in. When the diaphragm and rib muscles relax, the diaphragm moves upward and the rib cage moves down and inward.

ANALOGIES

57. c (is removed by the)

58. d (diffuses)

59. b (are different types of)

60. c (is measured with a[n])

61. e (causes)

62. e (transports air to the)

63. a (causes)

64. b (causes)

Chapter 38

TRUE/FALSE Read each statement. If the statement is true, write T in the space provided. If the statement is false, write F in the space provided.

1.______ According to most authorities, a well-balanced diet has more fat than protein.

2.______ Carbohydrates are obtained mostly from oils, margarine, and butter.

3.______ Excess calories that the body takes in are metabolized to produce heat energy.

4.______ Calcium is an example of a trace element that is required by the body.

5.______ Bile salts are manufactured and metabolized in the small intestine.

6.______ The synthesis of hydrochloric acid in the stomach is regulated by the hormone ulcerin.

7.______ The rectum is the final section of the digestive tract.

8.______ The inner surface of the large intestine contains villi specialized for the absorption of water.

9.______ The skin can be considered an excretory organ.

10.______ Urea is a highly toxic nitrogenous waste.

11.______ Dialysis is a permanent solution to kidney failure.

MULTIPLE CHOICE Circle the letter of the term or phrase that best completes each statement or best answers each question.

12. Nutrients provide the body with the energy and materials it needs for
  a. growth.
  b. maintenance.
  c. repair.
  d. All of the above

13. Vitamins and minerals
  a. are both proteins.
  b. do not supply the body with energy.
  c. supply the body with some energy.
  d. None of the above

14. Essential amino acids
  a. must be obtained from the food we eat.
  b. are made by our body.
  c. occur only in plants.
  d. None of the above

15. Vitamins are compounds that
  a. function as parts of enzymes.
  b. provide energy.
  c. form cell membranes.
  d. are not obtained from food.

16. Most of the body's energy needs come from
  a. carbohydrates.
  b. fats.
  c. vitamins.
  d. proteins.

17. Which of the following filtrates is *not* reabsorbed in significant quantities back into the bloodstream by the nephron?
   a. glucose   c. urea
   b. ions   d. water

18. The function of the digestive system is to
   a. chemically digest foods.   c. absorb nutrient materials.
   b. mechanically break down foods.   d. All of the above

19. Chemical digestion begins in the
   a. mouth.   c. esophagus.
   b. throat.   d. stomach.

20. Fat molecules are broken down into fatty acids by
   a. emulsifiers.   c. sphincters.
   b. amylases.   d. lipases.

21. The pharynx is
   a. located in the colon.   c. the voice box.
   b. in the back of the throat.   d. None of the above

22. The digestive enzyme in saliva begins the chemical digestion of
   a. fat.   c. starch.
   b. protein.   d. vitamins.

23. The wavelike contraction of muscle that moves food through the digestive system is called
   a. peristaltic contraction.   c. mechanical digestion.
   b. voluntary contraction.   d. tracking.

24. The first portion of the small intestine is the
   a. colon.   c. duodenum.
   b. esophagus.   d. rectum.

25. Bile
   a. breaks globules of fat into tiny droplets.
   b. is a detergent.
   c. is produced by the liver.
   d. All of the above

26. The villi of the small intestine allow for an increase in
   a. absorptive area.   c. acid production.
   b. mechanical digestion.   d. bile production.

27. Ammonia is converted to urea because
   a. urea is less toxic to the body.
   b. urea can be used as a nutrient.
   c. the nitrogen wastes in urea can be recycled and do not need to be excreted.
   d. All of the above

28. The kidneys play a major role in maintaining
   a. the heartbeat.
   b. respiration rate.
   c. homeostasis.
   d. the concentration of digestive enzymes in the blood.

29. Urine leaves the body through the
   a. ureter.   c. bladder.
   b. urethra.   d. intestine.

30. A kidney dialysis machine

a. reduces blood volume.
b. increases the volume of blood.
c. removes wastes from the blood.
d. oxygenates the patient's blood.

31. Urea is formed in the

a. cells.
b. lungs.
c. kidneys.
d. liver.

The following question(s) refer(s) to the diagram of the large molecule.

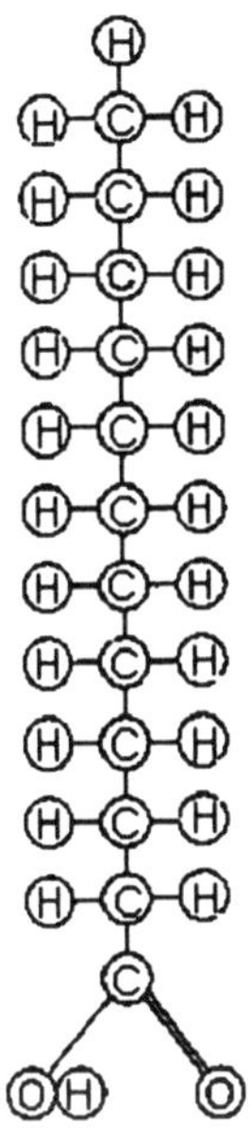

32. The structure shown is most likely a portion of a(n)

a. fat molecule.
b. carbohydrate.
c. protein molecule.
d. amino acid.

33. Most of the energy in the large molecule is stored in the

a. carbon-oxygen bonds.
b. carbon-hydrogen bonds.
c. oxygen-hydrogen bonds.
d. carbon-oxygen double bond.

The following question(s) refer(s) to the diagram of the structure.

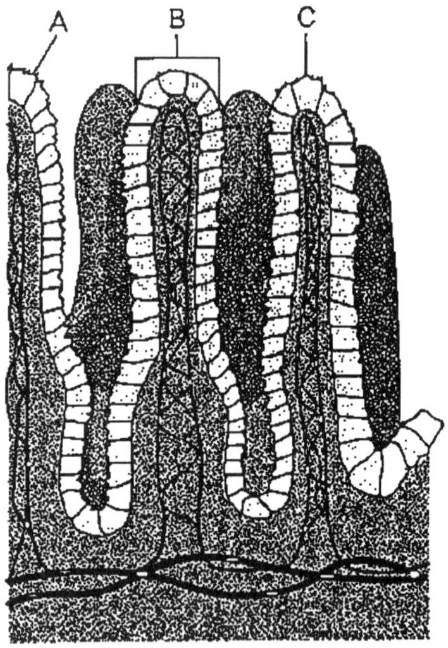

34. Structure B is a
    a. villus.
    b. nephron.
    c. ureter.
    d. urethra.

35. This structure is found in the
    a. kidney.
    b. esophagus.
    c. small intestine.
    d. tongue.

36. This structure allows for an increase in
    a. absorptive area.
    b. mechanical digestion.
    c. acid production.
    d. bile production.

37. Structure C is
    a. a passageway for bile to flow into the stomach.
    b. a capillary.
    c. a nephron.
    d. found only in the duodenum.

The following question(s) refer(s) to the diagram of the structure.

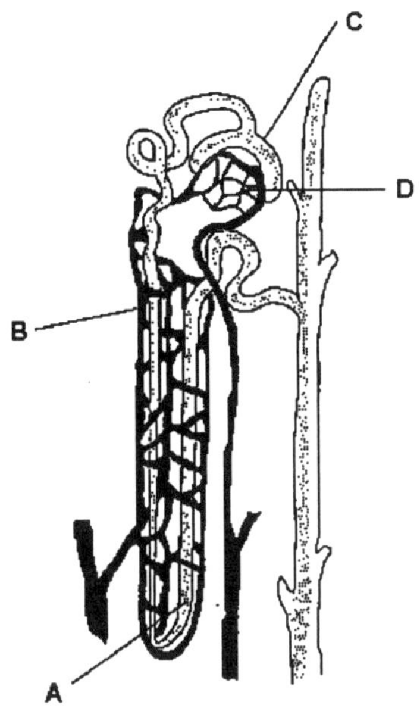

38.The structure shown in the diagram is known as a
   a. villus. c. ureter.
   b. nephron. d. urethra.

39.This structure is the basic functional unit of the
   a. esophagus. c. kidney.
   b. small intestine. d. liver.

40.At label A in the diagram,
   a. filtration is taking place.
   b. water and solutes are moving back into the blood.
   c. red blood cells are moving out of the blood.
   d. red blood cells are being forced back into the blood.

SHORT ANSWER Complete each statement by writing the correct term or phrase in the space provided.

41.A ______________________ is a unit of energy that indicates the amount of energy contained in food.

42.A person who is more than 20 percent heavier than the average person of the same sex and height is said to be _______.

43.The major building blocks of body tissues are supplied by foods containing ______________________.

44.The eight amino acids that humans are unable to manufacture are called ______________________ amino acids.

45.Proteases and hydrochloric acid in the stomach begin the digestion of ______________________.

46.Too much acid in the stomach may result in the formation of a(n) ______________________.

47. Starches are broken down into sugar molecules by enzymes called ______________________.

48. ______________________ is the process whereby fat globules are broken into tiny fat droplets.

49. The kidneys play a major role in maintaining ______________________.

50. The semisolid mixture of food, acid, and enzymes in the stomach is known as ______________________.

51. Hydrochloric acid is secreted by cells in the ______________________ found on the stomach's inner wall.

52. The first part of the small intestine is called the ________________.

53. The large intestine is known as the ______________________.

54. Urine produced in the kidneys passes into the bladder through tubes called ______________________.

55. The ______________________ sends enzymes through a duct into the first part of the small intestine.

56. From the liver, blood is re-collected into the ____________________ vein, which carries the blood back toward the heart.

ESSAY Read and answer each question or statement on a separate sheet of paper.

57. The structure of proteins makes it difficult for the body to digest proteins. How does the body solve the problem of digesting proteins?

58. Although the amount of material that flows into the digestive system is large, the amount that leaves the body as waste is small. Offer an explanation for this fact.

59. Explain how the kidneys play a role in maintaining homeostasis in the body.

ANALOGIES An analogy is a relationship between two pairs of words or phrases written as: a:b::c:d. The symbol : is read "is to," and the symbol :: is read "as." Write the letter of the pair of terms that best completes the analogy shown.

60. carbohydrates : energy::
   a. proteins : insulation for nerve tissue.
   b. amino acids : enzymes.
   c. fats : protein.
   d. fats : muscle.
   e. fats : bone.

61. vitamins : minute amounts::
a. B vitamins : scurvy
b. vitamin D deficiency : pellagra
c. vitamin C deficiency : rickets
d. trace elements : minute amounts
e. trace elements : large amounts

---

62. small intestine : large intestine::
a. large intestine : small intestine
b. stomach : large intestine
c. esophagus : stomach
d. small intestine : esophagus
e. esophagus : mouth

---

63. grinding up food in mouth : digestion in the stomach::
a. concentration of solids : digestion and absorption
b. concentration of solids : digestion in the stomach
c. absorption by small intestine : digestion in the stomach
d. concentration of solids : grinding up food in the mouth
e. absorption by small intestine : concentration of solids

---

64. pancreas : enzymes for small intestine::
a. stomach : saliva
b. stomach : proteins from amino acids
c. liver : bile salts
d. liver : HCl
e. large intestine : chyme

---

65. carbon dioxide : lungs::
a. blood : kidneys
b. sweat : kidneys
c. excess salt : lungs
d. carbon dioxide : skin
e. urea : kidneys

---

66. glomerulus : baseball::
a. loop of Henle : stop sign
b. renal tubule : pitcher's mound
c. renal tubule : baseball scoreboard
d. Bowman's capsule : baseball glove
e. Bowman's capsule : baseball bat

---

67. urea, glucose, and sodium : glomerulus, into Bowman's capsule::
   a. urine and sweat : collecting duct into the tissues of the body
   b. nitrogen wastes : Bowman's capsule into the glomerulus
   c. nitrogen wastes : renal tubule into capillaries
   d. glucose and ions : renal tubule into collecting duct
   e. glucose and ions : renal tubule into capillaries

____

68. ureter : urinary bladder::
   a. urinary bladder : ureter
   b. urethra : urinary bladder
   c. urinary bladder : urethra
   d. urethra : urinary bladder
   e. urethra : ureter

____

69. kidney transplant : kidney failure::
   a. infection : kidney failure
   b. long-term diabetes : kidney failure
   c. untreated long-term high blood pressure : kidney failure
   d. hemodialysis : kidney failure
   e. autoimmune kidney disease : kidney failure

____

TEST ANSWERS

TRUE/FALSE

1. T
2. F
3. F
4. F
5. F
6. F
7. T
8. F
9. T
10. F
11. F

MULTIPLE CHOICE

12. d
13. b
14. a
15. a
16. a
17. c
18. d
19. a
20. d
21. b
22. c
23. a
24. c

25. d

26. a

27. a

28. c

29. b

30. c

31. c

32. a

33. b

34. a

35. c

36. a

37. b

38. b

39. c

40. b

SHORT ANSWER

41. Calorie

42. obese

43. protein

44. essential

45. proteins

46. ulcer

47. amylases

48. Emulsification

49. homeostasis

50. chyme

51. gastric pits

52. duodenum

53. colon

54. ureters

55. pancreas

56. hepatic

ESSAY

57. Almost all proteins are chains of amino acids that are either folded into tight balls or wound together into tough fibers. Enzymes cannot get at the individual protein chains. The human body solves this problem by carrying out protein digestion in two steps. First, hydrochloric acid in the stomach is used to unfold large proteins into single polypeptide strands; then enzymes called proteases attack the individual protein strands, cutting them into smaller fragments.

58. Most of the fluids and solids that flow into the digestive system are absorbed back into the body during their passage through the small intestine.

59. The kidneys regulate the amount of substances such as salts, minerals, and other chemicals that are retained in the blood or excreted in the urine. In addition, the kidneys determine the concentration of substances in the blood by adjusting the total amount of water in the blood in order to hold the total blood volume nearly constant. Kidneys also remove from the body urea and other waste products that could damage cells.

ANALOGIES

60. b (are used to make)

61. d (are required in)

62. c (delivers food to the)

63. e (comes before)

64. c (produces)

65. e (is excreted by the)

66. d (is shaped like a)

67. e (pass out of the)

68. c (empties into the)

69. d (is a treatment for)

Chapter 39

TRUE/FALSE Read each statement. If the statement is true, write T in the space provided. If the statement is false, write F in the space provided.

1.______ Secretions of sweat and oil glands make the skin extremely basic, allowing it to be an effective barrier to infection.

2.______ Natural killer cells attack cells that have been infected by microbes, but not the microbes themselves.

3.______ Killer T cells and B cells are activated by interleukin-2, which is secreted by helper T cells.

4.______ B cells function by attacking and destroying body cells that have been infected by viruses.

5.______ Killer T cells are able to recognize and attack virus-infected cells because the infected cells have been coated with a protein called interleukin-2 by helper T cells.

6.______ The body possesses millions of different types of T cells, each of which bears unique receptor molecules capable of recognizing millions of different foreign proteins.

7.______ The first exposure to a pathogen results in a much faster immune response than the second exposure to the same pathogen.

8.______ Type I diabetes is an autoimmune disease affecting the thyroid gland.

9.______ AIDS patients often succumb to infections or cancers that are rare in normal individuals.

10.______ Any person who is HIV-positive has the disease called AIDS.

MULTIPLE CHOICE Circle the letter of the term or phrase that best completes each statement or best answers each question.

11. The body is protected from pathogens by lines of defense, which include
 a. preventing infection in the first place.
 b. fighting minor infections at localized breaks in the skin.
 c. the action of the immune system against major infections.
 d. All of the above

12. The first line of defense against infection is
 a. the skin.
 b. neutrophils.
 c. killer T cells.
 d. antibodies.

13. All of the following possess mucous membranes, *except* the
 a. digestive tract.
 b. surface of the skin.
 c. nasal passages.
 d. reproductive tract.

14. Mucus is produced by cells lining the walls of the bronchi and bronchioles
   a. only when a person has a severe respiratory infection.
   b. to allow oxygen to diffuse into the blood more efficiently.
   c. as a lubricant for the expulsion of food that might "go down the wrong tube."
   d. to protect against microbes that might be inhaled.

15. The stomach is involved in defense against infection by
   a. regurgitating any pathogen that might be swallowed.
   b. secreting mucus that is carried away by cilia.
   c. possessing acid that destroys potential pathogens that are swallowed.
   d. sending potential pathogens to the liver for destruction.

16. White blood cells that ingest invading microbes and cellular debris resulting from attacks by invaders are called
   a. macrophages.
   b. neutrophils.
   c. natural killer cells.
   d. complement cells.

17. Neutrophils are responsible for
   a. ingesting individual microbes.
   b. destroying viruses.
   c. secreting toxic chemicals that kill bacteria.
   d. producing antibodies.

18. Proteins in blood plasma that kill microbes by attacking their cell membranes and forming holes in them are called
   a. neutrophilins.
   b. antibodies.
   c. antibiotics.
   d. complement.

19. The redness and swelling associated with an inflammatory response is caused by
   a. secretion of antibodies.
   b. expansion of local blood vessels.
   c. complement activity.
   d. natural killer cells destroying bacteria.

20. Pus associated with some infections consists of
   a. dead or dying neutrophils.
   b. broken-down tissue cells.
   c. dead pathogens.
   d. All of the above

21. All of the following are white blood cells that are involved in immune responses, *except*
   a. B cells.
   b. T cells.
   c. macrophages.
   d. neutrophils.

22. Which of the following pairs is *incorrectly* associated?
   a. killer T cells-attack and kill infected cells
   b. helper T cells-activate killer T cells and B cells
   c. B cells-engulf cells that are infected with microbes
   d. macrophages-consume pathogens and infected cells

23. The role of helper T cells in immune responses is to
   a. secrete interleukin-1.
   b. stimulate macrophages to initiate an "alarm signal."
   c. initiate the activities of neutrophils.
   d. activate two different types of immune system cells.

24. Killer T cells recognize cells that have been infected by viruses
   a. only after the infected cells have been ingested by macrophages.
   b. because the infected cells have viral proteins on their surfaces.
   c. when the infected cells have been coated with complement.
   d. at the same time that neutrophils release their toxins into the damaged tissue.

25. When B cells encounter a pathogen, they
   a. secrete interleukin-2, which stimulates killer T cells.
   b. divide and produce large amounts of antibody.
   c. initiate an inflammatory response.
   d. attack the cell by making a hole in its membrane.

26. A few of the B cells that have encountered a pathogen
   a. become killer T cells.
   b. are ingested by macrophages.
   c. have viral protein on their cell membrane surface.
   d. become memory cells.

27. Secondary exposure to a pathogen
   a. results in very rapid production of antibodies.
   b. stimulates memory cells to divide quickly.
   c. may result in destruction of the pathogen before a person knows he or she is infected.
   d. All of the above

28. Vaccines are effective in preventing disease because they
   a. contain proteins from pathogens, but lack the ability to cause disease.
   b. are antibodies directed against specific pathogens.
   c. contain specific B cells and T cells.
   d. contain pathogens to which the person is already immune.

29. An autoimmune disease in which the immune system attacks the myelin sheath of motor nerves is
   a. multiple sclerosis.
   b. rheumatoid arthritis.
   c. Graves' disease.
   d. lupus erythematosus.

30. An immune response directed against a harmless substance is called
   a. an autoimmune disease.
   b. a secondary immune reaction.
   c. an allergic reaction.
   d. a vaccination reaction.

31. The release of histamine from cells in nasal passages
   a. occurs during an allergic reaction.
   b. causes nearby capillaries to swell.
   c. may cause increased secretion by mucous membranes.
   d. All of the above may occur.

32. Immunization against influenza virus does not provide long-term protection against the disease because
   a. the genes for the cell surface proteins of the virus mutate very rapidly.
   b. antibodies do not work against viruses.
   c. the vaccine used to immunize people against influenza causes the virus to change.
   d. the virus destroys the vaccine before it has a chance to work.

33. HIV causes AIDS by attacking and destroying
   a. helper T cells.
   b. B cells.
   c. neutrophils.
   d. antibodies.

34. The debilitating effects of AIDS are due to inability of the immune system to
   a. activate B cells and killer T cells.
   b. produce antibodies against pathogens.
   c. recognize and destroy cancer cells.
   d. All of the above

35. HIV can be transmitted
   a. through sexual intercourse with an infected person.
   b. by passage of infected macrophages across a mucous membrane.
   c. by sharing hypodermic syringes and needles.
   d. All of the above are ways that HIV can be transmitted.

The following question(s) refer(s) to the illustration of an immune response.

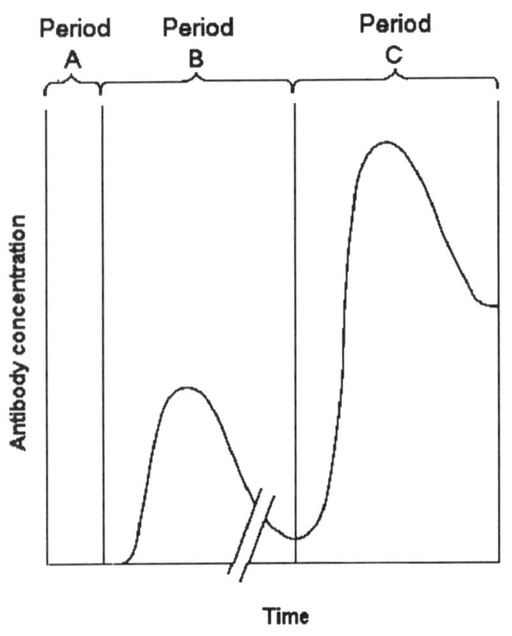

36. During which time period shown would the first antibodies to the pathogen be produced?
   a. period A
   b. period B
   c. period C
   d. None of the above

37. Which period of time would be characterized by the most rapid division of B cells?
   a. period A
   b. period B
   c. period C
   d. None of the above

SHORT ANSWER Complete each statement by writing the correct term or phrase in the space provided.

38. Sweat is protective because it contains the enzyme ________________, which attacks the cell walls of many bacteria.

39. Epithelial layers that line the internal surfaces of the body through which pathogens could pass and that produce mucus as a protective substance are called ________________________.

40. Immune surveillance by ________________________ cells is one of the body's most potent defenses against cancer.

41. ________________________ are white blood cells that travel throughout the body, killing bacteria one at a time by ingesting them.

42. A membrane attack complex is formed when proteins called _______________ are activated by contact with cell walls of invading bacteria or fungi.

43. An "alarm signal" is emitted by macrophages in the form of a protein called _______________, which activates helper T cells.

44. Interleukin-2 is produced by _______________ cells.

45. Proteins, produced by B cells, that can mark pathogens for destruction or can be released into the bloodstream to directly attack pathogens are called _______________.

46. _______________ is the process by which a dead or disabled pathogen (or proteins from that pathogen) are introduced into the body so that an immune response results without an actual infection.

47. After a primary exposure to a pathogen, the bloodstream contains _______________ cells that are specifically directed against that particular pathogen.

48. A condition in which the immune system loses the ability to distinguish between pathogens and the body's own cells is called a(n) _______________.

49. _______________ is a chemical that is released from mast cells during an allergic reaction.

50. The virus that causes AIDS is called _______________.

ESSAY Read and answer each question or statement on a separate sheet of paper.

51. Imagine that a potential pathogen has been able to get through the skin, the first line of body defense. What four steps does the body use as a second line of defense to prevent the pathogen from initiating a major infection?

52. Briefly describe how a cell that has been infected by a virus can be recognized and destroyed.

53. Explain how the human immunodeficiency virus (HIV) disables the immune system.

ANALOGIES An analogy is a relationship between two pairs of words or phrases written as: a:b::c:d. The symbol : is read "is to," and the symbol :: is read "as." Write the letter of the pair of terms that best completes the analogy shown.

54. neutrophils : releasing chemicals::
   a. macrophages : releasing chemicals
   b. natural killer cells : releasing chemicals
   c. natural killer cells : puncturing their membranes
   d. neutrophils : puncturing their membranes
   e. neutrophils : ingesting them

____

55. temperature response against bacteria : body temperature::
   a. inflammatory response : blood flow to infection site
   b. inflammatory response : infection
   c. fever : microbial growth
   d. fever : infection
   e. high fever : activity of cellular enzymes

____

56. pathogenic bacteria and viruses : dangerous enemies to humans::
   a. B cells and T cells : mucous membrane cells
   b. helper T cells and killer T cells : skin cells
   c. killer T cells and macrophages : pathogens
   d. killer T cells and B cells : white blood cells
   e. interleukin-1 and interleukin-2 : viruses

____

57. macrophage : helper T cells::
   a. killer T cell : macrophages
   b. helper T cell : killer T cells and B cells
   c. B cell : killer T cells and macrophages
   d. mucous membrane cell : helper T cells and B cells
   e. interleukin-1 : macrophages

____

58. macrophages destroy pathogens : B cells mark pathogens::
   a. B cells mark pathogens : memory cells are produced
   b. interleukin-1 is released : interleukin-2 is released
   c. body cells get infected : recognition by killer T cells
   d. vaccination occurs : memory cells are in bloodstream
   e. memory cells remain in bloodstream : vaccination

____

59. asthma attacks : narrowing of breathing passages::
   a. antibodies : release of antihistamines
   b. mast cells and antibodies : release of histamines
   c. mast cells : release of macrophages
   d. antihistamines : capillary swelling
   e. histamines : reduction of mucus production

____

60. pneumocystic pneumonia : AIDS patients::
   a. mast cell : HIV
   b. helper T cell : HIV
   c. HIV : helper T cells
   d. Kaposi's sarcoma : macrophages
   e. HIV : myelin nerve covering

____

61. HIV : HIV-contaminated hypodermic needles::

a. HIV : unprotected sexual contact
b. HIV : shaking hands
c. HIV : drinking from same water fountain as infected person
d. HIV : contact with saliva, tears, and urine
e. HIV : a mosquito bite

____

62. redness and swelling : the inflammatory response::

a. increased blood flow : AIDS
b. inflammatory response : membrane attack complex.
c. neutrophils : autoimmune disease
d. message to brain : temperature response.
e. macrophages : immune system failure

____

TEST ANSWERS

TRUE/FALSE

1.F

2.T

3.T

4.F

5.F

6.T

7.F

8.F

9.T

10.F

MULTIPLE CHOICE

11.d

12.a

13.b

14.d

15.c

16.a

17.c

18.d

19.b

20.d

21.d

22.c

23.d

24.b

25. b

26. d

27. d

28. a

29. a

30. c

31. d

32. a

33. a

34. d

35. d

36. b

37. c

SHORT ANSWER

38. lysozyme

39. mucous membranes

40. natural killer

41. Macrophages

42. complement

43. interleukin-1

44. helper T

45. antibodies

46. Vaccination

47. memory

48. autoimmune disease

49. Histamine

50. human immunodeficiency virus (HIV)

ESSAY

51. There are four steps that serve as a second line of defense against infection. (1) Cells, such as macrophages, neutrophils, and natural killer cells, can destroy invading microbes or cells that have been invaded by them. (2) Complement consists of a series of proteins that interact when they encounter the cell membranes of bacteria or fungi, causing a puncturing of the microbes' cell membranes, thus destroying them. (3) Localized infection initiates an inflammatory response that suppresses infection and speeds healing. (4) Fever is initiated by macrophages that send a message to the brain that results in an elevation of body temperature. This interferes with the ability of bacteria to grow and reproduce.

52. Macrophages contact infected cells and release an "alarm signal" called interleukin-1. This protein activates helper T cells that secrete another protein, interleukin-2. This substance, in turn, stimulates killer T cells, which bind to the infected cell using surface receptor molecules that recognize traces of viral protein on the surface of the infected cell. This interaction causes damage to the cell membrane of the infected cell, resulting in its destruction.

53. AIDS is caused by infection with HIV. HIV, carried throughout the body by macrophages, attacks helper T cells, causing a severe reduction in their numbers. Without a sufficient number of helper T cells, B cells and killer T cells cannot be activated efficiently. This leaves the infected individual susceptible to a variety of pathogens and cancers.

ANALOGIES

54. c (kill bacteria by)

55. a (increases)

56. d (are two kinds of)

57. b (activates)

58. e (after)

59. b (cause)

60. c (attacks)

61. a (can be transmitted by)

62. d (is/are part of)

Chapter 40

TRUE/FALSE Read each statement. If the statement is true, write T in the space provided. If the statement is false, write F in the space provided.

1.______ In an inhibitory synapse, the receptor is a gated potassium channel.

2.______ In some neurons, a form of supporting cell called a sheath cell wraps around the axon.

3.______ Motor nerves that conduct impulses to skeletal muscles under our conscious control make up the limbic system.

4.______ Emotions are controlled by the sympathetic division of the autonomic nervous system.

5.______ Most of the activity of the cerebrum occurs in its gray matter.

6.______ The fovea, the most sensitive part of the eye, consists of millions of rod cells.

7.______ A taste bud is stimulated when a chemical dissolved in saliva binds to small hairs in the taste buds, which are located on the surface of the tongue.

8.______ Nociceptors are located throughout all tissues and organs except the brain.

9.______ The brain adjusts to prolonged exposure to nicotine by making fewer receptor proteins to which nicotine can bind.

10.______ Addiction to psychoactive drugs is a physiological response because addiction involves interactions of drug molecules and nerve cell membranes.

11.______ Alcohol is able to change the structure of a neuron's membrane bilayer, producing changes in the shape of receptor proteins.

MULTIPLE CHOICE Circle the letter of the term or phrase that best completes each statement or best answers each question.

12. A nerve impulse may be described as
   a. an electrical message.
   b. a chemical message.
   c. an electromagnetic message.
   d. a chemical change occurring in the brain.

13. Which statement describes the chemical composition of the axon that is not transmitting an impulse?
   a. There are many times more sodium ions outside the axon membrane than inside.
   b. Sodium ions are in balance inside and outside the axon membrane.
   c. There are fewer potassium ions inside the axon membrane than outside.
   d. Potassium and sodium ions are equal on both sides of the axon membrane.

14. Nodes
   a. strengthen axons.
   b. slow the nerve impulse.
   c. occur in the diseased axons.
   d. are gaps in the myelin sheath.

15. The junction at which an impulse travels from one neuron to another is called a
a. synapse. c. nerve message.
b. stimulus. d. node.

16. The sodium-potassium pump
a. rebuilds axon fibers.
b. restores the resting potential.
c. creates a stimulus.
d. is found only in the peripheral nervous system.

17. When a nerve impulse reaches a synapse, neurotransmitters
a. become enzymes in the space between the neurons.
b. bind to the membrane receptor proteins on the membrane of the next neuron.
c. cover the membrane of the axon.
d. cause the cell body in the next neuron to enlarge.

18. The myelin sheath
a. transmits impulses from one neuron to another.
b. insulates the synapses.
c. nourishes neurons.
d. insulates the axons.

19. Short, slender input channels that extend from one end of a neuron's body are called
a. axons. c. synapses.
b. cell bodies. d. dendrites.

20. Morphine works as a painkiller because
a. it insulates sense organs.
b. it increases the thickness of the myelin sheath.
c. it slows down nerve impulses by filling in nodes.
d. it acts on synapses and binds to receptors in the brain.

21. Electrical changes in an axon create
a. a stimulus. c. an action potential.
b. an electrical shock. d. light and sound.

22. Some neurotransmitters cross a synapse and open sodium channels in the membrane of the next neuron, causing
a. inhibition of impulses in the neuron.
b. the death of the neuron.
c. initiation of an impulse in the neuron.
d. the formation of protein receptors in the neuron.

23. The central nervous system consists of
a. the brain and spinal cord. c. the brain stem and cerebellum.
b. spinal nerves only. d. the cerebrum and spinal column.

24. The peripheral nervous system
a. is not linked to the central nervous system.
b. provides pathways to and from the central nervous system.
c. consists of the cerebellum and spinal cord.
d. is composed of only motor neurons.

25. Motor neurons transmit messages
a. from the environment to the brain.
b. from the environment to the spinal cord.
c. from the spinal cord to the brain.
d. from the central nervous system to a muscle or gland.

26. The gray matter of the brain consists of
a. unmyelinated neurons.
b. only synapses.
c. myelin.
d. nodes.

27. The autonomic nervous system controls
a. reflexes.
b. voluntary movement.
c. involuntary functions of glands and internal organs.
d. locomotion.

28. Specialized hearing receptors are found in the
a. cornea.
b. semicircular canals.
c. cochlea.
d. cerebellum.

29. The layer of light receptors and neurons at the back of the eye is called the
a. retina.
b. cochlea.
c. iris.
d. optic nerve.

30. Sensory receptors essential for balance are located in the
a. sclera.
b. middle ear.
c. cochlea.
d. semicircular canals.

31. Drugs may cause addiction by
a. competing with protein receptors for binding of neurotransmitters.
b. stimulating a reduction of protein receptors in synapses, decreasing the sensitivity of the nerve pathway.
c. stimulating enzymes that destroy neurotransmitters.
d. None of the above

32. Narcotics and enkephalins have very similar molecular structures, therefore
a. narcotics are able to bind to the same receptors as enkephalins.
b. enkephalins are addictive drugs.
c. enkephalins are derived from cocaine.
d. narcotics are natural pain relievers.

33. Prolonged use of narcotics
a. causes an increase in the number of opiate receptors in spinal neurons.
b. stimulates the limbic center in the brain.
c. desensitizes nerve cells.
d. All of the above

34. While narcotics can stimulate the pleasure center of the brain, removal of the effect of these narcotics can
a. desensitize the neurons.
b. make the nerve cells become hypersensitive.
c. cause sodium channels to open, speeding nerve impulse transmission.
d. result in an increase in opiate receptors on the membranes of neurons.

The following question(s) refer(s) to the three-part diagram of a nerve impulse.

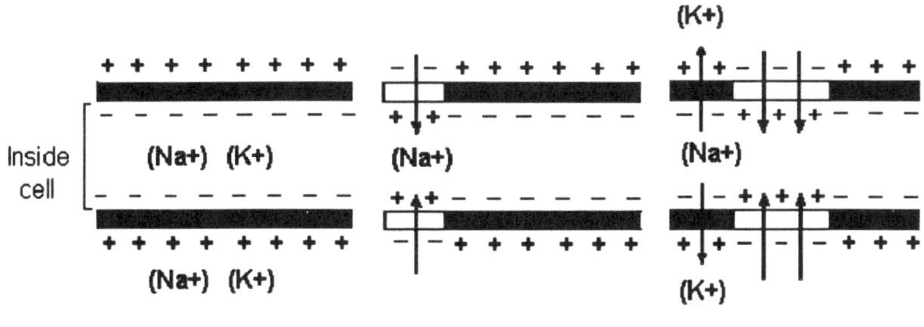

35.The diagrams indicate that a nerve impulse
a. moves from the inside to the outside of an axon.
b. moves from the outside to the inside of an axon.
c. is the movement of an action potential along a neuron.
d. moves slowly.

36.When a neuron is at rest,
a. sodium ions are found on the outside of the cell.
b. potassium ions are found on the outside of the cell.
c. the outside of the cell is positively charged.
d. All of the above

37.Electrical changes in a neuron create
a. a stimulus.
b. an electrical shock.
c. an action potential.
d. light and sound.

38.When an impulse moves down an axon,
a. sodium ions first rush out of the cell.
b. a small part of the axon momentarily reverses its polarity.
c. the resting potential of the cell does not change.
d. potassium ions are pumped into the axon.

The following question(s) refer(s) to the diagram of a human brain.

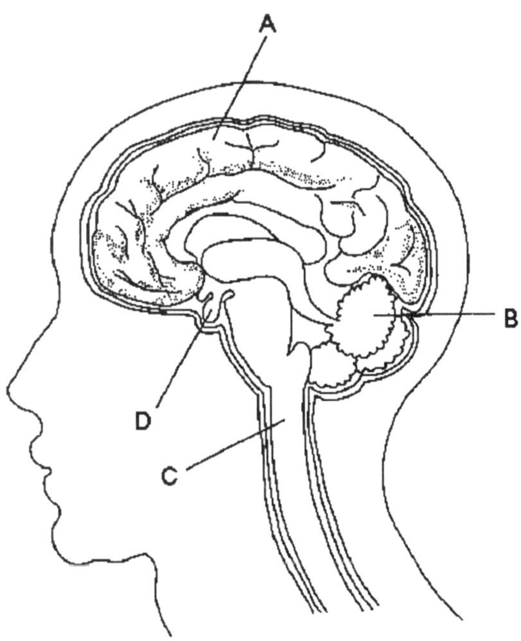

39.Structure B is the
a. reticular formation.
b. brain stem.
c. cerebellum.
d. cerebrum.

40.Body temperature and blood pressure are regulated by structure
a. A.
b. B.
c. C.
d. D.

41.The reticular formation runs through structure
a. A.
b. B.
c. C.
d. D.

42. The gray matter of the brain is found in structure

a. A.
b. B.
c. C.
d. D.

SHORT ANSWER Complete each statement by writing the correct term or phrase in the space provided.

43. A neuron transmits a nerve impulse as a wave of ______________________.

44. Cytoplasmic extensions called ___________________________ allow a neuron to receive information simultaneously from many different sources.

45. Destruction of large patches of myelin sheath can result in a disease known as ___________________________.

46. A sudden, involuntary movement in response to a stimulus is called a ___________________________.

47. The main site of sensory processing in the brain is the ______________.

48. Nerves that control breathing, swallowing, heartbeat, and the diameter of the blood vessels are found in the ___________________________.

49. A division of the autonomic nervous system known as the ___________________________ division can conserve energy by slowing the heartbeat and breathing rate.

50. Tiny pieces of calcium carbonate in the ear called ____________________ help the body maintain its balance.

51. When light enters the eye, it first passes through the ___________________________.

52. The ___________________________ is the light-sensing portion of the eye.

53. The amount of light entering the eye is controlled by the ___________________________.

54. A ___________________________ is a globular cluster of cells specialized to detect the four basic types of chemicals found in foods.

55. Narcotics that are used to relieve pain and induce sleep are known as ___________________________.

ESSAY Read and answer each question or statement on a separate sheet of paper.

56. How is a signal from one neuron transferred to another neuron?

57. Explain why you cannot hold your breath indefinitely.

58. How is light turned into an image?

59. Describe the action of cocaine at the synapse.

ANALOGIES An analogy is a relationship between two pairs of words or phrases written as: a:b::c:d. The symbol : is read "is to," and the symbol :: is read "as." Write the letter of the pair of terms that best completes the analogy shown.

60. unmyelinated axon : slow nerve impulses::
   a. neuron : being composed of many axons
   b. nerve impulse : not traveling through axons
   c. myelinated axon : fast nerve impulses
   d. dendrite : nerve impulses traveling through it
   e. multiple sclerosis : increased muscle strength

____

61. multiple sclerosis : destruction of large patches of myelin::
   a. resting potential : pumping of potassium ions out of cell
   b. resting potential : pumping of sodium ions out of the cell
   c. action potential : pumping of sodium ions out of the cell
   d. action potential : pumping of calcium into the cell
   e. action potential : pumping of cocaine out of the cell

____

62. resting potential : more positive charge on the outside::
   a. depolarization : sodium flowing out, potassium out
   b. repolarization : sodium flowing in, potassium out
   c. action potential : reversal in the charge across the membrane
   d. positive charge on the outside : more sodium inside the cell
   e. positive charge on inside : more sodium outside the cell

____

63. synapse : 2 neurons::
   a. neuron : 2 cell bodies
   b. cell body : 2 axons
   c. synapse : cell body and axon
   d. axon : cell body and synapse
   e. axon : dendrites and the cell body

____

64. limbic system : processing emotions like pain and hunger::
   a. cerebellum : balance and posture
   b. brain stem : nerve connections for consciousness
   c. hypothalamus : connecting the brain to the spinal cord
   d. reticular formation : activities related to the body temperature and blood pressure
   e. spinal cord : connecting neurons to each other in the cerebrum

____

65. sensory nerve : body to the central nervous system::
   a. motor nerve : thalamus to the hypothalamus
   b. motor nerve : heart to the spinal cord
   c. motor nerve : glands to the cerebrum
   d. motor nerve : brain and spinal cord to the body
   e. motor nerve : lungs to the brain stem

____

66. tortoise : slow::
   a. nerve impulse along a myelinated nerve : slow
   b. autonomic nervous system activities : quick
   c. action potential : slow
   d. reflex : slow
   e. sensory nerve to motor nerve response : quick

---

67. sympathetic division : parasympathetic division::
   a. hot : cold
   b. bring : brought
   c. letter : message
   d. alphabet : letters
   d. pudding : food

---

68. iris : amount of light entering the eye::
   a. pupil : amount of light entering the eye
   b. cornea : shape of the lens
   c. lens : point of focus on retina
   d. retina : movement of iris muscles
   e. cone : response to motion

---

69. narcotics : binding to the opiate receptors in the brain::
   a. opiate receptors : producing opium
   b. opiate receptors : producing morphine
   c. potassium ions : stimulating pain messages
   d. sodium ions : decreasing the number of opiate receptors
   e. enkephalins : binding to the opiate receptors in the brain

---

70. enkephalin : potassium channels open::
   a. cocaine : dopamine in synapse
   b. narcotic : sensitivity to pain at normal levels
   c. cocaine : effect of addiction low
   d. cocaine : increasing the number of dopamine receptors
   e. synapse : sensitivity

TEST ANSWERS

TRUE/FALSE

1. T
2. F
3. F
4. F
5. T
6. F
7. T
8. T
9. T
10. T
11. T

MULTIPLE CHOICE

12. a
13. a
14. d
15. a
16. b
17. b
18. d
19. d
20. d
21. c
22. c
23. a
24. b

25. d

26. a

27. c

28. c

29. a

30. d

31. b

32. a

33. d

34. b

35. c

36. d

37. c

38. b

39. c

40. d

41. c

42. a

SHORT ANSWER

43. depolarization

44. dendrites

45. multiple sclerosis

46. reflex

47. thalamus

48. brain stem

49. parasympathetic

50. otoliths

51. cornea

52. retina

53. iris

54. taste bud

55. depressants

ESSAY

56. When a nerve impulse gets to the end of an axon, its message must cross the synapse if the message is to continue. Messages are carried across synapses by chemical transmitters that are called neurotransmitters. Neurotransmitters are found in tiny sacs at the end of the axon. When a nerve impulse reaches the end of the axon, it causes the sacs to release neurotransmitters into the synapses. The neurotransmitters diffuse across the synapse and bind to receptors in the membrane of the cell on the other side, permitting an action potential to start in the second cell.

57. The autonomic nervous system carries messages to muscles and glands that usually work without our noticing, enabling the nervous system to govern most of the body's homeostasis. We have voluntary control over some functions, such as breathing, that are regulated by the autonomic nervous system. Any voluntary control of the autonomic nervous system that endangers life disturbs homeostasis of the brain tissue, causing unconsciousness. Then the autonomic nervous system takes over and restores normal functioning—in this case, breathing.

58. A photoreceptor in the eye is able to detect a single photon of light. The primary sensing event of vision is the absorption of a photon of light by pigment complexes called rhodopsins and photopsins. When they absorb a photon of light, the pigments change shape and initiate a chain of events that leads to an action potential. The nerve impulse travels along the optic nerve to the brain, which translates the information into a meaningful image.

59. Cocaine prevents the reabsorption of dopamine from the synapse. The trapped dopamine repeatedly stimulates neurons. Neurons adjust to the presence of cocaine by decreasing their number of dopamine receptors. This causes neurons in the limbic system to become less sensitive, requiring more and more cocaine for stimulation.

ANALOGIES

60. c (has the characteristic of)

61. b (results from)

62. c (is caused by)

63. d (is located between)

64. a (is responsible for)

65. d (sends information from the)

66. e (is)

67. a (is the opposite of)

68. c (adjusts)

69. e (relieve pain by)

70. a (keeps)

Chapter 41

TRUE/FALSE Read each statement. If the statement is true, write T in the space provided. If the statement is false, write F in the space provided.

1._______ Although hormones are slow-acting, their effects tend to persist for a long time.

2._______ Endocrine glands deliver their products directly to where they are needed through ducts.

3._______ Peptide hormones are lipid-soluble, while steroids are water-soluble.

4._______ A prostaglandin is a peptide hormone that is produced exclusively by brain cells and can act in the same manner as a neurotransmitter during emergency situations.

5._______ When sufficient levels of testosterone are present in a male's bloodstream, production is inhibited by a negative feedback system.

6._______ Steroid hormones directly influence the activities of the genes in their target cells.

7._______ The hypothalamus is controlled by the pituitary gland.

8._______ A goiter is the result of futile attempts by the thyroid gland to make thyroxine when the person is suffering from an iodine deficiency.

9._______ The adrenal cortex produces steroid hormones, and the adrenal medulla produces peptide hormones.

10._______ The pancreas is both an endocrine and an exocrine gland.

MULTIPLE CHOICE Circle the letter of the term or phrase that best completes each statement or best answers each question.

11.Hormones are
a. chemicals that stimulate nerve cells during times of stress.
b. the same as electrical nerve impulses.
c. transported to their targets by the bloodstream.
d. neurons along which messages travel.

12.Compared to nerve messages, hormonal messages
a. are transmitted faster. c. have a shorter duration.
b. are essentially the same. d. last longer.

13.A disadvantage of nervous signals, which is not true of hormonal messages, is that
a. nerve transmissions reach many cells at once.
b. nerve cells send messages at a much slower rate than hormones.
c. the effects of nervous signals last for only a short time.
d. nervous signals work on only one specific type of cell.

14.Which of the following organs contain cells that have an endocrine function?
a. brain c. small intestine
b. stomach d. All of the above

15. Since hormones must often travel to distant organs in the body, they
   a. are each produced by several endocrine organs simultaneously.
   b. are very stable molecules that do not decompose quickly.
   c. stimulate nerve cells to send impulses to target organs.
   d. all act on only one target cell.

16. All of the following are nonendocrine chemical signal molecules, *except*
   a. steroids.
   b. neuropeptides.
   c. neurotransmitters.
   d. prostaglandins.

17. Neurotransmitters
   a. are produced by endocrine glands throughout the body.
   b. are the same as neuromodulators.
   c. are usually derived from amino acids.
   d. are a type of peptide hormone.

18. Modified lipids that tend to accumulate in areas of tissue disturbance or injury are
   a. endorphins.
   b. enkephalins.
   c. neuromodulators.
   d. prostaglandins.

19. Biologists often refer to the "neuroendocrine system" because
   a. the nervous system makes all of the body's hormones.
   b. the nervous and endocrine systems work closely together.
   c. the endocrine system controls the body through electrical messages.
   d. both systems are structurally and functionally identical.

20. The hypothalamus has a dual function in that it
   a. sends nerve impulses and also makes hormones.
   b. is both a nervous and an excretory organ.
   c. belongs to both the nervous and circulatory systems.
   d. is found connected to both kidneys.

21. Hormones produced by the pituitary gland
   a. are regulated by secretions from the hypothalamus.
   b. control the activity of other endocrine glands.
   c. are produced as the result of stimulation by releasing hormones.
   d. All of the above

22. In order for a hormone to work,
   a. it must reach its target cell.
   b. it must bind to a receptor protein.
   c. its message must cross a plasma membrane.
   d. All of the above must occur.

23. When a peptide hormone attaches to a target cell,
   a. it binds to a receptor in the plasma of the organism.
   b. it passes through the plasma membrane.
   c. a series of chemical reactions forms a "second messenger."
   d. the hormone is converted into a steroid.

24. Which of the following would *not* be caused by cyclic AMP?
   a. activation of enzymes within a target cell
   b. inhibition of the chemical activities within a target cell
   c. increases in the gene activity of a target cell
   d. amplification of the effect of a peptide hormone

25. Which of the following is a steroid hormone?
   a. glucagon
   b. prostaglandin
   c. testosterone
   d. insulin

26. A substance that functions by affecting the activities of genes in a target cell is a
a. prostaglandin.
b. steroid hormone.
c. peptide hormone.
d. second messenger.

27. Steroids are able to enter their target cells
a. after combining with a receptor protein.
b. because they are made of DNA.
c. in order to produce cyclic AMP.
d. because they are fat-soluble.

28. Thyroxine
a. stimulates cell metabolism and growth.
b. slows growth of its target cells.
c. stimulates synthesis of DNA.
d. binds to receptor molecules on the target cell's surface.

29. Since steroid hormones are lipids, they
a. attach only to lipid receptor molecules.
b. cannot enter the target cell.
c. activate only fat cells.
d. pass through the plasma membrane.

30. Until recently, the pituitary gland was called the "master gland." That name is no longer used because
a. it was recently discovered that the pituitary gland does not actually do anything.
b. the hypothalamus is responsible for controlling the activity of the pituitary gland.
c. the pituitary gland controls the hypothalamus, which in turn controls the other glands.
d. hormones of the other endocrine glands of the body activate the pituitary gland.

31. The posterior lobe of the pituitary gland
a. secretes releasing hormones that stimulate the anterior lobe of the pituitary gland.
b. produces and secretes certain steroid hormones.
c. stores and releases hormones made in the hypothalamus.
d. is responsible for producing and secreting seven peptide hormones.

32. Hypothyroidism can cause
a. nervousness.
b. weight loss.
c. increased blood sugar.
d. lack of energy.

33. Parathyroid hormone is important for survival because it
a. stimulates the body's metabolic rate.
b. regulates the amount of calcium in the blood.
c. causes the heart to contract and pump blood.
d. increases the excretion of calcium by the kidneys.

34. The adrenal medulla is different from other endocrine glands in that it
a. is stimulated to release its hormones by nerves from the hypothalamus.
b. produces only steroid hormones.
c. is located around the inner adrenal cortex.
d. does not release its hormones into the bloodstream.

35. The islets of Langerhans in the pancreas are responsible for
  a. production of epinephrine and norepinephrine.
  b. making hormones that regulate blood sugar levels.
  c. regulating calcium levels in the blood and in the bones.
  d. controlling the amount of iodine that reaches the thyroid gland.

36. Individuals with Type II diabetes
  a. require daily injections of insulin.
  b. suffer from an autoimmune disorder that attacks the islets of Langerhans.
  c. have an abnormally low number of insulin receptors on their cell membranes.
  d. exhibit especially low levels of insulin in their blood.

The following question(s) refer(s) to the diagram showing the organs of the human endocrine system.

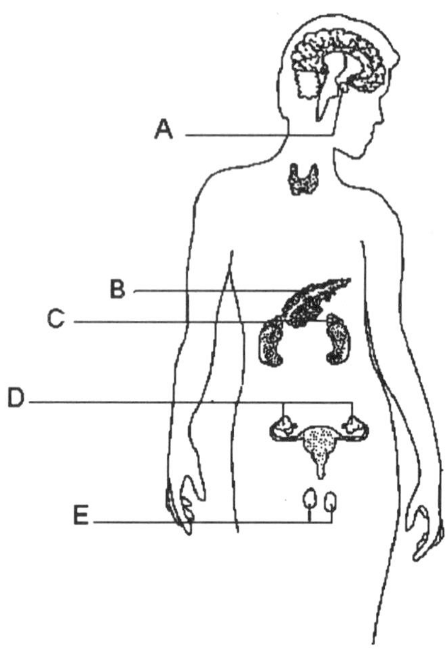

37. The pituitary gland is indicated as
  a. gland A. c. gland C.
  b. gland B. d. gland D.

38. Development and maintenance of female characteristics are stimulated by secretions of
  a. gland A. c. gland D.
  b. gland B. d. gland E.

39. The gland that produces the hormones insulin and glucagon is
  a. gland A. c. gland C.
  b. gland B. d. gland D.

40. The gland that is stimulated during emergency situations, causing the "fight or flight" response, is
  a. gland A. c. gland C.
  b. gland B. d. gland D.

41. Testosterone is produced by
  a. gland A. c. gland D.
  b. gland C. d. gland E.

SHORT ANSWER Complete each statement by writing the correct term or phrase in the space provided.

42. Chemical messengers that are produced in one part of the body and are transported by the bloodstream to another part of the body where they cause an effect are called ________________.

43. Hormones are produced by ________________, which release their products directly into the bloodstream.

44. Fat-soluble lipid hormones, similar in structure to cholesterol, are ________________.

45. Peptide hormones are water-soluble hormones consisting of chains of ________________.

46. ________________ are nonendocrine molecules, such as endorphins and enkephalins, that affect cells near the nerve cells that produce them.

47. Chemicals that alter a cell's response to a neurotransmitter are called ________________.

48. The ________________, a portion of the brain, controls much of the endocrine activity of the body by regulating the secretions of the pituitary gland.

49. ________________ hormones remain outside of their target cells, while ________________ hormones carry out their function from within their target cells.

50. A common "second messenger" that is produced by cells that have been stimulated by a peptide hormone is ________________.

51. A steroid hormone combines with a receptor protein to form a(n) ________________, which enters the nucleus of the cell and binds to DNA.

52. ________________ is a steroid hormone that is produced by the adrenal glands in response to stress.

53. A hormone that increases the body's metabolic rate is ____________.

54. High levels of calcium in the blood stimulate the production of ________________, a hormone that causes more calcium to be deposited in bone tissue, thus lowering the blood-calcium level.

55. The frequently prescribed medication prednisone, used as an anti-inflammatory agent, is a derivative of the hormone ____________.

56. A disorder in which cells of an individual cannot obtain glucose from the blood is called ________________.

57. A hormone that enables the cells of certain tissues to take in glucose molecules is ________________.

ESSAY Read and answer each question or statement on a separate sheet of paper.

58. Describe the advantages that hormonal messages have over messages of the nervous system.

59. Glucagon, a hormone produced by the pancreas, cannot enter its target cells. Yet, it causes an increase in glucose in the bloodstream. Explain how this hormone works.

60. Compare the action mechanisms of peptide and steroid hormones.

61. Explain why the cells of a patient with diabetes mellitus starve from lack of glucose even though blood glucose levels are higher than normal.

ANALOGIES An analogy is a relationship between two pairs of words or phrases written as: a:b::c:d. The symbol : is read "is to," and the symbol :: is read "as." Write the letter of the pair of terms that best completes the analogy shown.

62. nerve messages : fast-acting::
   a. nerve messages : transported by the blood system
   b. nerve messages : released from glands
   c. chemical messages : transported by nerves
   d. chemical messages : transported by bloodstream
   e. chemical messages : fast-acting

____

63. peptide hormone : outside::
   a. receptor protein : inside
   b. cyclic AMO : outside
   c. steroid hormone : inside
   d. peptide hormone : complex
   e. steroid hormone : outside

____

64. endocrine glands : hormones::
   a. neurons : neurotransmitters
   b. neurons : hormones
   c. all cells : neurotransmitters
   d. all cells : hormones
   e. endocrine glands : neurotransmitters

____

65. neurotransmitters : one neuron to the next::
   a. prostaglandins : one neuron the next
   b. small blood vessels : hypothalamus to the pituitary
   c. nerves : hypothalamus to the pituitary
   d. nerves : one neuron to the next
   e. prostaglandins : hypothalamus to the pituitary

____

66. delivery person : deliver a package::
   a. cyclic AMP : produce a peptide hormone
   b. cyclic AMP : amplify effects of peptide hormone
   c. cyclic AMP : bind to a protein receptor
   d. cyclic AMP : dissolve in fat
   e. cyclic AMP : go into a cell nucleus

____

67. mosquito : a screen to reach the inside of a house::
   a. receptor protein : a cell membrane to reach the cell nucleus
   b. receptor protein : the thyroid gland to reach a cell nucleus
   c. thyroxine : the cell nucleus to the cell membrane
   d. thyroxine : a cell membrane to the cell nucleus
   e. thyroxine : a cell membrane to the thyroid gland

___

68. parathyroid gland : PTH::
   a. thyroid gland : thyroxine
   b. pancreas : estrogen
   c. pituitary gland : insulin
   d. adrenal gland : FSH
   e. ovary : aldosterone

___

69. increase in blood sugar level : glucagon breakdown::
   a. calcitonin production : low blood calcium level
   b. calcitonin production : underproduction of thyroxine
   c. hyperthyroidism : overproduction of thyroxine
   d. hypothyroidism : overproduction of thyroxine
   e. calcitonin production : overproduction of thyroxine

___

70. adrenaline : reaction to stress::
   a. norepinephrine : changes in blood pressure
   b. aldosterone : readiness to "fight"
   c. aldosterone : readiness to "flight"
   d. PTH : blood-calcium level
   e. cortisol : "fight" or "flight" response

___

71. aldosterone and PTH : retrieval of fluids from the body::
   a. cortisol and adrenaline : loss of sodium and potassium
   b. cortisol and PTH : loss of potassium
   c. adrenaline and norepinephrine : response to stress
   d. adrenaline and norepinephrine : potassium ions to be excreted in urine
   e. potassium and sodium loss : PTH

___

TEST ANSWERS

TRUE/FALSE

1.T

2.F

3.F

4.F

5.T

6.T

7.F

8.T

9.T

10.T

MULTIPLE CHOICE

11.c

12.d

13.c

14.d

15.b

16.a

17.c

18.d

19.b

20.a

21.d

22.d

23.c

24.c

25. c

26. b

27. d

28. a

29. d

30. b

31. c

32. d

33. b

34. a

35. b

36. c

37. a

38. c

39. b

40. c

41. d

SHORT ANSWER

42. hormones

43. endocrine glands

44. steroids

45. amino acids

46. Neuropeptides

47. neuromodulators

48. hypothalamus

49. Peptide; steroid

50. cyclic AMP

51. hormone-receptor complex

52. Cortisol

53. thyroxine

54. calcitonin

55. cortisol (or hydrocortisone)

56. diabetes mellitus

57. insulin

ESSAY

58. Hormones spread to many cells via the bloodstream, rather than being directed to individual cells as in the case of nerve impulses. Although hormones are spread throughout the body, they also exhibit specificity, in that they have a shape that allows them to interact with specific receptors on particular target cells. Third, since the same receptor molecule may be found on different kinds of cells in different organs, hormonal effects exhibit flexibility, which is the ability to achieve different effects in different tissues.

59. Glucagon attaches to a target cell in the liver by binding to a protein receptor on the cell's surface. This binding causes an enzyme to produce cyclic AMP, a second messenger that activates certain enzymes within the target cell. These enzymes break down glycogen to form glucose, which then moves from the cell to the bloodstream.

60. Peptide hormones attach to receptor proteins on the surface of target cells, causing the production of cyclic AMP, a second messenger, within the cell. Cyclic AMP activates enzymes within the cell; thus, the cell is "turned on." Steroid hormones enter the cell, where they combine with receptor proteins in the cytoplasm. The hormone-receptor complex enters the cell's nucleus, where it activates specific genes on the DNA molecules.

61. In a diabetes patient, the pancreas does not make enough insulin. Insulin stimulates cells to take up glucose from the bloodstream. When insulin is absent, there may be plenty of glucose in the blood, but the cells are unable to absorb it. Therefore, the cells cannot conduct cellular respiration, so they starve.

ANALOGIES

62. d (are)

63. c (works on the)

64. a (produce)

65. b (connect)

66. b (is used to)

67. d (passes through)

68. a (secretes)

69. c (is caused by)

70. d (regulates the body's)

71. c (regulate)

Chapter 42

TRUE/FALSE Read each statement. If the statement is true, write T in the space provided. If the statement is false, write F in the space provided.

1.______ Sperm are produced and mature in the testes.

2.______ The adrenal glands regulate the functioning of the testes.

3.______ In humans, both the male and female gamete-producing organs are located within the abdominal cavity.

4.______ Fertilization usually takes place in the fallopian tubes.

5.______ An ovum is a mature egg cell.

6.______ Prescription pills that contain luteinizing hormone (LH) and follicle stimulating hormone (FSH) are used to disrupt the ovarian cycle.

7.______ The uterus is a pear-shaped organ made of three muscular layers.

8.______ Immediately after fertilization, a human embryo rapidly increases in size.

9.______ The process through which a hollow dorsal nerve cord forms from mesoderm is called neurulation.

10.______ Ectoderm, mesoderm, and endoderm form during gastrulation.

11.______ In its early stage, genital herpes can be cured by the use of antibiotics.

12.______ One of the most common causes of infertility in women is pelvic inflammatory disease.

MULTIPLE CHOICE Circle the letter of the term or phrase that best completes each statement or best answers each question.

13. The testes
   a. produce sperm.
   b. produce male hormones.
   c. are suspended in the scrotum.
   d. All of the above

14. Semen is composed of
   a. sperm.
   b. fluid from the seminal vesicles.
   c. bulbourethral gland secretions.
   d. All of the above

15. A vasectomy is an operation in which
   a. the testes are removed.
   b. the seminal vesicles are removed.
   c. the epididymis is coiled.
   d. each vas deferens is cut and the ends are tied.

16. The process by which sperm leave the male's body is called
   a. secretion.
   b. ejaculation.
   c. erection.
   d. fertilization.

17. The period of time during which a female reaches reproductive maturity is called
a. puberty.
b. maturation.
c. growth.
d. menopause.

18. The fallopian tubes
a. secrete estrogen.
b. produce eggs.
c. are tubes through which each egg travels.
d. All of the above

19. Menopause is the time at which
a. adult sex characteristics first appear.
b. eggs are produced.
c. menstruation begins.
d. females cease to release eggs.

20. During implantation,
a. the follicle matures.
b. the embryo attaches itself to the uterine wall.
c. the sperm reaches the egg.
d. menstruation occurs.

21. The strong ring of muscles at the entrance to the uterus is known as the
a. vagina.
b. endometrium.
c. cervix.
d. diaphragm.

22. "Spermicidal"
a. refers to the testes.
b. means "sperm production."
c. refers to the release of sperm.
d. means "sperm-killing."

23. The embryo is formed
a. in the ovary.
b. during the third trimester of pregnancy.
c. at fertilization.
d. when the fertilized egg divides.

24. The structure that exchanges substances between the mother and fetus is the
a. yolk sac.
b. fallopian tube.
c. placenta.
d. mature follicle.

25. Embryonic cells are organized into the major organ systems
a. during the sixth month.
b. by the end of the third trimester.
c. during the second trimester.
d. just before birth.

26. Cells produced by division of the fertilized egg give rise to the
a. embryo.
b. placenta.
c. umbilical cord.
d. All of the above

27. Sexually transmitted diseases such as gonorrhea and chlamydia, which are caused by bacteria
a. are almost always incurable.
b. are a direct cause of viral infections in a fetus.
c. can be treated with antibiotics.
d. None of the above

28. The most common sexually transmitted disease in the United States is
a. gonorrhea.
b. AIDS.
c. syphilis.
d. chlamydia.

29. Sexually transmitted diseases may be
   a. caused by bacteria.
   b. caused by a virus.
   c. passed from the mother's blood to the blood of her baby.
   d. All of the above

30. One type of herpes virus causes genital herpes, while another herpes virus causes
   a. cold sores.
   b. gonorrhea.
   c. syphilis.
   d. AIDS.

31. Gonorrhea can be acquired
   a. by sharing glassware and dishes with infected persons.
   b. through sexual contact with infected persons.
   c. by swimming in infected swimming pools.
   d. through poor personal hygiene habits.

The following question(s) refer(s) to the diagram of the male reproductive system.

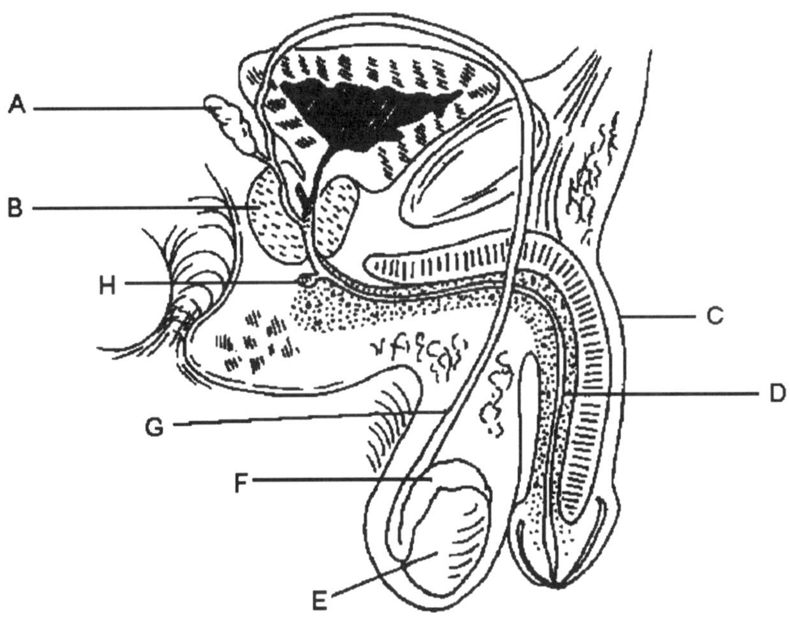

32. The structure that connects the epididymis to the urethra is labeled
   a. A.
   b. E.
   c. F.
   d. G.

33. The structure labeled B is
   a. the urinary bladder.
   b. the scrotal sac.
   c. the prostate gland.
   d. the bulbourethral gland.

34. The tube that carries urine during excretion and semen during ejaculation is labeled
   a. A.
   b. H.
   c. G.
   d. D.

The following question(s) refer(s) to the diagram of the female reproductive system.

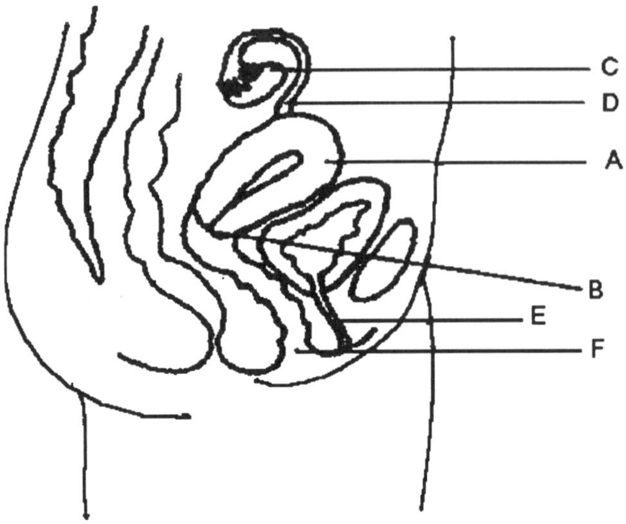

35. The structure labeled A is
   a. a fallopian tube. c. the uterus.
   b. the urethra. d. a ureter.

36. Fertilization usually occurs in the structure labeled
   a. A. c. C.
   b. D. d. F.

37. Eggs mature in the structure labeled
   a. C. c. D.
   b. G. d. A.

The following question(s) refer(s) to diagram of one part of the human reproductive system.

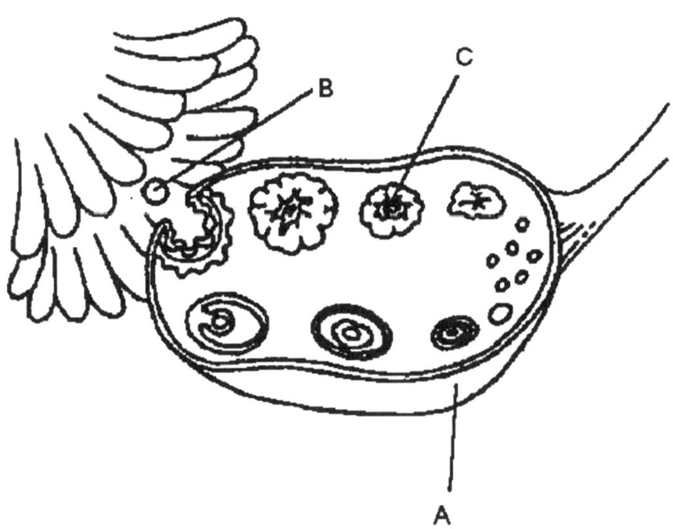

38. The diagram represents the
   a. menstrual cycle. c. testicular cycle.
   b. ovarian cycle. d. menopausal cycle.

39. The structure indicated by B is
a. a sperm cell.
b. a follicle cell.
c. an egg cell.
d. the cervix.

40. The structure indicated by A is
a. a follicle.
b. an egg cell.
c. the uterus.
d. an ovary.

41. The structure indicated by C is
a. an immature follicle.
b. a follicle beginning to break down.
c. a mature egg cell.
d. an immature egg cell.

SHORT ANSWER Complete each statement by writing the correct term or phrase in the space provided.

42. The testes are suspended in the ________________________.

43. In the testes, sperm are produced in tubes called ________________________.

44. In males, the ________________________ gland, which is located just below the bladder, secretes a fluid that counteracts the acids produced by the female reproductive tract.

45. The immature eggs found in a female at birth are arrested in the ________________________ stage of the first meiotic division.

46. Cutting or tying the fallopian tubes to prevent pregnancy is called ________________________.

47. The inner wall of the uterus is called the ________________________.

48. Clusters of cells that surround an immature egg are called ________________________.

49. The ovarian and ________________________ cycles of the female are usually about 28 days long.

50. The period of rapid division of an egg immediately after fertilization is known as ________________________.

51. Pregnancy is often divided into three 3-month periods known as ________________________.

52. Following implantation, a membrane known as the ________________________ encloses the embryo.

53. The body organs of an embryo develop in a process called ________________________.

54. ________________________ is the most common bacterial STD in the United States.

55. Syphilis usually begins with the appearance of an ulcer called a ________________________

ESSAY Read and answer each question or statement on a separate sheet of paper.

56. Describe the structure of a sperm cell and explain how it is adapted for its function.

57. Explain why the first trimester of pregnancy is such a crucial time of development.

58. Describe the development that occurs in a fetus from the end of the first trimester to the end of the third trimester.

ANALOGIES An analogy is a relationship between two pairs of words or phrases written as: a:b::c:d. The symbol : is read "is to," and the symbol :: is read "as." Write the letter of the pair of terms that best completes the analogy shown.

59. sperm production : testes::
   a. scrotum : testes
   b. alkaline fluid production : testes
   c. increased body temperature : scrotum
   d. FSH production : testes
   e. testosterone production : testes

____

60. ovary : egg production::
   a. ovary : sperm production
   b. female reproductive system : sperm production
   c. testis : sperm production
   d. ovary : fertilization
   e. abdominal cavity : testes

____

61. cleavage : rapid cell division of zygote::
   a. gastrulation : cleavage of cells
   b. gastrulation : beginning of cell formation
   c. neurulation : development of nerve cord
   d. neurulation : distribution of cells into 3 layers
   e. organogenesis : development of nerve cord

____

62. miscarriage : the disruption of proper course of events during pregnancy::
   a. gastrulation : birth defects
   b. rubella : fetal alcohol syndrome
   c. alcohol use : fetal alcohol syndrome
   d. birth defect : alcohol use by pregnant females
   e. organogenesis : fetal alcohol syndrome

____

63. AIDS patients : infections and cancer::
   a. HIV patients : painful blisters
   b. HIV patients : cancer
   c. genital herpes sufferers : painful blisters
   d. genital herpes sufferers : cancer
   e. genital herpes sufferers : cold sores

____

64. gonorrhea infection : no symptoms::

a. syphilis infection : cancer
b. gonorrhea infection : chancre
c. syphilis infection : pus discharge
d. chlamydia infection : no symptoms
e. chlamydia infection : fever blister

____

65. pneumocystic pneumonia : AIDS patients::

a. mast cell : HIV
b. helper T cell : HIV
c. HIV : helper T cells
d. Kaposi's sarcoma : macrophages
e. HIV : myelin nerve covering

____

TEST ANSWERS

TRUE/FALSE

1.F

2.F

3.F

4.T

5.T

6.F

7.F

8.F

9.F

10.T

11.F

12.T

MULTIPLE CHOICE

13.d

14.d

15.d

16.b

17.a

18.c

19.d

20.b

21.c

22.d

23.d

24.c

25. b

26. d

27. c

28. d

29. d

30. a

31. b

32. d

33. c

34. d

35. c

36. b

37. a

38. b

39. c

40. d

41. b

SHORT ANSWER

42. scrotum

43. seminiferous tubules

44. prostate

45. prophase

46. tubal ligation

47. endometrium

48. follicles

49. menstrual

50. cleavage

51. trimesters

52. amnion

53. organogenesis

54. Chlamydia

55. chancre

ESSAY

56. A sperm cell consists of a head with very little cytoplasm and a long tail. Digestive enzymes in the head enable a sperm cell to penetrate an egg. Located between the head and tail are mitochondria that supply the tail with ATP containing the energy that a sperm cell needs to swim through a female's reproductive system to an egg.

57. Most of the major developmental events (for example, organ formation) occur during early pregnancy. Most miscarriages occur during this period. Fetal alcohol syndrome is due to alcohol use by pregnant women, especially during early pregnancy.

58. By the end of the first trimester, all of the major body organs have differentiated, and the sex of the fetus has been established. During the second and third trimesters, the fetus grows rapidly as its organs are completed and become functional. By the end of the third trimester, the fetus is able to exist outside of its mother's body.

ANALOGIES

59. e (occurs in the)

60. b (is the location of)

61. c (consists of the)

62. d (can result from)

63. c (suffer from)

64. d (often shows)

65. c (attacks)

Introduction to Performance-Based Assessment

Teacher Information

The Test Generator contains the following Performance-Based Assessment Activities:

- Sample Unit: *Introduction to Performance-Based Assessment*
- Unit 1: *Effect of Acid Rain on Trout Populations*
- Unit 2: *Identifying Dominant and Recessive Traits*
- Unit 3: *Diffusion Through a Cell Membrane*
- Unit 4: *Testing Pond Water for Carbon Dioxide*
- Unit 5: *Fermentation by Yeast*
- Unit 6: *Testing Enzyme Activity in Seeds*
- Unit 7: *Exploring Invertebrates*
- Unit 8: *The Challenge of Water Retention*
- Unit 9: *Testing Human Digestive Enzymes*

The Performance-Based Assessment Activities are correlated with the content of the units of *Biology: Principles and Explorations*. Each Performance-Based Assessment Activity is designed for group or independent assessment and can usually be completed in one 50-minute class period. The assessment activities assess student mastery of the following skills:

- Identifying a problem that can be solved by conducting an experiment in the laboratory
- Forming a hypothesis
- Designing and conducting an experiment
- Making observations and measurements
- Organizing the data in a table
- Making a graph of the data
- Interpreting the data
- Evaluating the hypothesis based on the data

Each assessment activity consists of two parts. **Part I** is an introduction to the materials and concepts that students will be working with in Part II of the activity. **Part II** consists of two or three alternative experiments in which students are assigned similar tasks. For each assessment activity, individual students (or teams) should complete both a Part I and one of the alternative versions of Part II. Because students will be conducting different experiments, they should realize that they may not get useful information by observing others. Thus, students are encouraged to work independently.

Performance-Based Assessment Activities may be new to your students. To introduce your students to performance-based assessment, *Introduction to Performance-Based Assessment* leads students through the steps of the scientific process and familiarizes them with the expectations and format of these Performance-Based Assessment Activities. The assessment activities become increasingly challenging with each unit. In earlier assessments, you may want to concentrate on helping students understand the scientific process. You may decide to use the activities for Units 1 and 2 as team assessments, so that students will have an opportunity to develop their problem-solving strategies with the support of peers. By the third or fourth assessment, students should be ready to handle more complex problems. These Performance-Based Assessment

Activities are sequenced with this strategy in mind. After each assessment activity, encourage students to discuss their approach to completing the assessment. Begin individual assessment when you feel your students are ready.

Each assessment activity begins with a practice activity, so that students can practice the laboratory skills required in an experiment before they are expected to independently conduct an experiment. Students are always provided a list of appropriate safety precautions before they are instructed to work with any laboratory materials. A portion of each assessment score covers students' observance of safety precautions and appropriate use of the laboratory materials.

Before beginning a Performance-Based Assessment Activity, you may wish to modify your laboratory environment to ensure that students complete assessments independently and without distraction from others. It may be helpful to divide work stations into two sections with inexpensive partitions, made from cardboard or tag board. In addition, it is a good idea to designate an area of the laboratory for a central supply table. In this area, students should be able to find all the materials necessary for conducting experiments. When chemical disposal is necessary, you will want to provide a disposal area for your students, such as a ventilated hood, lab bench, or lab sink located in the front of the classroom.

Each Performance-Based Assessment Activity includes **Teacher Notes** and a **Scoring Rubric**. **Teacher Notes** describe the concepts and vocabulary that students should review before beginning the assessment activity, as well as strategies and instructions on how to set up the laboratory for the assessment activity. The materials required for each activity are divided into two lists: a list of the items you will need to prepare materials for an assessment activity and a list of the materials each student or team completing an assessment activity will require. Step-by-step directions explain how to prepare any solutions and materials and how to avoid potential problems.

Each **Scoring Rubric** lists preferred student responses in italics. Suggested point values for each step in the activity appear in parentheses next to the responses. The total points earned by a student for each step in the activity can be recorded in the blanks to the right of the steps. The Scoring Rubrics can be modified to accommodate your specific circumstances. You may wish to use this rubric only as a guide in scoring the assessment activity. If possible, make a copy of the rubric for each student or team, and return a copy of the rubric showing the scores earned by the students when you return their graded papers.

When using the Test Generator to edit or print a Performance-Based Assessment Activity, it is important to note that each activity consists of three to five questions. There is one question for each Part I and Part II of the assessment activity. The Answer Key for each question contains a Scoring Rubric for that question. The final question for each unit contains the Teacher Notes. Figures that are referred to in any question or answer of a Performance-Based Assessment Activity are found in the Answer Key for the Teacher Notes. Print these figures, and cut and paste them in place before copying an assessment for your class.

Unit Sample
Performance-Based Assessment Activity
Introduction to Performance-Based Assessment
Part I–Sample Assessment

This activity is designed to introduce you to the Holt Performance-Based Assessment Program. Completing this investigation will help you understand how the scientific process can be applied to studying a problem. This activity will also prepare you for future performance-based assessments.

Follow the guidelines below during this assessment activity.

- Remember not to talk to other students taking the test. If you are working with a team, talk only with those in your group and talk very quietly, so that you do not disturb other teams.
- When you conduct your experiment you will be able to use only the materials and equipment on the supply table. Some of the materials on the table will not be needed.
- Most of the other students involved in today's activity have different problems to solve. If you observe other students using materials different from those that you are using, this may be because their problems are not the same as yours.
- Remember that a problem often has several different acceptable solutions.

Objectives

After reading the Background, you will be expected to do the following.

- Perform a simple experiment and make observations to identify a scientific problem.
- Formulate a hypothesis based on the observations you made.
- Design and conduct an experiment to test the hypothesis.
- Organize the data from your experiment in a table.
- Evaluate the hypothesis based on your data.

Background

Enzymes are a type of molecule made by living cells. Enzymes act as catalysts for the chemical reactions that are necessary for life, such as breaking large molecules into smaller ones. In your mouth, for example, enzymes in saliva help break apart long starch molecules into thousands of the smaller molecules that make up starch—the simple sugar glucose. Catalase is another enzyme that is produced by the cells of some living tissues. Catalase assists the reaction that breaks apart hydrogen peroxide molecules (H_2O_2) and forms molecules of water (H_2O) and oxygen (O_2).

= an oxygen atom

Sample Figure 1a

= a hydrogen atom

Sample Figure 1b

This chemical reaction can be illustrated as follows.

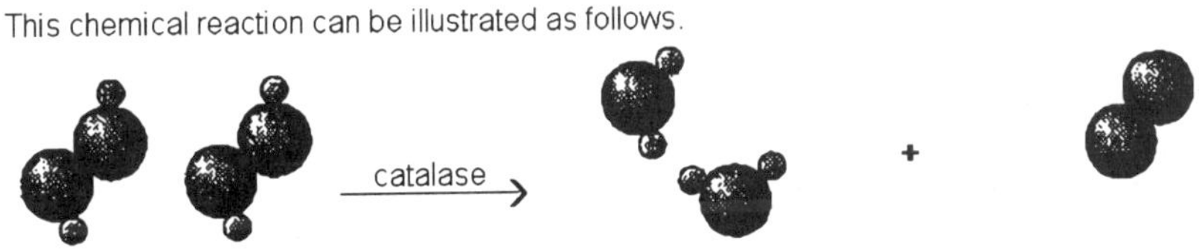

Sample Figure 2a
two H_2O_2 molecules

Sample Figure 2b
two H_2O molecules

Sample Figure 2c
one O_2 molecule

Chemists summarize the reaction illustrated above with the following equation.

$$2H_2O_2 \xrightarrow{\text{catalase}} 2H_2O + O_2$$

This equation is read as follows: two molecules of hydrogen peroxide are broken apart in the presence of catalase, and two molecules of water and one molecule of oxygen are formed.

Safety Precautions

The following safety precautions should be observed during Parts I and II of this activity.

- Wear safety goggles, protective gloves, and a lab apron.
- Confine loose hair and clothing.
- Do not touch or taste any chemicals. Wash off any solutions that get on skin or clothing at the sink while calling to the teacher. If you get a solution in your eyes, immediately flush it out at the eyewash station while calling to the teacher.
- Be careful not to spill any solutions. Call the teacher in the event of a spill.
- Exercise care when working with boiling liquids. Be sure that test tubes in a hot water bath are not pointing toward you or anyone else. Handle hot test tubes with a test-tube holder. Apply ice to burns or scalds.
- Exercise caution when working with a hot plate. Electrical equipment is an electric shock hazard. Be sure the area under and around a hot plate is dry, and be sure that your hands are dry when using a hot plate. Do not allow electrical cords to dangle from your work station. Do not use a hot plate with a kinked or frayed cord, but return it to your teacher.
- Dispose of broken glass in a separate container. Never put broken glass in a regular waste container.
- Clean up the lab and all equipment after use. Dispose of solutions according to proper disposal methods. After all lab work is finished, wash hands thoroughly before leaving the lab.

Practice

As you work through this activity, you will learn the steps in the **scientific process**. Learn the name and definition of each of these steps so that you can apply them to the scientific investigation of other problems.

Step 1: Observations

Careful observation is the foundation of scientific investigation. Observation involves perceiving and noting what happens in nature and in laboratory experiments. Observations are made before an investigation is planned and throughout all subsequent experiments. Before continuing, perform the following activity and record your observations.

1. **Materials** You should find the following materials at your work station:

   dropper bottle of 3% hydrogen peroxide solution
   2 glass microscope slides
   container of sand
   container of potassium iodide
   chemical spatula or scoop

2. **Procedure** Place a small amount of sand (10-15 grains) on the left side of a glass slide. Place a similar amount of potassium iodide on the right side of the slide. On each of the two samples, place one drop of the 3% hydrogen peroxide solution.

3. Data Describe what you observed below.

____

____

____

The bubbles you observed are oxygen gas. Potassium iodide acts as a catalyst for the reaction that breaks apart hydrogen peroxide molecules and forms water and oxygen. Sand contains no ingredients that can help break apart hydrogen peroxide molecules. With this knowledge, you can now investigate a simple problem in the laboratory.

Step 2: Problem

Before a laboratory investigation can begin, a problem must be identified. A problem should be clearly defined and is usually stated as a question, such as the one that follows.

The enzyme catalase is found in neither potatoes nor wood.

The problem in a scientific investigation is based on a researcher's observations or on information gathered from reading about the observations and experiments of other researchers.

4. Using the Background information given above and the observations you have made so far, suggest two other problems that could be investigated in a laboratory.

____

____

Step 3: Hypothesis

The next step in the scientific process is to offer a possible answer to the question that describes the problem you want to investigate. Scientists call this possible answer a hypothesis. One hypothesis about the problem stated above is as follows.

The enzyme catalase is found in neither potatoes nor wood.

5. In the space provided, write at least two other possible hypotheses for the problem in Step 2.

a. ____

b. ____

Often, many different hypotheses are offered for a single problem. When selecting among many appropriate hypotheses, a researcher uses past experience and understanding to decide which one to test first.

6. From the three hypotheses stated above, select the hypothesis that you think is most likely to be the answer for the problem given in Step 2. Write this hypothesis in the space below.

____

Step 4: Experimental Design

The next step in the scientific process is to plan an experiment that will test your hypothesis. In this step, a scientist writes out the procedure, or series of steps, that will be used to test a hypothesis. It is often helpful to draw a diagram of how equipment will be set up. You will be asked to do this in future performance-based assessment activities. Here, a simple description of the experiment is adequate. Before you plan an experiment to test the hypothesis you selected in Step 3, visit the materials table and examine the available materials. *Do not remove* any materials at this time. Think about how you would design an experiment to test your hypothesis using the available materials.

7. **Materials** In the space below, list the materials that you will use in your experiment.

8. **Procedure** In the space below, describe how you will set up and conduct an experiment that will test your hypothesis.

____

____

____

9. **Data** List the observations or measurements you will make during the experiment.

____

____

____

Step 5: Control

An experimental design should have at least two parts. One part is called the *experimental setup*. The other is called the *control setup*. Control tests are an important part of every experimental design. They answer questions about the materials and procedures used and enable researchers to draw appropriate conclusions about results. For example, if you observed no production of bubbles when hydrogen peroxide was placed on living tissue, you might conclude that the tissue does not contain catalase. Another possibility is that the hydrogen peroxide you used is of poor quality. Hydrogen peroxide molecules break down very slowly by the reaction described previously, even when a catalyst is not present. Therefore, hydrogen peroxide solutions have a limited shelf life. As a control, you should test the reactive ability of the hydrogen peroxide you are using in your experiment by exposing it to a substance that is known to catalyze the breakdown of hydrogen peroxide into water and oxygen gas.

10. In the space below, describe a control test that would verify the quality of a hydrogen peroxide solution.

____

____

One additional control test could be useful for this experiment. Placing a liquid on a dry material sometimes results in the appearance of a few air bubbles, which might be interpreted as oxygen bubbles. When oxygen gas is produced by the breakdown of hydrogen peroxide, a large number of bubbles is produced.

11. In the space below, describe a control test that would distinguish the appearance of a few air bubbles from the amount of bubbles that would indicate oxygen gas production.

____

____

Step 6: Conduct the Experiment as Planned
Before conducting your experiment, compare the following experimental design with your design.

Experimental Setup:
- Place a small piece of potato on one end of a glass slide or some other suitable surface. Place a similar sized piece of wood, such as a piece of a toothpick, on the other end of the glass slide.
- Add 1 drop of 3% hydrogen peroxide to each sample.
- Watch for the formation of oxygen bubbles in both samples. The formation of bubbles is an indication that the enzyme catalase is present in the sample.

Control Setup:
It is well known that potassium iodide always causes good quality hydrogen peroxide to break down into oxygen gas and water. To test the quality of hydrogen peroxide, do the following.
- Place a small amount of potassium iodide on one end of a glass slide.
- Add one drop of 3% hydrogen peroxide to the potassium iodide and observe.

Washed sand is known to produce no reaction of hydrogen peroxide. Any small bubbles that result when hydrogen peroxide is mixed with sand are probably air. A very large quantity of bubbles is produced when hydrogen peroxide breaks down into water and oxygen gas. To compare the quantity of bubbles produced when a reaction does not occur, do the following.
- Place a small amount of sand on the other end of the glass slide.
- Add one drop of 3% hydrogen peroxide to the sand and observe.

12. Gather the materials you need to conduct the experiment above. Notice that you do not need all of the items on the materials table. This will usually be the case in future performance-based assessments. The extra items are included to provide practice in determining which equipment and supplies are essential for an experiment and which are not.

13. Conduct the experiment above, following the appropriate **safety precautions**.

Step 7: Record Experimental Results
Experimental results should be neatly recorded and clearly labeled. When appropriate, the data collected during an experiment should be organized into a chart like the example that follows.

14. Data Record the results of your experiment in the blanks below.

a. Experimental Test Results ____

____

b. Control Test Results____

____

15. Record your data in the data chart below.

Experimental Tests	Results	Control Tests	Results
potato		potassium iodide	
wood		washed sand	

Step 8: Interpret Results
When you have finished collecting your data, write a few sentences that summarize what the data mean with respect to the problem being investigated. It is important to distinguish the meaning of *data*, which are the records of what happened in an experiment, from an *interpretation* of the data, which describes what the data mean.

16. Interpretation In the space below, write your interpretation of the data you recorded.

____

____

____

Step 9: Conclusion
In a scientific investigation, the conclusion is a statement of whether or not the hypothesis was a possible answer to the problem being investigated. To determine a conclusion, researchers compare their hypothesis with their interpretation of their data. If the hypothesis and the data agree, scientists say "the hypothesis is *supported* by the data." If the interpretation of the data is an interpretation that would be impossible or unlikely if the hypothesis were true, scientists conclude that "the hypothesis is *refuted,* or *contradicted,* by the data." Sometimes the interpretation of the data does not allow a researcher to conclude that the hypothesis is either supported or refuted. Many scientists would then say "the hypothesis is *not supported* by the data." Other scientists would say "the hypothesis is *neither supported nor refuted* by the data."

17. Conclusion Examine your hypothesis and the interpretation of your data, and then state a conclusion in the space below.

____

____

Unit Sample
Performance-Based Assessment Activity
Introduction to Performance-Based Assessment
Part II–Sample Assessment Application

In this part of the sample assessment, you will apply what you have learned about using the scientific process to solve a lab problem. You will turn in your response for teacher evaluation.

Problem

Consider the following **problem.**

Does boiling a piece of potato destroy the ability of the enzyme catalase to assist the breakdo[wn] of hydrogen peroxide into water and oxygen gas?

1. **Hypothesis** In the space below, write a **hypothesis** that you think provides the best answer to this question.

____

____

Experiment

Design an experiment that you can conduct to test the above hypothesis. You will be able to use only the materials and equipment from the materials table or at your testing station. Not all of the materials at the materials table will be needed. If you observe other students using materials different from those you are using, this may be because they are testing a different hypothesis. Also remember that a problem often has different acceptable solutions.

2. **Materials** Go to the equipment and materials table, and examine the available materials. While at the materials table, decide which of the items available would be useful in your experiment. List those materials in the space below.

3. **Experimental Design** In the space below, explain how you will set up an experiment that would test your hypothesis.

   a. Experimental Setup______________________________

____

____

____

b. In the space below, draw a diagram of how your experiment test will be set up.

c. Control Setup ____

____

____

____

4. **Observations** In the space below, explain what observations or measurements you will make during the experiment.

____

____

5. **Safety Precautions** Review the safety precautions listed on page 2, and follow them during the experiment.

6. **Procedure** To conduct your experiment, gather *only* the materials and equipment you need from the materials table. Pour into separate glassware the amount of each solution you need for your experiment, and transfer the solutions to your work station. Do not take an entire quantity of any solution. Others may need to use the same solutions. When you have gathered the needed materials, set up and conduct your experiment as planned.

Data

7. **Records** Record your data in the table provided below.

8. **Interpretation** In the space below, write your interpretation of the data you recorded.

____

____

9. **Conclusion** In the space below, state whether your data support or refute your hypothesis.

____

10. **Analysis** List any factors that could have affected the accuracy of the data you obtained.

____

Cleanup

After you turn in your paper, dispose of your materials according to the instructions provided by your teacher. Clean up your work area, and wash your hands before leaving the lab.

Unit Sample
Performance-Based Assessment Activity
Introduction to Performance-Based Assessment
Teacher Notes and Scoring Rubric

Teacher Notes

Assessment Time Required (Four 15- to 40-minute class periods)
The following is a suggested schedule for using the Introduction to Performance-Based Assessment.

Day 1: Have students perform Part I–Sample Assessment. (30 minutes)
Day 2: Review Part I–Sample Assessment, the names and descriptions of each step in the scientific process, the dynamics of the hydrogen peroxide reaction, and the procedure for assigning points. (30 minutes)
Day 3: Have students perform Part II–Sample Assessment Application. (30-40 minutes)
Day 4: Review answers for Part II–Sample Assessment Application. (15 minutes)

Concepts and Vocabulary
Before beginning this assessment, students should be familiar with the following concepts:

- control
- enzyme
- enzyme activity
- glucose
- starch
- scientific process

Introduction
This introductory Performance-Based Assessment Activity is designed to introduce students to the steps in the scientific process. It will also provide students with an opportunity to practice the skills needed for performance-based assessments. This activity consists of two parts. Part I is a sample assessment that leads students through the steps of a scientific investigation. Part II of the sample assessment is an activity that allows students to apply what they learned about the process of scientific investigation. In both parts, students design and perform experiments to test for the presence of the enzyme catalase in various materials. In later assessments, students will be given different problems to solve.

Strategies
Because this introductory assessment activity may be your students' first experience with performance-based assessment, administer it early in the year. This activity is designed to be used to teach and assess students either in teams or as individuals. You might, for example, use Part I—Sample Assessment as a team assignment, allowing students to work in teams of two, three, or four, and use Part II—Sample Assessment Application to assess students individually. You may require students working in teams to submit one paper per team, or you may require them to submit individual papers. Tell students that they may be assessed individually later in the year.

Tell students that a Performance-Based Assessment Activity is a test and that the standard test-taking rules apply. Allow talking among team members, but stress that talking must be done very quietly and must be relevant to the assessment activity. Advise students not to base their responses on what they see other students doing. During future Performance-Based Assessment Activities, other students may be working on different problems. When all students have completed an assessment, conduct a class discussion of the students' approaches to the Performance-Based Assessment Activity.

Begin this activity by reviewing the reaction of hydrogen peroxide and the action of catalase. Tell students that it is essential to understand that bubble production indicates the presence of an active enzyme and that the absence of bubbles indicates either the absence of the enzyme or the presence of an inactive enzyme. Before assigning Part II-Sample Assessment Application, review the completed Part I-Sample Assessment to be sure that all students understand the definition of each step in the scientific process and that they can apply the steps to solving a problem. Review how the assessments will be graded. One option is to have students grade Part I of this assessment together and talk about how points were assigned.

After Part II has been completed and reviewed, you might want to pose the following problems for discussion to further review the application of the scientific process. Ask: At what temperatures is the enzyme catalase most active? Can the enzyme catalase work equally well in an acidic or basic environment? What is the relationship between the size of the piece of potato and the length of time it takes to break down all of the hydrogen peroxide?

Student Safety Precautions

The safety precautions for this assessment activity are listed in Part I. Before allowing students to begin the activity, make sure they have read these safety precautions. Advise students that they will be given points for properly following the safety precautions during this activity.

Use only electrical equipment equipped with a three-wire cord and a three-prong plug. Each electrical socket in the laboratory must be of the three-hole type with a GFI (ground-fault interrupter) circuit. Be sure that hot plates have an "on-off" switch and indicator light. Students should not be allowed to leave operating hot plates unattended. Be sure electrical equipment is turned off before plugging it in or unplugging it. Do not allow electrical cords to dangle from work stations; dangling cords are a tripping and shock hazard. Do not use electrical equipment with frayed or kinked cords. The area under and around electrical equipment should be dry. Cords should not lie in puddles of spilled liquid. Students' hands should be dry when using electrical equipment. Turn off and unplug all electrical equipment before leaving the laboratory.

Materials

For each student or team, you will need the following materials in the quantities indicated:

- 3% hydrogen peroxide (available at drug stores)
- 1 small dropper bottle for the hydrogen peroxide
- 1 piece of fresh raw potato
- 1 toothpick
- potassium iodide (just a pinch)
- washed beach or construction sand (just a pinch)
- 2 glass microscope slides
- 1 chemical spatula or scoop
- 1 250-mL beaker
- 1 test-tube holder
- 1 test-tube rack
- 1 pair of forceps (for transferring hot potato piece from boiling water)
- 1 hot plate

Laboratory Setup

For this activity, some materials should be placed at students' work stations and others should be placed at a central supply table to prepare students for this arrangement in future assessments.

Place the following materials on a centrally located supply table:

- 1 container of fresh raw potato pieces, cut into 1/8 to 1/4 in. cubes
- 1 box of toothpicks
- 1 beaker or flask of liquid marked "Starch Solution" (decoy)
- 1 beaker or flask of liquid marked "Protein Solution" (decoy)
- 1 test-tube holder per student or team
- 1 test-tube rack per student or team
- 1 250-mL beaker per student or team
- 1 pair of forceps per student or team

Place the following at each testing station:

- 1 hot plate
- 1 small container of potassium iodide
- 1 small container of washed beach or construction sand
- 1 dropper bottle of 3% hydrogen peroxide
- 2 glass microscope slides
- 1 chemical spatula or scoop

Cleanup and Disposal

The hydrogen peroxide, potassium iodide, sand, potato pieces, and toothpicks used during this activity can be safely disposed of in the classroom's trash can. Instruct students to first use a paper towel to sweep into the trash the materials on their glass microscope slides, and then wash and dry the glassware they use and return it to the materials table before they leave the lab.

Scoring Rubric

The following pages are a Scoring Rubric that provides possible answers to this Performance-Based Assessment Activity. Preferred student responses are shown in italics. Suggested point values for each step in the activity appear in parentheses next to the responses. The total points earned by a student for each step in the activity can be recorded in the blanks to the right of the steps. The Scoring Rubric can be modified to accommodate your specific circumstances. You may wish to use this rubric only as a guide in scoring the assessment activity. If possible, make a copy of the rubric for each student or team, and return a copy of the rubric showing the scores earned by the students when you return their graded papers.

Activity and Scoring Rubric Art

These teacher notes are found in question 3 of the Unit Sample File. In the answer key for this question, you will find the art for Sample Figure 1 and Sample Figure 2, which are found in Part I of this assessment. Cut and paste these figures in the places indicated before photocopying the assessment for your students.

Unit Sample Performance-Based Assessment Activity Introduction to Performance-Based Assessment (50 points) Scoring Rubric

______________ Total Score

Possible answers to this Performance-Based Assessment Activity are provided below. Preferred student responses are shown in italics. Suggested point values for each step in the activity appear in parentheses next to the responses. The total points earned for each step in the activity are recorded in the blanks to the right of the steps.

Part I–Sample Assessment (25 points total)

______________ Part I Score

Step 1: Observation (1 point total)

3. Data
Bubbles can be observed in the H_2O_2 placed on potassium iodide. Only a few bubbles or an absence of bubbles are observed in the H_2O_2 placed on sand. (1 point) ______________

Step 2: Problem (2 points total)

4. Problems to be investigated:
Any two appropriate problems that are consistent with the background provided and the observations made so far. For example, a problem for this experiment might be *Is the enzyme catalase found in saliva?* (2 points) ______________

Step 3: Hypothesis (3 points total)

5. Possible hypotheses: (1 point each for any 2 of the following) ______________
The enzyme catalase is found in both potatoes and wood.
The enzyme catalase is found in neither potatoes nor wood.
The enzyme catalase is found in potatoes but not in wood.
The enzyme catalase is found in wood but not in potatoes.

6. Selected hypothesis:
Any one of the four hypotheses is acceptable. (1 point) ______________

Step 4: Experimental Design (3 points total)

7. Materials The student made appropriate use of the materials. (1 point) ______________

8. Procedure The student provided an appropriate description of an experiment. Descriptions of appropriate tests are found in **Step 6** on page 5. (1 point) ______________

9. Data The student provided an appropriate list of observations and measurements. (1 point) ______________

Step 5: Control (2 points total)

10. The student described an appropriate control test. (1point) The test using hydrogen peroxide and potassium iodide described in **Step 1** is an appropriate control in this case. ______________

11. The student described an appropriate control test. (1 point) ______________
The test using hydrogen peroxide and sand described in **Step 1** is an appropriate control in this case.

Step 6: Conduct the Experiment as Planned (2 points total)

12. Materials The student made appropriate use of materials. (1 point) ________

13. The student followed appropriate **safety precautions**. (1 point) ________

Step 7: Record Results (8 points total)

14. Data (4 points total)

a. Experimental Test Results:

The potato produced large quantities of bubbles. (1 point) ________

The wood produced few or no bubbles. (1 point) ________

b. Control Test Results:

The potassium iodide produced large quantities of bubbles. *(1 point)* ________

The sand produced few or no bubbles. *(1 point)* ________

15. Data Table (4 points total)

Experimental Tests	Results	Control Tests	Results
potato	*large quantities of bubbles* (1 point)	potassium iodide	*large quantities of bubbles* (1 point)
wood	*few or no bubbles* (1 point)	washed sand	*few or no bubbles* (1 point)

Step 8: Interpret Results (3 points total)

16. Interpretation

Potatoes contain catalase. (1 point) ________

Wood does not contain catalase. (1 point) ________

Sand does not contain catalase. (1 point) ________

Step 9: Conclusion (1 point total)

17. Conclusion

The hypothesis that pototoes contain the enzyme catalase is ***supported*** *by the data.* (1 point) ________

or

All other hypotheses are refuted, or contradicted, by the data. (1 point)

Part II–Sample Assessment Application (25 points total)

Part II Score

Problem (1 point)

1. **Hypothesis**

   Boiling a piece of potato destroys the ability of catalase to assist the conversion of H_2O_2 to H_2O and O_2. (1 point)

   or

   Boiling a piece of potato does not destroy the ability of catalase to assist the conversion of H_2O_2 to H_2O and O_2. (1 point)

Experiment (14 points total)

2. **Materials** The student listed appropriate materials. (1 point)

3. **Experimental Design** (10 points total)

   a. Experimental Setup (4 points total) Obtain a small piece of raw potato, and place it into boiling water for five or more minutes. (2 points)
   (If potato is left for less than 5 minutes, assign 1 point.)

   With forceps, remove the potato, place it on a glass slide or similar container, add one drop of hydrogen peroxide, and observe. (1 point)

   b. Student's experimental setup diagram should show the following:
   - 250 mL beaker half-filled with water on top of a hot plate. (1 point)
   - Potato piece inside of beaker of boiling water. (1 point)

   c. Control Setup (5 points)
   Obtain a small piece of raw potato, place it on a glass slide or similar container, add one drop of hydrogen peroxide, and observe. (1 point)

   Also place a small amount of potassium iodide on a slide, and test it with hydrogen peroxide. (1 point)*

   *Place a small amount of sand on a slide, and test it with hydrogen peroxide.**
   (1 point)

   *You may assign these points if students indicate that this step was performed during the sample assessment and offer the results of these tests.

4. **Observations**

   When hydrogen peroxide is added, watch for the presence or absence of large amounts of bubbles. (1 point)

5. **Safety Precautions** The student followed appropriate safety precautions. (2 points)

6. **Procedure** The student made appropriate use of materials. (2 points)

Data (10 points total)

7. **Records** (8 points total)

Experimental Tests	Results	Control Tests	Results
raw potato (1 point)	*large quantities of bubbles* (1 point)	*potassium iodide* (1 point)	*large quantities of bubbles* (1 point)
boiled potato (1 point)	*few or no bubbles* (1 point)	*washed sand* (1 point)	*few or no bubbles* (1 point)

8. **Interpretation**

Boiling a raw potato destroys the ability of the enzyme catalase to convert hydrogen peroxide to oxygen and water. (1 point)

9. **Conclusion**

The hypothesis that boiling a piece of potato destroys the ability of catalase to assist the breakdown of H_2O_2 into H_2O and O_2 is supported by the data. (1 point)

or

The hypothesis that boiling a piece of potato destroys the ability of catalase to assist the breakdown of H_2O_2 into H_2O and O_2 is contradicted by the data. (1 point)

Unit 1
Performance-Based Assessment Activity
Effect of Acid Rain on Trout Populations
Part I–Identifying a Problem

Follow the guidelines below during this assessment activity.

- Remember not to talk to other students taking the test. If you are working with a team, talk only with those in your group and talk very quietly, so that you do not disturb other teams.
- When you conduct your experiment, you will be able to use only the materials and equipment on the supply table. Some of the materials on the table will not be needed.
- Most of the other students involved in today's activity have different problems to solve. If you observe other students using materials different from those that you are using, this may be because their problems are not the same as yours.
- Remember that a problem often has several different acceptable solutions.

Objective

After reading the following paragraph, you will be expected to identify a scientific problem.

Background

The Pacific Northwest is known for its fine sport fishing. People have fished the lakes and streams in the foothills of the Cascade mountain range for years. In many of the lakes and streams of one area, however, the number of a certain species of trout has dropped drastically. Some lakes and streams in this area no longer have a single trout. In this area, the wind blows from the west, off the Pacific Ocean. On the way east, moisture-laden clouds pass over a large city where air pollution has become an increasing problem in recent years. The clouds give up the moisture as rain or snow on the western slopes of the Cascade Range. Waste gases from the city's industries react with the water in the air to produce acid rain and snow. Acid precipitation that falls on the western slopes of the mountains runs into the creeks and streams that supply Lakes A, B, and C. A reduction in trout population has been observed in each of these lakes.

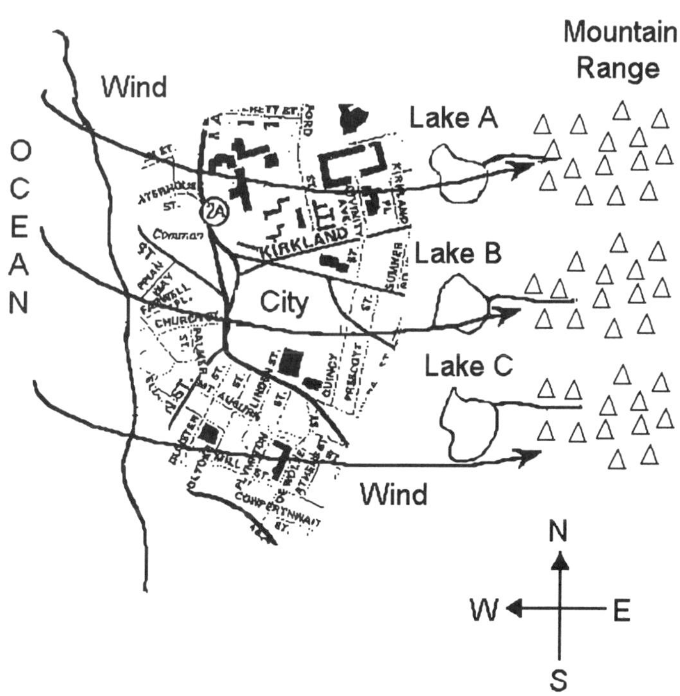

Unit 1 - Figure 1

1. Based on the Background, state a **problem** that might be investigated in the laboratory.

____

____

Safety Precautions

The following safety precautions should be observed during Parts I and II of this activity.

- Wear safety goggles, protective gloves, and a lab apron.
- Confine loose hair and clothing.
- Do not touch or taste any chemicals. Wash off any solutions that get on skin or clothing at the sink while calling to the teacher. If you get a solution in your eyes, immediately flush it out at the eyewash station while calling to the teacher.
- Be careful not to spill any solutions. Call the teacher in the event of an acid or base spill.
- Dispose of any broken glass in a separate container. Never put broken glass in a regular waste container.
- Clean up the lab and all equipment after use. Dispose of solutions according to proper disposal methods. Wash hands thoroughly before leaving the lab after all lab work is finished.

Practice

To see how a pH test is done, do the following simple exercise.

2. Go to the supply table, and obtain about 2–3 mL of each of the following solutions: distilled water, lemon juice, baking soda, and tap water.

3. The acidity of a solution is measured in pH units. A dispenser of pH test paper should be on the supply table. Remove four 2.5 cm (1.0 in.) strips of pH paper. Dip a strip of pH test paper into one of the solutions, and remove it immediately. Note the color of the pH paper. Compare it to the color scale on the pH test dispenser to determine the pH of the solution.

4. Determine the pH of each of the other three solutions and record the pH of each of the four solutions in the table below. Be sure to add the necessary heads to the table.

5. The pH scale ranges from 1 to 14. Solutions with a pH number between 0 and 7 are acidic. Solutions with a pH number between 7 and 14 are basic. A solution with a pH of 7 is neutral. Add another column to your table and indicate which solutions are acidic, basic, or neutral.

6. Dispose of your materials according to the instructions provided by your teacher, and rinse each container thoroughly before continuing.

Unit 1
Performance-Based Assessment Activity
Effect of Acid Rain on Trout Populations
Part II–Experiment

Objectives

After reading the Additional Background, you will be expected to do the following

- Formulate a hypothesis about the effects of acid rain on a trout population in a given lake.
- Design and conduct an experiment to test your hypothesis.
- Organize the data from your experiment in a table.
- Evaluate your hypothesis based on your data.

Problem

Consider the following **problem** based on the information in Part I.

Could acid precipitation be responsible for the decline of the trout population in Lake ______? (Record the letter of the lake assigned to you.)

Additional Background

After a female trout lays eggs, the eggs develop in the water over a 28-day period before hatching. A biologist from the University of Washington simulated the effects of acid precipitation on trout eggs. First, the biologist placed trout eggs in dishes of stream water. Then, at four different stages during the trout eggs' development, an acidic solution was added to make the pH of the stream water in one set of dishes 4.4. In the control group, stream water was added instead of an acidic solution. Later the eggs that hatched in each set of dishes were counted. The biologists' findings are summarized in the following table.

Effect of Acidic pH on Trout Egg Development

Number of days after eggs were laid	1–4 days	5–10 days	11–15 days	16–28 days
Percentage of eggs that later hatched in test group (acidic solution added)	98%	4%	90%	99%
Percentage of eggs that later hatched in control group (stream water added)	99%	97%	98%	99%

1. What do the data indicate about the time of exposure of trout eggs to pH 4.4 and the percentage of eggs that eventually hatch?

____

____

2. In this experiment, what was the test group? ______________________________

What was the control group? ______________________________

Studies show that the time of egg laying in trout varies from year to year. Biologists' findings over a three-year period are summarized in the following table.

Dates Egg Laying Occurred Over Three Years

Lake	**Current Year**	**Last Year**	**2 Years Ago**
Lake A	May 1	May 4	May 10
Lake B	April 1	April 30	May 4
Lake C	May 2	May 5	May 14

3. During what dates of the current year would trout egg development be most affected by a pH of 4.4 in Lake

A?________________ Lake B?_______________ Lake C?_______________

Experiment

4. **Hypothesis** In the space below, write a hypothesis for the **problem** that is stated at the beginning of Part II.

____

____

5. **Materials** Go to the equipment and materials table, and examine which materials are available. Think about how you would design an experiment to test your hypothesis using these materials. Do not remove any materials from the table yet. Make a list of the materials that you will use in your experiment.

6. **Procedure**

   a. Describe how you will test your hypothesis.

   ____

   ____

   ____

   b. List the observations or measurements you will make during the experiment.

   ____

   ____

   c. Be sure to observe the **safety precautions** listed in Part I during your experiment.

   d. To conduct your experiment, gather *only* the materials and equipment you need from the materials table. Pour into separate glassware the amount of each solution you need for your experiment, and transfer the solution or solutions to your work station. Do not take an entire quantity of any solution. Others may also need to use the same solutions. When you have gathered the needed materials, set up and conduct your experiment as planned.

Data

7. **Records** Record your data below in a neat table.

8. **Interpretation** Write your interpretation of the data you recorded.

____

____

____

9. **Conclusion** State whether your data support or refute your hypothesis.

____

10. **Analysis** List any factors that could have affected the accuracy of the data you obtained during your experiment.

____

____

____

Cleanup

After you turn in your paper, dispose of your materials according to the instructions provided by your teacher. Clean up your work area, and wash your hands before leaving the lab.

Unit 1
Performance-Based Assessment Activity
Effect of Acid Rain on Trout Populations
Teacher Notes and Scoring Rubric

Teacher Notes

Assessment Time Required (One 50-minute class period)

Concepts and Vocabulary
Before beginning this assessment, students should be familiar with the following concepts:
- acid
- acid precipitation
- base
- pH
- scientific processes

Introduction
The Performance-Based Assessment Activity for Unit 1 consists of two parts. Part I involves reading a paragraph of background information and determining a scientific problem that can be tested in a lab. Part II of the assessment activity consists of three separate problems in which students are asked to test the pH of lake water in order to determine whether acid precipitation may have affected a trout population within the lake.

Strategies
All students should complete Part I. For Part II, assign each student one of the three lakes (Lake A, Lake B, or Lake C) mentioned in the Background section of Part I. If Part I is completed the day before Part II is to be done, students' papers for Part I should be returned to them for use during Part II of the assessment.

This assessment activity may be the first performance-based lab assessment for your students. Because this assessment is not as difficult as those in later units, it is a good introduction to performance-based assessment. Administer this assessment early in the year, and allow students to work in teams of two, three, or four. You may require students working in teams to submit one paper per team, or you may require them to submit individual papers. Tell students that they may be assessed individually later in the year.

Remind students that this activity is a test and that the standard rules of test-taking apply. Allow talking among team members, but stress that talking must be done very quietly and must be relevant to the assessment activity. When all students have completed the assessment, conduct a class discussion of the students' approaches to the Performance-Based Assessment Activity.

Student Safety Precautions
The safety precautions for this assessment activity are listed in Part I. Before allowing students to begin the activity, make sure they have read these safety precautions. Advise students that they will be given points for properly adhering to the safety precautions during this activity.

Materials
You will need the following materials to prepare the solutions needed for this assessment activity:
- 0.1 M HCl (1 mL is enough to prepare 1 L each of the acidic solutions)
- baking soda (sodium bicarbonate) (10 g is enough to prepare 1 L of each of the basic solutions)
- distilled or demineralized water (pH of close to 7.0)
- 17 large (200–500 mL) beakers or flasks with labels (one for each of the test solutions)

For each student or team, you will need the following materials in the quantities indicated:

- 4 small glass containers
- 8 strips of masking or labeling tape 5 cm (2 in.) in length
- 8 pieces of pH test paper 2.5 cm (1 in.) in length (Use pH test paper with a range of at least 3.5–9.0, color coded in 0.5 pH unit increments.)
- 2–3 mL distilled or demineralized water (pH of close to 7.0)
- 2–3 mL lemon juice (pH of about 2.2)
- 2–3 mL baking soda, or sodium bicarbonate, solution (pH of about 8.0)
- 2–3 mL tap water (pH will vary—determine in advance)
- 2–3 mL water from Lake A, B, or C, April 7
- 2–3 mL water from Lake A, B, or C, May 7
- 2–3 mL water from Lake A, B, or C, June 7
- 2–3 mL water from Lake A, B, or C, July 7

Preparation

1. Prepare solutions of the following pH: approximately 4.0, 6.1, 6.8, 7.0, 7.1, 7.2, and 8.0. As you prepare each solution, use a pH meter or narrow range pH paper to determine the pH of each solution.
2. **Caution** Do not use concentrated HCl to prepare the solutions for this activity. Purchase 0.1 M HCl for use in preparing solutions. When you prepare solutions containing acids or bases, wear safety goggles, a face shield, impermeable gloves, and a lab apron. In case of an acid spill, dilute first with water. Then mop up the spill with wet cloths designated for spill cleanup.
3. Prepare the acidic solutions as follows:
   pH 4.0—Add 1 mL of 0.1 M HCl to distilled or demineralized water and dilute to 1 L.
   pH 6.1—Dilute 7.9 mL of the pH 4.0 solution to 1 L.
   pH 6.8—Dilute 1.6 mL of the pH 4.0 solution to 1 L.
4. Prepare the basic solutions as follows:
   pH 8.0—Mix 7 g baking soda ($NaHCO_3$) with 1 L of distilled or demineralized water.
   pH 7.1 and 7.2—Mix pinches of baking soda ($NaHCO_3$) with distilled or demineralized water until proper pH is reached.
5. You may need to adjust the pH of distilled or demineralized water to make it about 7.0. If you use tap water that does not have a pH of 7.0 to prepare the solutions, the pH of the solutions will also have to be adjusted to the proper level. To adjust the pH of a solution, add small increments of 0.1 M HCl to tap water to lower its pH, or add small increments of baking soda (sodium bicarbonate) to water to raise its pH. Buffered capsules for various pHs can also be used and are available from a chemical or biological supply house.
6. Place a container of water identified as "Ocean Water" among the other solutions to be tested. This solution is a decoy that will help students become accustomed to the presence of materials that are *not* needed for their experiments.
7. Very small glass containers that hold only about 5 mL of liquid are recommended for use by students in transporting and testing solutions. If larger containers are used, students might waste solutions, and you may run out of solutions before all students complete the activity.

Laboratory Setup

Place the following materials on a centrally located supply table:

(Parts I and II)

- sign that reads, "Take no more than 2–3 mL of any solution"
- 4–8 dispensers of pH test paper
- 4–8 rolls of masking tape or labeling tape
- 4 small glass containers per student or team in a class
- tap water identified as "Ocean Water"

- Solutions* (Part I only)
  - distilled or demineralized water marked "Distilled Water"
  - lemon juice identified as "Lemon Juice"
  - baking soda solution identified as "Baking Soda Solution"
  - tap water identified as "Tap Water"
- Solutions* (Part II only)
  - water identified as "Water From Lake A, April 7" (pH 7.2)
  - water identified as "Water From Lake A, May 7" (pH 4.0)
  - water identified as "Water From Lake A, June 7" (pH 6.8)
  - water identified as "Water From Lake A, July 7" (pH 7.0)
  - water identified as "Water From Lake B, April 7" (pH 7.2)
  - water identified as "Water From Lake B, May 7" (pH 6.8)
  - water identified as "Water From Lake B, June 7" (pH 7.1)
  - water identified as "Water From Lake B, July 7" (pH 7.0)
  - water identified as "Water From Lake C, April 7" (pH 4.0)
  - water identified as "Water From Lake C, May 7" (pH 6.1)
  - water identified as "Water From Lake C, June 7" (pH 6.8)
  - water identified as "Water From Lake C, July 7" (pH 7.1)

*100 mL of each solution is more than enough for 20 students or teams. (Note: Only one-third of your students will use each set of lake-water solutions.)

Cleanup and Disposal

Before discarding leftover solutions in the sink, collect them in a single container, and neutralize them to a pH of 7 with 0.1 M NaOH. For student use during the assessment activity, provide a container marked "Solution Disposal Container." Avoid having students handle acid spills themselves. Instruct students to wash and dry the glassware they use and return it to the materials table before they leave the lab.

Scoring Rubric

The following pages are a Scoring Rubric that provides possible answers to this Performance-Based Assessment Activity. Preferred student responses are shown in italics. Suggested point values for each step in the activity appear in parentheses next to the responses. The total points earned by a student for each step in the activity can be recorded in the blanks to the right of the steps. The Scoring Rubric can be modified to accommodate your specific circumstances. You may wish to use this rubric only as a guide in scoring the assessment activity. If possible, make a copy of the rubric for each student or team, and return a copy of the rubric showing the scores earned by the students when you return their graded papers.

Activity and Scoring Rubric Art

These teacher notes are found in question 3 of the Unit 1 file. In the answer key for this question, you will find the art for Unit 1-Figure 1, which is found in Part I of this assessment. Cut and paste this figure in the place indicated before photocopying the assessment for your students.

Unit 1
Performance-Based Assessment Activity
Testing the Effects of Acid Rain (50 points)
Scoring Rubric

______ Total Score

Possible answers to this Performance-Based Assessment Activity are provided below. Preferred student responses are shown in italics. Suggested point values for each step in the activity appear in parentheses next to the responses. The total points earned for each step in the activity are recorded in the blanks to the right of the steps.

Part I (17 points total)

______ Part I Score

Background (2 points total)

1. **Problem** (2 points total)
   Could acid precipitation be causing the decline of trout populations in the lakes and streams? (2 points) ______
   or
   Other appropriately stated problems that are relevant to the background information are also acceptable. (1 point) ______

Practice (15 points total)

Steps 4-5. (1 point per table cell) ______

Substance Tested	*pH*	*Interpretation*
Distilled water	*7.0*	*neutral*
Lemon juice	*2.5–3.5*	*acidic*
Baking soda	*8–8.2*	*basic*
Tap water	*	*

*The pH of tap water will vary in different localities. Use the pH value for your local tap water.

Part II (33 points total)

______ Part II Score

Additional Background (6 points total)

1. *The data indicate that trout egg development is disrupted by exposure to a pH of 4.4 during the period 5–10 days after they are laid.* (1 point) ______

2. *The test group was the one to which acid was added.* (1 point) ______

   The control group was the one to which stream water was added. (1 point) ______

3. *Lake A—May 5 to May 10* (1 point) ______

   Lake B—April 5 to April 10 (1 point) ______

   Lake C—May 6 to May 11 (1 point) ______

Experiment (15 points total)

4. **Hypothesis** (3 points total)
Acid precipitation is (or is not) causing the decline of the trout populations in Lake A, B, or C. (3 points)
or
Acid precipitation is lowering the pH in Lake A, B, or C between the 5th and 10th days after trout eggs are laid, and this decreases the hatching rate. (3 points) __________
or
Other appropriately stated hypotheses that are relevant to the background may also be acceptable but are not indicated by the data. (2 points) __________

5. **Materials** (2 points total)
Students should include only materials that are found on the supply table and that are needed for their particular experiment. (2 points) __________

6. **Procedure** (10 points total)
   a. Students should plan to test the pH of water samples taken from their assigned lake and compare their pH measurements to those provided in the introduction. (2 points) __________
   b. For Lake A or C: Test samples taken from Lake A or C on May 7. For Lake B: Test samples taken from Lake B on April 7. (2 points) __________
   (If students plan to test other dates, this is acceptable but not necessary.)
   c. **Safety Precautions** The student followed appropriate safety precautions (4 points) __________
   d. **Materials** The student made appropriate use of materials (2 points) __________

Data (12 points total)

Lake A

7. **Records** (6 points total)

Date (1 point)	*pH* (1 point)
Apr. 7	7.2
May 7 (2 points)	*4.0* (2 points)
June 7	6.8
July 7	7.0

8. **Interpretation** (3 points total)
The pH of Lake A was low enough to cause the death of trout embryos (2 points) __________
during the time they were vulnerable to a low pH (May 6–11). (1 point) __________
or
Any other interpretation that is consistent with the data (1 point) __________

9. **Conclusion** (2 points total)
The hypothesis that acid precipitation is causing the decline of the trout population in Lake A is supported by the data (2 points) __________
or
Other appropriate conclusion when hypothesis and interpretation are compared (1 point) __________

10. **Analysis**
Acceptable factors include any possible constraining factor, such as unclean glassware or contaminated samples of lake water. (1 point) __________

Lake B

7. **Records** (6 points total)

Date (1 point)	*pH* (1 point)
Apr. 7 (2 points)	*7.2* (2 points)
May 7	6.9
June 7	7.1
July 7	7.0

8. **Interpretation** (3 points total)
The pH of Lake B was not low enough to cause the death of trout embryos
(2 points) ______________
during the time they were vulnerable to a low pH (April 6–11).
(1 point) ______________
or
Any other interpretaion that is consistent with the data (1 point) ______________

9. **Conclusion** (2 points total)
The hypothesis that acid precipitation is causing the decline of the trout population in Lake B is refuted by the data. (2 points) ______________
or
Other appropriate conclusion when hypothesis and interpretation are compared
(1 point) ______________

10. **Analysis**
Acceptable factors include any possible constraining factor, such as unclean glassware or contaminated samples of lake water. (1 point) ______________

Lake C

7. **Records** (6 points total)

Date (1 point)	*pH* (1 point)
Apr. 7	4.3
May 7 (2 points)	*5.8* (2 points)
June 7	6.7
July 7	7.1

8. **Interpretation** (3 points total)
The pH of Lake C was not low enough to cause the death of trout embryos
(2 points) ______________
during the time they were vulnerable to a low pH (May 7–15).
(1 point) ______________
or
Any other interpretation that is consistent with the data (1 point) ______________

9. **Conclusion** (2 points total)
The hypothesis that acid precipitation is causing the decline of the trout population in Lake C is refuted by the data. (2 points) ______________
or
Other appropriate conclusion when hypothesis and interpretation are compared
(1 point) ______________

10. **Analysis**
Acceptable factors include any possible constraining factor, such as unclean glassware or contaminated samples of lake water. (1 point) ______________

Unit 2
Performance-Based Assessment Activity
Identifying Dominant and Recessive Traits
Part I–Identifying a Problem

Follow the guidelines below during this assessment activity.

- Remember not to talk to other students taking the test. If you are working with a team, talk only with those in your group and talk very quietly, so that you do not disturb other teams.
- When you conduct your experiment, you will be able to use only the materials and equipment on the supply table. Some of the materials on the table will not be needed.
- Other students involved in today's activity have different problems to solve. If you observe other students using materials different from those that you are using, this may be because their problems are not the same as yours.
- Remember that a problem often has several different acceptable solutions.

Objective

After reading the following paragraph, you will be expected to identify a scientific problem.

Background

A certain species of plant produces either bright red flowers or pure white flowers. In working out the inheritance of a trait with contrasting forms such as flower color, it is important to determine which symbols will be assigned to the alternative forms (alleles) of the genes for the trait (in this case, red and white). The first letter for one of the alternative forms of the trait involved is generally used to represent the alleles for a trait. A capital letter is usually assigned to the dominant allele, and a lowercase letter is assigned to the recessive allele. For example, if **R** is used to represent red, then **r** would represent white. Or if **W** is used to represent white, then **w** would represent red. The pair of genes that determines a trait is called a genotype. The genotype is represented by the pair of letters that symbolizes the alleles present. When both genes in a pair are the same, the genotype is said to be **homozygous**. When the genes in a pair are different, the genotype is said to be **heterozygous**.

1. In this activity, you will perform simulated crosses between red-flowered plants and white-flowered plants. Based upon the preceding paragraph, identify and state a problem that needs to be solved before the inheritance of flower color in this plant species can be studied further.

____

____

2. To recall how the results of a cross are predicted, complete the following exercise. In cats, black coat color, **B**, is dominant to white coat color, **b.** To predict the possible coat colors of the kittens that will result from a cross between two heterozygous black cats (**Bb** × **Bb**), complete the following Punnett square.

Unit 2
Performance-Based Assessment Activity
Identifying Dominant and Recessive Traits
Part II–Experiment A

Objectives

After reading the Additional Background, you will be expected to do the following:

- Formulate a hypothesis for a scientific problem.
- Conduct an experiment to test the hypothesis.
- Record and interpret data to determine if the hypothesis is a suitable answer for the problem.

Safety Precautions

The following safety precautions should be observed during Part II of this activity.

- Wear safety goggles, protective gloves, and a lab apron.
- Confine loose hair and clothing.
- Do not touch or taste any chemicals. Wash off any solutions that get on skin or clothing at the sink while calling to the teacher. If you get a solution in your eyes, immediately flush it out at the eyewash station while calling to the teacher.
- Be careful not to spill any solutions. Call the teacher in the event of an acid or base spill.
- Dispose of any broken glass in a separate container. Never put broken glass in a regular waste container.
- Clean up the lab and all equipment after use. Dispose of solutions according to proper disposal methods. After all lab work is finished, wash hands thoroughly before leaving the lab.

Additional Background

In order to simulate crosses between plants with red or white flowers, you will use solutions to represent flower color. The red solutions represent plants with red flowers. The clear solutions represent plants with white flowers. Assume that plants with a particular flower color may or may not be carrying a gene for the alternate flower color. When you mix two of the solutions, the color of the resulting mixture represents a flower color that *could* be observed among the offspring of two plants if you were to actually cross plants with flowers the same colors as the solutions.

1. **Caution** *The solutions you will use in the following experiment contain chemicals that could damage your skin, eyes, or clothing.* Follow the suggested safety precautions.

2. **Materials** Obtain the following materials and 5 mL of each of the solutions indicated from the materials table. Transfer to your work station only the needed amounts of the materials and each solution. Do not take all of any solution. Others may need access to the same solutions. When you have gathered the needed materials, return to your work station and continue.

5 test tubes	masking tape	solution #4
1 test-tube rack	solution #1	solution #5
3 beakers or similar containers	solution #2	solution #7

3. After returning from the materials table, record the colors for each solution in the blanks below.

solution #1 ____________ solution #5 ____________

solution #2 ____________ solution #7 ____________

solution #4 ____________

Practice

4. Procedure

a. Cross 1 Mark one empty container "Cross 1." Then pour 5 mL of solution #1 and 5 mL of solution #2 into the empty container marked "Cross 1." Record the colors of the parents and the resulting offspring in the boxes below.

Parent #1 × Parent #2

↓

F_1 Offspring

b. Cross 2 Mark another empty container "Cross 2." Then pour 5 mL of solution #4 and 5 mL of solution #5 into the empty container marked "Cross 2." Record the colors of the parents and the resulting offspring in the boxes below.

Parent #4 × Parent #5

↓

F_1 Offspring

Problem

Based on the background information, consider the following **problem**.

Which gene symbols for flower color mentioned in the opening paragraph should be assigned to represent the genotypes of these parents and offspring?

Experiment

5. Hypothesis Write a hypothesis for the **problem** that is stated above by filling in the boxes below with the appropriate gene symbols.

a. Cross 1

Parent #1 × Parent #2

↓

#3 F_1 Offspring

b. Cross 2

Parent #4	×	Parent #5
	↓	
	#6 F_1 Offspring	

c. Which is the dominant form of the trait (flower color), red or white?

____

d. **Cross 3** Consider the following cross, and predict the genotype and flower color of #8.

Parent #6	×	Parent #7
	↓	
	#8 F_1 Offspring	________ color

6. **Procedure**
   a. Perform Cross 3 above by mixing equal amounts of solutions #6 and #7.
   b. Suggest an experiment that would enable you to determine the exact genotype of Parent #7.

Data

7. **Records** Record the results obtained in the blank. _______________

8. **Conclusion** State whether your data support or refute your hypothesis.

____

9. **Analysis** Discuss the strengths and weaknesses of the model used in this assessment activity to represent crosses involving plants with different flower colors.

____

____

____

____

Cleanup

After turning in your paper, dispose of your materials according to the instructions provided by your teacher. Clean up your work area, and wash your hands before leaving the lab.

Unit 2
Performance-Based Assessment Activity
Identifying Dominant and Recessive Traits
Part II–Experiment B

Objectives

After reading the Additional Background, you will be expected to do the following:

- Formulate a hypothesis for a scientific problem.
- Conduct an experiment to test the hypothesis.
- Record and interpret data to determine if the hypothesis is a suitable answer for the problem.

Safety Precautions

The following safety precautions should be observed during Part II of this activity.

- Wear safety goggles, protective gloves, and a lab apron.
- Confine loose hair and clothing.
- Do not touch or taste any chemicals. Wash off any solutions that get on skin or clothing at the sink while calling to the teacher. If you get a solution in your eyes, immediately flush it out at the eyewash station while calling to the teacher.
- Be careful not to spill any solutions. Call the teacher in the event of an acid or base spill.
- Dispose of any broken glass in a separate container. Never put broken glass in a regular waste container.
- Clean up the lab and all equipment after use. Dispose of solutions according to proper disposal methods. After all lab work is finished, wash hands thoroughly before leaving the lab.

Additional Background

In order to simulate crosses between plants with red or white flowers, you will use solutions to represent flower color. The red solutions represent plants with red flowers. The clear solutions represent plants with white flowers. Assume that plants with a particular flower color may or may not be carrying a gene for the alternate flower color. When you mix two of the solutions, the color of the resulting mixture represents a flower color that *could* be observed among the offspring of two plants if you were to actually cross plants with flowers the same colors as the solutions.

1. **Caution** *The solutions you will use in the following experiment contain chemicals that could damage your skin, eyes, or clothing.* Follow the suggested safety precautions.

2. **Materials** Obtain the following materials and 5 mL of each of the solutions indicated from the materials table. Transfer to your work station only the needed amounts of the materials and each solution. Do not take all of any solution. Others may need access to the same solutions. When you have gathered the needed materials, return to your work station and continue.

   4 test tubes
   1 test-tube rack
   3 beakers or similar containers
   masking tape
   solution #1
   solution #2
   solution #4
   solution #5

3. After returning from the materials table, record the colors for each solution in the blanks below.

   solution #1 ______________
   solution #2 ______________
   solution #4 ______________
   solution #5 ______________

Practice

4. Procedure

a. Cross 1 Mark one empty container "Cross 1." Then pour 5 mL of solution #1 and 5 mL of solution #2 into the empty container marked "Cross 1." Record the colors of the parents and the resulting offspring in the boxes below.

Parent #4 × Parent #5

↓

#6 F_1 Offspring

b. Cross 2 Mark another empty container "Cross 2." Then pour 5 mL of solution #4 and 5 mL of solution #5 into the empty container marked "Cross 2." Record the colors of the parents and the resulting offspring in the boxes below.

Parent #1 × Parent #2

↓

#3 F_1 Offspring

Problem

Based on the background information, consider the following **problem**.

Which gene symbols for flower color mentioned in the opening paragraph should be assigned to represent the genotypes of these parents and offspring?

Experiment

5. Hypothesis Write a hypothesis for the **problem** that is stated above by filling in the boxes below with the appropriate gene symbols.

a. Cross 1

Parent #4 × Parent #5

↓

#6 F_1 Offspring ____________ color

b. Cross 2

[ ] × [ ]

Parent #1 — Parent #2

↓

[ ]

#3 F_1 Offspring ____________ color

c. Which is the dominant form of the trait (flower color), red or white?

____

d. Cross 3 Consider the following cross, and predict the genotype and flower color of #9.

[ ] × [ ]

Parent #6 — Parent #6

↓

[ ]

#9 F_1 Offspring ____________ color

6. Procedure

a. Perform Cross 3 above by mixing equal amounts of solution #6.

b. Suggest an experiment that would enable you to determine the exact genotype of Parent #2.

Data

7. Records Record the results obtained in the blank. ______________________

8. Conclusion State whether your data support or refute your hypothesis.

____

9. Analysis Discuss the strengths and weaknesses of the model used in this assessment activity to represent crosses involving plants with different flower colors.

____

____

____

____

Cleanup

After turning in your paper, dispose of your materials according to the instructions provided by your teacher. Clean up your work area, and wash your hands before leaving the lab.

Unit 2
Performance-Based Assessment Activity
Identifying Dominant and Recessive Traits
Teacher Notes and Scoring Rubric

Teacher Notes

Assessment Time Required (One 50-minute class period)

Concepts and Vocabulary

Before beginning this assessment, students should be familiar with the following concepts:

- dominant trait
- recessive trait
- genotype
- scientific processes
- monohybrid cross

Introduction

The Performance-Based Assessment Activity for Unit 2 consists of two parts. Part I involves reading a paragraph of background information and determining a scientific problem that can be tested in a lab. Part II of the assessment activity consists of two separate problems in which students are asked to simulate crosses among individuals of a plant species that has either red or white flowers.

Strategies

All students should complete Part I. For Part II, half of the students in the class should be assigned Experiment A, and the other half should be assigned Experiment B. If Part I is completed the day before Part II is to be done, students' papers for Part I should be returned to them for use during Part II of the assessment.

Because this Performance-Based Assessment Activity may be administered early in the year, you may want this activity to be group assessment rather than an individual assessment. If this is the case, limit the group size to two students. You may require students working in teams to submit one paper per team, or you may require them to submit individual papers. If students work in teams, tell them that they may be assessed individually later in the year.

Remind students that this activity is a test and that standard test-taking rules apply. Allow talking among team members, but stress that talking must be done very quietly and must be relevant to the assessment activity. When all students have completed an assessment, conduct a class discussion of the students' approaches to the Performance-Based Assessment Activity.

Student Safety Precautions

The safety precautions for this assessment activity are listed in Part II. Before allowing students to begin Part II of the activity, make sure they have read these safety precautions. Advise students that they will be given points for properly adhering to the safety precautions during this activity.

Materials

You will need the following materials to prepare the solutions needed for this assessment activity:

- 1 M HCl (1 mL is enough to prepare 1 L each of the acidic solutions)
- 1 M NaOH (10 g is enough to prepare 1 L of each of the basic solutions)
- phenolphthalein powder (1 g is enough to prepare 160 mL of phenolphthalein solution)
- 100% ethyl alcohol (40 mL is enough to prepare 160 mL of phenolphthalein solution)
- distilled or demineralized water (pH of close to 7.0)
- tap water (pH of close to 7.0)
- 10 large (200–500 mL) beakers or flasks with labels (two for each of the test solutions)

For each student or team, you will need the following materials in the quantities indicated:

- 4 or 5 test tubes
- test-tube rack
- 3 beakers
- 7 or 8 strips of masking or labeling tape 5 cm (2 in.) in length
- 5 mL* Solution #1
- 5 mL* Solution #2
- 5 mL* Solution #4
- 5 mL* Solution #5
- 5 mL* Solution #7 (Experiment A only)

*To determine the total amount of the test solutions you will need, multiply [the number of work stations used] × [number of classes] × [7 mL] (instead of 5 mL to allow for some waste).

Preparation

1. To prepare a phenolphthalein solution, dissolve 1 g of phenolphthalein powder in 40 mL of 100% ethyl alcohol, then dilute this solution with 120 mL of distilled or demineralized water.
2. **Caution** Teachers should mix all NaOH and HCl solutions. Do not use concentrated HCl to prepare the solutions for this activity. Purchase 1 M HCl for use in preparing solutions. You may also purchase 1 M NaOH or prepare a 1 M NaOH solution from concentrated NaOH by using the instructions in step 3 below. When you prepare solutions containing acids or bases, wear safety goggles, a face shield, impermeable gloves, and a lab apron. In case of an acid spill, dilute first with water. Then mop up the spill with wet cloths designated for spill cleanup.
3. Place 20 g of concentrated NaOH into 500 mL of distilled or demineralized water. Use a spatula to carefully measure concentrated NaOH without touching it with your hands. Stir and heat until NaOH dissolves. When dissolved, pour into stock bottle labeled 1 M NaOH.
4. Prepare the test solutions as follows:
   Solution #1 (red)—Add 10 mL of phenolphthalein solution and 5 mL of 1 M NaOH to 1 L of tap water, and mix thoroughly.
   Solution #2 (clear or white)—Add 20 mL of 1 M HCl to 1 L of tap water, and mix thoroughly.
   Solution #4 (clear or white)—Add 15 mL of phenolphthalein solution to 1 L of tap water, and mix thoroughly.
   Solution #5 (clear or white)—Add 10 mL of 1 M NaOH to 1 L of tap water, and mix thoroughly.
   Solution #7 (clear or white)—Add 25 mL of 1 M HCl to 1 L of tap water, and mix thoroughly.
5. You may need to adjust the pH of distilled or demineralized water to make it about 7.0. If you use tap water that does not have a pH of 7.0 to prepare the solutions, the pH of the solutions will also have to be adjusted to the proper level. To adjust the pH of a solution, add small increments of 1 M HCl to tap water to lower its pH, or add small increments of 1 M NaOH to water to raise its pH.

Laboratory Setup (Part II only)

Place the following materials on a centrally located supply table:

- Solutions (2 containers of each to reduce the time it will take students to collect solutions)
  - Solution #1 (appropriately labeled)
  - Solution #2 (appropriately labeled)
  - Solution #4 (appropriately labeled)
  - Solution #5 (appropriately labeled)
  - Solution #7 (appropriately labeled)
- 10 25-mL graduated cylinders (2 for each solution, appropriately labeled)
  - 4 to 8 rolls of masking or labeling tape
  - 5 10-mL test tubes per student or team
  - 1 test-tube rack per student or team
  - 3 100-mL beakers or similar-sized transparent containers per student or team

Cleanup and Disposal

Before discarding leftover solutions in the sink, collect them in a single container, and neutralize them to a pH of 7 with 1 M HCl or 1 M NaOH. For student use during the assessment activity, provide a container marked "Solution Disposal Container." Avoid having students handle acid or base spills themselves. Instruct students to wash and dry the glassware they use and return it to the materials table before they leave the lab.

Scoring Rubric

The following pages are a Scoring Rubric that provides possible answers to this Performance-Based Assessment Activity. Preferred student responses are shown in italics. Suggested point values for each step in the activity appear in parentheses next to the responses. The total points earned by a student for each step in the activity can be recorded in the blanks to the right of the steps. The Scoring Rubric can be modified to accommodate your specific circumstances. You may wish to use this rubric only as a guide in scoring the assessment activity. If possible, make a copy of the rubric for each student or team, and return a copy of the rubric showing the scores earned by the students when you return their graded papers.

Unit 2
Performance-Based Assessment Activity
Identifying Dominant and Recessive Traits (50 points)
Scoring Rubric

______________ Total Score

Possible answers to this Performance-Based Assessment Activity are provided below. Preferred student responses are shown in italics. Suggested point values for each step in the activity appear in parentheses next to the responses. The total points earned for each step in the activity are recorded in the blanks to the right of the steps.

Part I (10 points total)

______________ Part I Score

Background

1. **Problem**
   Which characteristic is dominant, red or white? (2 points) ______________

2. (8 points, 1 point per table cell) ______________

	B (1 point)	***b*** (1 point)
B (1 point)	*BB* (1 point)	*Bb* (1 point)
b (1 point)	*Bb* (1 point)	*bb* (1 point)

Part II–Experiment A (40 points total)

Part II-A Score ______

Additional Background (8 points total)

1. **Safety Precautions** The student followed appropriate safety precautions. (2 points) ______

2. **Materials** The student made appropriate use of materials. (1 point) ______

3. Color of Solutions (5 points total)
   solution #1 *red or pink* (1 point) ______

   solution #2 *white or clear* (1 point) ______

   solution #4 *white or clear* (1 point) ______

   solution #5 *white or clear* (1 point) ______

   solution #7 *white or clear* (1 point) ______

Practice (8 points total)

4. **Procedure**
   a. **Cross 1** (4 points total) ______

red (1 point)	×	*white* (1 point)
Parent #1	↓	Parent #2
	white (2 points)	
	F_1 Offspring	

   b. **Cross 2** (4 points total) ______

white (1 point)	×	*white* (1 point)
Parent #4	↓	Parent #5
	red (2 points)	
	F_1 Offspring	

Experiment (18 points total)

5. **Hypothesis**
   a. **Cross 1** (5 points total) ______

ww (1 point)	×	*WW or Ww* (1 point each)
Parent #1	↓	Parent #2
	Ww (2 points)	
	#3 F_1 Offspring	

b. Cross 2 (4 points total)

Parent #4	×	Parent #5
Ww (1 point)		*Ww* (1 point)

↓

#6 F_1 Offspring: *ww* (2 points)

c. *white* (1 point)

d. Cross 3 (6 points total)

Parent #6	×	Parent #7
ww (1 point)		*WW or Ww* (1 point each)

↓

#8 F_1 Offspring: *Ww* (2 points)

white color (1 point)

6. Procedure

b. *You should be able to determine whether the genotype of Parent #7 is homozygous or heterozygous by mixing equal amounts of solution #7.* (2 points)

Data (6 points total)

7. Records *white* (1 point) (*Ww*) (1 point)

8. Conclusion

The hypothesis that the white Parent #7 is either WW or Ww and that #8, an F_1 offspring of Cross 3, will be white and have the genotype Ww is supported by the data. (2 points)

9. Analysis

Students should recognize that this model cannot indicate all the possible outcomes of a cross, unless the parents are known to be homozygous. (2 points)

Part II–Experiment B (40 points total)

Part II-B Score ____________

Additional Background (7 points total)

1. Safety Precautions The student followed appropriate safety precautions. (2 points) ____________

2. Materials The student made appropriate use of materials. (1 point) ____________

3. Color of Solutions

solution #1 *red or pink* (1 point) ____________

solution #2 *white or clear* (1 point) ____________

solution #4 *white or clear* (1 point) ____________

solution #5 *white or clear* (1 point) ____________

Practice (8 points total)

4. Procedure

a. Cross 1 (4 points total) ____________

white (1 point)	×	*white* (1 point)
Parent #4		Parent #5

↓

red (2 points)

F_1 Offspring

b. Cross 2 (4 points total) ____________

red (1 point)	×	*white* (1 point)
Parent #1		Parent #2

↓

white (2 points)

F_1 Offspring

Experiment (19 points total)

5. Hypothesis

a. Cross 1 (5 points total) ____________

Ww (1 point)	×	*Ww* (1 point)
Parent #4		Parent #5

↓

ww (2 points)

#6 F_1 Offspring

red

color (1 point)

b. Cross 2 (6 points total) ____________

ww (1 point)	×	*WW or Ww* (1 point each)
Parent #1	↓	Parent #2

Ww (2 points)	
#3 F_1 Offspring	*white* color (1 point)

c. *white* (1 point) ____________

d. Cross 3 (5 points total) ____________

ww (1 point)	×	*ww* (1 point)
Parent #6	↓	Parent #6

ww (2 points)	
#9 F_1 Offspring	*red* color (1 point)

6. Procedure

b. *You should be able to determine whether the genotype of Parent #2 is homozygous or heterozygous by mixing equal amounts of solution #2.* (2 points) ____________

Data (6 points total) ____________

7. Records *red* (1 point) (*ww*) (1 point)

8. Conclusion

The hypothesis that the two red parents with the genotype ww in Cross 3 will produce an F_1 offspring that will be red and have the genotype ww is supported by the data. (2 points) ____________

9. Analysis

Students should recognize that this model cannot indicate all the possible outcomes of a cross, unless the parents are known to be homozygous. (2 points) ____________

Unit 3
Performance-Based Assessment Activity
Diffusion Through a Cell Membrane
Part I–Identifying a Problem

Follow the guidelines below during this assessment activity.

- Remember not to talk to other students taking the test. If you are working with a team, talk only with those in your group and talk very quietly so that you do not disturb other teams.
- When you conduct your experiment, you will be able to use only the materials and equipment on the supply table. Some of the materials on the table will not be needed.
- Other students involved in today's activity have different problems to solve. If you observe other students using materials different from those that you are using, this may be because their problems are not the same as yours.
- Remember that a problem often has several different acceptable solutions.

Objectives

After reading the Background, you will be expected to do the following:

- Identify a scientific problem that can be investigated in a laboratory.
- Perform laboratory tests for the presence of carbon dioxide and starch.
- Make a simulated cell out of dialysis tubing.

Background

As cells evolved, they developed characteristics that enabled them to become the basic structural and functional units of life. Certain fundamental characteristics that evolved in the earliest cells, such as the ability to replicate, are found in all cells today. One characteristic of all cells is the plasma membrane that separates the inside of the cell from the outside environment. In order for materials to enter and leave a cell, the molecules of those materials must pass through the cell's membrane. Thus, the plasma membranes of living cells are considered to be selectively permeable because they allow only certain materials to pass through. The selectively permeable plasma membrane is a great evolutionary advantage to all organisms because it enables cells to take in only the materials they need, compartmentalize certain materials, and rid themselves of wastes.

Some molecules are able to pass through a plasma membrane by diffusion. Water is one such material. Water molecules are small enough to pass through tiny spaces between the molecules in the plasma membranes of cells. Other materials can pass through a cell's plasma membranes only through protein channels, some of which require the cell to expend energy. Besides water, all cells must be able to take in oxygen, food molecules, and a variety of minerals, and they must be able to rid themselves of carbon dioxide and other waste products. Plant cells must also be able to take in carbon dioxide, which they use to make starch during the process of photosynthesis. The starch made in the cells of leaves is then exported to other parts of the plant where there are cells that are not able to photosynthesize.

1. Based on the preceding paragraphs, identify and state a **problem** that can be tested in a laboratory.

____

____

Safety Precautions

The following safety precautions should be observed during Parts I and II of this activity.

- Wear safety goggles, protective gloves, and a lab apron.
- Confine loose hair and clothing.
- Do not touch or taste any chemicals. Wash off any solutions that get on skin or clothing at the sink while calling to the teacher. If you get a solution in your eyes, immediately flush it out at the eyewash station while calling to the teacher.
- Be careful not to spill any solutions. Call the teacher in the event of a spill.
- Dispose of any broken glass in a separate container. Never put broken glass in a regular waste container.
- Clean up the lab and all equipment after use. Dispose of solutions according to proper disposal methods. After all lab work is finished, wash hands thoroughly before leaving the lab.

Practice

Simple tests for the presence of carbon dioxide, CO_2, and starch can be performed in a laboratory. You can also make a simulated cell out of dialysis tubing.

2. Go to the the materials table, and obtain the following materials.
   - small flask
   - 100 mL tap water
   - test tube with 2 mL tap water
   - test tube with 2 mL starch solution
   - test-tube rack
   - plastic straw
   - small funnel
   - large test tube
   - 15 cm piece of dialysis tubing
   - 2 pieces of thread or string
   - dropper bottle of bromothymol blue (BTB) indicator
   - dropper bottle of Lugol's iodine solution
   - masking or labeling tape

3. Place about 100 mL of tap water into a small flask. Add 10 drops of bromothymol blue (BTB) indicator. If BTB remains blue, no CO_2 is present in the water. If BTB turns yellow, CO_2 is present in the water.

   a. Is CO_2 present in tap water?

   Now use the straw to gently blow into the tap water containing BTB for one or more minutes or until you observe no further change.

   b. Describe what happened and what this means.

   ____

4. Add 2 drops of Lugol's iodine solution to the test tube with 2 mL of tap water.

   a. What is the resulting color? ______________________________

   Now add 2 drops of Lugol's iodine solution to the test tube with 2 mL of starch solution.

   b. What is the resulting color? ______________________________

Dialysis tubing is made of a material that has some of the same properties as the plasma membranes of living cells. This material contains tiny pores that allow only molecules that are smaller than the pores to pass through. Thus, some of the same kinds of molecules that pass into and out of cells by diffusion can also pass through dialysis tubing by diffusion. However, dialysis tubing will not allow the passage of other kinds of molecules that cells are able to take in through the protein channels in their plasma membranes.

5. To make a simulated cell, perform the following steps.

   a. Thoroughly wet a 15 cm (6 in.) piece of dialysis tubing, and then securely tie one end of the tubing with a piece of thread or string as shown in the figure.

   b. Lower the tied end of the dialysis tubing into a large test tube, and gently fold the upper lip of the dialysis tubing over the opening of the test tube.

   c. Use a funnel to fill the tubing with tap water to within 5 cm (2 in.) of the opening.

   d. To close the opening and remove any air inside the tubing, twist the open end of the tubing just above the water level a couple of times before folding it over. Then tie this end securely with another piece of thread. Your simulated cell should resemble the one shown in the figure.

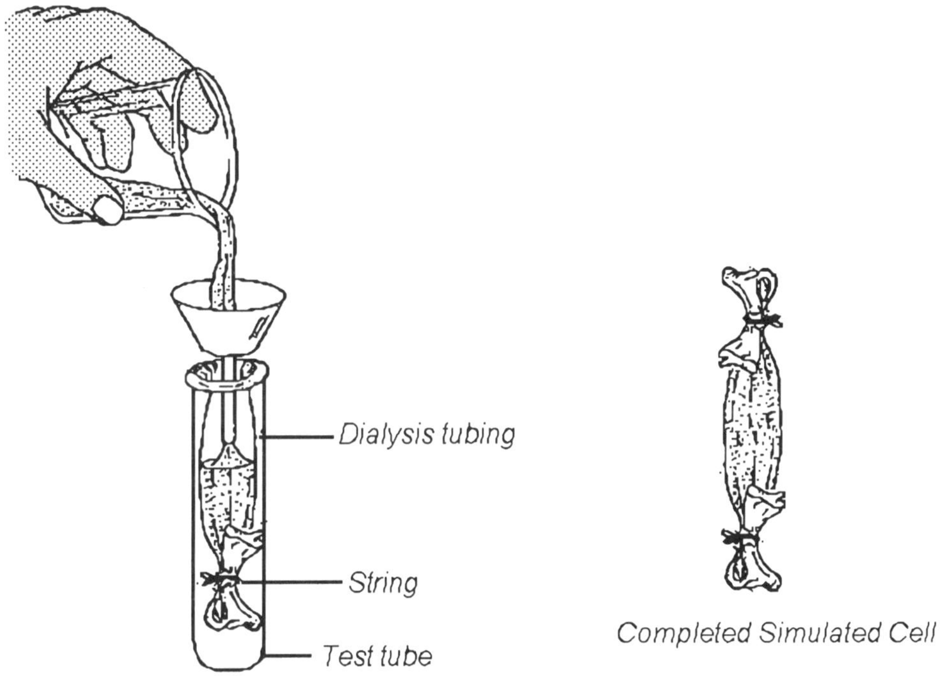

Unit 3 - Figure 1

6. Dispose of your materials according to the instructions provided by your teacher, and rinse each container thoroughly before continuing.

Unit 3
Performance-Based Assessment Activity
Diffusion Through a Cell Membrane
Part II–Experiment A

Objectives

During Part II, you will be expected to do the following:

- Formulate a hypothesis for a scientific problem.
- Conduct an experiment to test the hypothesis.
- Record and interpret data to determine if the hypothesis is a suitable answer for the problem.

Problem

Consider the following problem based on the background information in Part I.

Can CO_2 diffuse into a cell through the cell membrane?

Experiment

1. **Hypothesis** In the space provided below, write a hypothesis for the above problem.

____

____

2. **Materials** Go to the equipment and materials table, and examine the available materials. Think about how you would design an experiment to test your hypothesis using these materials. Do not remove any materials from the table yet. In the space below, make a list of the materials that you will use in your experiment.

3. **Revise Your Hypothesis** After returning from the materials table, rewrite your hypothesis in a form that takes into account the materials you will use.

   Revised Hypothesis: ____

   ____

4. **Procedure** In the space provided below, describe and illustrate an experiment that would test your hypothesis.
   a. List the steps of your experiment.

____

____

____

____

b. In the space provided, draw your experimental and control setups.

Experimental Test | Control Test

c. **Observations** List the observations or measurements you will make during your experiment.

---

d. During your experiment, be sure to observe the **safety precautions** listed in Part I.

e. To conduct your experiment, gather only the materials and equipment you need from the materials table. Pour into separate glassware the amount of each solution you need for your experiment, and transfer the solution or solutions to your work station. Do not take an entire quantity of any solution. Others may need to use the same solutions. When you have gathered the needed materials, set up and conduct your experiment as planned. Experiments of this type should be set up and data collected over a period of at least 7–10 minutes.

Data

5. **Records** Record your data below in a neat table.

6. **Interpretation** Write your interpretation of the data you recorded.

---

7. **Conclusion** State whether your data support or refute your hypothesis.

Cleanup

After you turn in your paper, dispose of your materials according to the instructions provided by your teacher. Clean up your work area, and wash your hands before leaving the lab.

Unit 3
Performance-Based Assessment Activity
Diffusion Through a Cell Membrane
Part II–Experiment B

Objectives

During Part II, you will be expected to do the following:

- Formulate a hypothesis for a scientific problem.
- Conduct an experiment to test the hypothesis.
- Record and interpret data to determine if the hypothesis is a suitable answer for the problem.

Problem

Consider the following problem based on the background information in Part I.

Can starch pass out of a cell through the cell membrane?

Experiment

1. **Hypothesis** In the space provided below, write a hypothesis for the above problem.

____

____

2. **Materials** Go to the equipment and materials table, and examine the available materials. Think about how you would design an experiment to test your hypothesis using these materials. Do not remove any materials from the table yet. In the space below, make a list of the materials that you will use in your experiment.

3. **Revise Hypothesis** After returning from the materials table, rewrite your hypothesis in a form that takes into account the materials you will use.

   Revised Hypothesis: _______________

   ____

4. **Procedure** In the space provided below, describe and illustrate an experiment that would test your hypothesis.

   a. In the blanks below, list the steps of your experiment.

   ____

   ____

   ____

   ____

b. In the space provided, draw your experimental and control setups.

Experimental Test | Control Test

c. Observations List the observations or measurements you will make during your experiment.

____

____

d. During your experiment, be sure to observe the **safety precautions** listed in Part I.

e. To conduct your experiment, gather only the materials and equipment you need from the materials table. Pour into separate glassware the amount of each solution you need for your experiment, and transfer the solution or solutions to your work station. Do not take an entire quantity of any solution. Others may need to use the same solutions. When you have gathered the needed materials, set up and conduct your experiment as planned. Experiments of this type should be set up and data collected over a period of at least 7–10 minutes.

Data

5. **Records** Record your data below in a neat table.

6. **Interpretation** Write your interpretation of the data you recorded.

____

____

7. **Conclusion** State whether your data support or refute your hypothesis.

____

Cleanup

After you turn in your paper, dispose of your materials according to the instructions provided by your teacher. Clean up your work area, and wash your hands before leaving the lab.

Unit 3
Performance-Based Assessment Activity
Diffusion Through a Cell Membrane
Teacher Notes and Scoring Rubric

Teacher Notes

Assessment Time Required (Two 30- to 50-minute class periods)

Concepts and Vocabulary
Before beginning this assessment, students should be familiar with the following concepts:

- dialysis tubing
- diffusion
- osmosis
- scientific processes
- selectively permeable membrane

Introduction
The Performance-Based Assessment Activity for Unit 3 consists of two parts. Part I involves reading three paragraphs of background information, determining a scientific problem that can be tested in a lab, and practicing the techniques needed for the assessment. Part II consists of two separate problems in which students are asked to make a simulated cell out of dialysis tubing and determine whether either carbon dioxide or starch will pass through a cell membrane.

Strategies
All students should complete Part I. If students have no experience in making simulated cells out of dialysis tubing or in using bromothymol blue and Lugol's iodine solution as indicators, you may want to have students complete Part I on the day before they complete Part II. For Part II, half of the students in the class should be assigned Experiment A, and the other half should be assigned Experiment B. If Part I is completed the day before Part II is to be done, students' papers for Part I should be returned to them for use during Part II of the assessment.

Because it may be difficult for individual students to make a simulated cell out of dialysis tubing, you may want this activity to be a group assessment rather than an individual assessment. If this is the case, limit the group size to two students. You may require students working in teams to submit one paper per team, or you may require them to submit individual papers. If students work in teams, tell them that they may be assessed individually later in the year.

Remind students that this activity is a test and that standard test-taking rules apply. Allow talking among team members, but stress that talking must be done very quietly and must be relevant to the assessment activity. When all students have completed an assessment, conduct a class discussion of the students' approaches to the Performance-Based Assessment Activity.

Student Safety Precautions
The safety precautions for this assessment activity are listed in Part I. Before allowing students to begin either part of the activity, make sure they have read these safety precautions. Advise students that they will be given points for properly adhering to the safety precautions during this activity.

Materials
You will need the following materials to prepare the solutions needed for this assessment activity:
- bromothymol blue powder (4 g is enough to make 1 L of a 0.4% solution)
- distilled or demineralized water (for preparing bromothymol blue solution)
- 0.1 M NaOH (amount needed will depend on the pH of your local tap water)
- potassium iodide, KI, crystals (100 g is enough to make 1 L of Lugol's iodine solution)
- iodine, I, crystals (50 g is enough to make 1 L of Lugol's iodine solution)
- soluble starch (0.03 g is enough to make 1 L of 0.3% starch solution)
- large (500 mL or 1 L) beakers (for mixing solutions)
- several dropper bottles (for solutions)
- hot plate

For each student or team, you will need the following materials in the quantities indicated:

- 1 dropper bottle of bromothymol blue indicator (BTB)
- 1 dropper bottle of Lugol's iodine indicator
- 1 plastic drinking straw
- 1 125-mL flask
- 1 test-tube rack
- 2 small test tubes
- 2 small (50–125 mL) beakers or similar glass containers (size is not critical)
- 3 15-cm (6-in.) pieces of dialysis tubing (35 mm to 43 mm flat width)
- 6 15-cm (6-in.) pieces of sturdy thread or string
- 2 large test tubes (slightly longer and larger in diameter than liquid-filled dialysis tubing)
- 1 small funnel 11 13
- 1 10-mL graduated cylinder
- 1 50-mL graduated cylinder
- 250–400 mL of distilled water or pH-adjusted tap water (depending on flask and beaker size)
- 50–100 mL of distilled water or pH-adjusted tap water (depending on dialysis tubing size)*
- 50–100 mL of starch solution (depending on dialysis tubing size)*

*Only half of the students or teams will use this solution.

Preparation

1. **Caution** When you prepare acid or base solutions, wear safety goggles, a face shield, impermeable gloves, and a lab apron. In case of an acid or base spill, dilute first with water. Then mop up the spill with wet cloths designated for spill cleanup.
2. To prepare a 0.4% bromothymol blue (BTB) solution, dissolve 4 g of bromothymol blue powder in a 1 L beaker with 200 mL of distilled water, and dilute the solution to 1 L with distilled water. Or you can purchase a 0.4% or 0.1% aqueous solution of bromothymol blue indicator from a biological supply house.
3. Check the pH of your local tap water with bromothymol blue indicator. If the water becomes yellow or green when BTB is added, it is too acidic. Slowly add drops of 0.1 M NaOH until the solution turns blue. Prepare several 1 L quantities of clear tap water to which enough 0.1 M NaOH has been added to ensure that bromothymol blue indicator will remain blue when it is added to the water. Label the adjusted tap water, and place it on the supply table. Instruct students to use this preparation for tap water in their experiments. If your tap water does not turn bromothymol blue indicator yellow or green, you may instruct students to obtain the tap water needed for this activity from the sink.
4. To prepare Lugol's iodine solution, dissolve 100 g potassium iodide, KI, in 1 L of tap water, and then add 50 g of iodine, I, crystals. Stir. Heat to dissolve if necessary. Pour into dropper bottles and label. Pour remainder of solution into a stock bottle and date. You can also purchase prepared Lugol's iodine solution from a biological supply house.
5. To prepare a 0.3 % starch solution, mix 0.03 g of soluble starch with about 50 mL of tap water, and heat to boiling while stirring. Remove starch solution from heat, and dilute to 1 L with tap water. Stir well and label. Pour into three separate beakers labeled as "Starch Solution."
6. Place a container labeled "Solution Disposal Container" at the front of the room. Instruct students to put leftover solutions in this container.

Laboratory Setup (Parts I and II)

Place the following materials on a centrally located supply table:

- 1 dropper bottle of bromothymol blue indicator (BTB) per student or team
- 1 dropper bottle of Lugol's iodine indicator per student or team
- 1 plastic drinking straw per student or team
- 1 125-mL flask per student or team
- 1 test-tube rack per student or team
- 2 small test tubes per student or team
- 2 small beakers or similar glass containers per student or team
- 3 15-cm (6-in.) pieces of dialysis tubing (35 mm to 43 mm flat width) per student or team
- 6 15-cm (6-in.) pieces of sturdy thread or string per student or team (You could substitute several rolls of thread or string and several pairs of scissors for these.)
- 2 large test tubes per student or team
- 1 small funnel per student or team
- 1 10-mL graduated cylinder per student or team
- 1 50-mL graduated cylinder per student or team
- several large containers of distilled water or pH-adjusted tap water
- 5 to 10 1-L containers of starch solution
- 5 or 6 thermometers (decoy)

Cleanup and Disposal

For student use during the assessment activity, provide a container marked "Solution Disposal Container." Solutions used in this activity can safely be disposed of in the laboratory sink. Instruct students to dispose of used dialysis tubing in the laboratory trash can. Avoid having students handle spills themselves. Instruct students to wash and dry the glassware they use and return it to the materials table before they leave the lab.

Scoring Rubric

The following pages are a Scoring Rubric that provides possible answers to this Performance-Based Assessment Activity. Preferred student responses are shown in italics. Suggested point values for each step in the activity appear in parentheses next to the responses. The total points earned by a student for each step in the activity can be recorded in the blanks to the right of the steps. The Scoring Rubric can be modified to accommodate your specific circumstances. You may wish to use this rubric only as a guide in scoring the assessment activity. If possible, make a copy of the rubric for each student or team, and return a copy of the rubric showing the scores earned by the students when you return their graded papers.

Activity and Scoring Rubric Art

These teacher notes are found in question 4 of the Unit 3 file. In the answer key for this question, you will find the art for Unit 3-Figure 1, which is found in Part I of this assessment. Cut and paste this figure in the place indicated before photocopying the assessment for your students.

Unit 3
Performance-Based Assessment Activity
Diffusion Through a Cell Membrane (50 points)
Scoring Rubric

_______ Total Score

Possible answers to this Performance-Based Assessment Activity are provided below. Preferred student responses are shown in italics. Suggested point values for each step in the activity appear in parentheses next to the responses. The total points earned for each step in the activity are recorded in the blanks to the right of the steps.

Part I (8 points total)

_______ Part I Score

Background (2 points total)

1. Problem _______

What materials are able to pass through a cell membrane by diffusion? (2 points)

or

Other appropriately stated problems that are relevant to the background information. (2 points)

Practice (6 points total)

3. Bromothymol Blue Tests (2 points total) _______

a. CO_2 Test (1 point) _______

no (if BTB remained blue)

or

yes (if BTB turned yellow)

b. Blowing your breath into the water adds CO_2 to the water and makes the BTB turn yellow. (1 point) _______

4. Lugol's Iodine Solution Tests (2 points total)

a. Lugol's iodine solution remains yellow, orange, or brown in tap water. (1 point) _______

b. Lugol's iodine solution turns blue-black, blue, or black in a starch solution. (1 point) _______

5. Student made an appropriate simulated cell out of dialysis tubing. (2 points) _______

Part II—Experiment A (42 points total)

Part II-A Score

Experiment (29 points total)

1. **Hypothesis** (1 point)
   Carbon dioxide can (or cannot) diffuse through the cell membrane.

2. **Materials**
   Students should include only materials that are found on the supply table and that are needed for their particular experiment. (2 points)

3. **Revised Hypothesis**
   Carbon dioxide can (or cannot) diffuse through the walls of dialysis tubing into a simulated cell. (1 point)

4. **Procedure** (25 points total)
   a. **Steps of Experiment** (7 points total)
   Experimental Setup:
   1. *Make a simulated cell out of dialysis tubing, thread, and tap water that does not contain CO_2 and to which BTB has been added.** (1 point)
      *The student may plan to add CO_2 to the water inside the dialysis tubing, but this is not recommended.
   2. *Place the simulated cell into a beaker containing tap water.*
      Note: The student could add BTB to the water in the beaker.* (1 point)
      *This is necessary only if in Step 1 the student plans to add CO_2 to the water inside the dialysis tubing.
   3. *Use a plastic straw to blow breath, which contains CO_2, into the water in the beaker.** (1 point)
      *Note: This step should not appear if the student plans in Step 1 to add CO_2 to the water inside the dialysis tubing.
   4. *Wait about 10 minutes, and record any evidence of a color change inside the simulated cell.** (1 point)
      *If the student plans in Step 1 to add CO_2 to the water inside the dialysis tubing, any color change would occur outside the simulated cell.

   Control Setup:
   Repeat steps 1, 2, and 4 in the experimental setup above. (3 points)

   b. Diagram of the experimental design must be consistent with the steps listed in part **4a**. The following elements should be indicated in student's drawing. (14 points total)

Experimental Setup	**Control Setup**
beaker (1 point)	*beaker* (1 point)
dialysis tubing "cell" (1 point)	*dialysis tubing "cell"* (1 point)
Inside dialysis tubing:	**Inside dialysis tubing:**
H_2O (1 point)	H_2O (1 point)
no CO_2* (1 point)	no CO_2 (1 point)
BTB (1 point)	BTB (1 point)
Outside dialysis tubing:	**Outside dialysis tubing:**
H_2O (1 point)	H_2O (1 point)
CO_2* (1 point)	no CO_2 (1 point)

*Some students may reverse location of CO_2. If so, BTB must be present outside the tubing.

c. Observations
In both the control and experimental setups, observe the color of BTB, and watch for changes in color inside the simulated cell.* (2 points) ______________

*If the student plans in Step 1 to add CO_2 to the inside of the dialysis tubing, the color change would occur outside the simulated cell.

d. Safety Precautions The student followed appropriate safety precautions. (1 point) ______________

e. Materials The student made appropriate use of materials. (1 point) ______________

Data (17 points total)

5. **Records** (13 points total) ______________

Tests for CO_2 (1 point)	*Color of BTB Inside* Simulated Cell—Start* (1 point)	*Color of BTB Inside* Simulated Cell—End* (1 point)
Experimental (1 point)	*blue*+ (2 points)	*yellow*+ (2 points)
Control (1 point)	*blue*+ (2 points)	*blue*+ (2 points)

*These words should be "Outside" if the student placed CO_2 inside the dialysis tubing.

+Accept test results other than those shown only if they are consistent with the experimental procedure. (1 point each instead of 2 points)

6. **Interpretation** (2 points total)
CO_2 passed through the walls of the dialysis tubing. (2 points) ______________
or
Interpretation is appropriate for student's data. (1 point) ______________
(Inaccurate data or procedures may result in a different interpretation—e.g., CO_2 did not pass through the walls of the dialysis tubing.)

7. **Conclusion** (2 points total)
The hypothesis that CO_2 is able to pass through the walls of dialysis tubing is supported by the data. (2 points) ______________
or
The hypothesis that CO_2 is not able to pass through the walls of dialysis tubing is refuted by the data. (2 points) ______________
or
Any other conclusion that is appropriate for the student's data. (1 point) ______________

Part II—Experiment B (42 points total)

Part II-A Score

Experiment

1. **Hypothesis** (1 point)
   Starch can (or cannot) diffuse through the cell membrane.

2. **Materials**
   Students should include only materials that are found on the supply table and that are needed for their particular experiment. (2 points)

3. **Revised Hypothesis**
   Starch can (or cannot) diffuse through the walls of dialysis tubing into a simulated cell. (1 point)

4. **Procedure** (25 points total)
   a. **Steps of Experiment** (7 points total)
      Experimental Setup:
      1. *Make a simulated cell out of dialysis tubing, thread, and a starch solution.* Note: Students could also add Lugol's iodine solution to starch solution inside of the dialysis tubing, but this is not necessary because they have already tested the starch solution in Part I. (1 point)
      2. *Rinse the outside of the simulated cell to remove any starch solution so that it will not contaminate the water in the beaker.* (1 point)
      3. *Place the simulated cell into a beaker containing tap water.* Note: At this time, students could also add Lugol's iodine solution to the tap water in the beaker. (1 point)
      4. *Wait about 10 minutes, add a few drops of Lugol's iodine solution to the water in the beaker* (only if Lugol's iodine solution was not added in Step 2), *and record any evidence of a color change in the solution in the beaker.* (1 point)

      Control Setup:
      Repeat Steps 1, 2, and 3 in the experimental setup above, but substitute tap water for starch solution in the simulated cell. (3 points)

   b. Diagram of experimental design must be consistent with the steps listed above in part **4a**. The following elements should be indicated in student's drawing. (14 points total)

Experimental Setup
beaker (1 point)
dialysis tubing "cell" (1 point)
Inside dialysis tubing:
starch (1 point)
H_2O (1 point)
Outside dialysis tubing:
H_2O (1 point)
no starch (1 point)
Lugol's iodine solution (1 point)

Control Setup
beaker (1 point)
dialysis tubing "cell" (1 point)
Inside dialysis tubing:
H_2O (1 point)
no starch (1 point)
Outside dialysis tubing:
water (1 point)
no starch (1 point)
Lugol's iodine solution (1 point)

c. **Observations**
In both the control and experimental setups, observe the color of Lugol's iodine solution, and watch for changes in the color of the solution in the beaker. (2 points) ____________

d. **Safety Precautions** The student followed appropriate safety precautions. (1 point) ____________

e. **Materials** The student made appropriate use of materials. (1 point) ____________

Data (17 points total)

5. **Records** (13 points total) ____________

Tests for Starch (1 point)	*Color of Lugol's Solution—Start* (1 point)	*Color of Lugol's Solution—End* (1 point)
Experimental (1 point)	*yellow or orange** (2 points)	*yellow or orange** (2 points)
Control (1 point)	*yellow or orange** (2 points)	*yellow or orange** (2 points)

*Accept test results other than those shown only if they are consistent with the experimental procedure. (1 point each instead of 2 points)

6. **Interpretation** (2 points total)
Starch did not pass through the walls of the dialysis tubing. (2 points) ____________
or
Interpretation is appropriate for student's data. (1 point) ____________
(Inaccurate data or procedures may result in a different interpretation—e.g., starch passed through the walls of the dialysis tubing.)

7. **Conclusion** (2 points total)
The hypothesis that starch is able to pass through the walls of dialysis tubing is refuted by the data. (2 points) ____________
or
The hypothesis that starch is not able to pass through the walls of dialysis tubing is supported by the data. (2 points) ____________
or
Any other conclusion that is appropriate for the student's data. (1 point) ____________

Unit 4
Performance-Based Assessment Activity
Testing Pond Water for Carbon Dioxide
Part I–Formulating Hypotheses

Follow the guidelines below during this assessment activity.

- Remember not to talk to other students taking the test. If you are working with a team, talk only with those in your group and talk very quietly, so that you do not disturb other teams.
- When you conduct your experiment, you will be able to use only the materials and equipment on the supply table. Some of the materials on the table will not be needed.
- Most of the other students involved in today's activity have different problems to solve. If you observe other students using materials different from those that you are using, this is because their problems are not the same as yours.
- Remember that a problem often has several different acceptable solutions.

Background

A group of ecology students wanted to determine if there was a relationship between a certain species of fish and plant A, and the amount of carbon dioxide present in the organisms' environment. The students set up four similar aquariums: one containing five fishes, one containing five fishes and 500 g of plant A, one containing no fishes and 500 g of plant A, and one containing neither fish nor plant A. Next, they collected water samples from each aquarium at sunrise and sunset over a three-week period. Then the students compared the average amounts of carbon dioxide present in each aquarium. Their calculations are shown in the following table.

#	Organisms present in the aquarium	Carbon dioxide content of aquarium water at sunrise	Carbon dioxide content of aquarium water at sunset
1	5 fishes	31 g	30 g
2	5 fishes, 500 g of plant A	45 g	31 g
3	500 g of plant A	16 g	1 g
4	no fish, no plant A	none	none

1. Compare the carbon dioxide content of the samples that were taken at sunrise and sunset. Which samples had substantially different amounts of carbon dioxide at sunrise and sunset?

____

2. Based on the data, write a hypothesis about the relationship between the presence of fish and the level of carbon dioxide.

____

3. Based on the data, write a hypothesis about the relationship between the presence of plant A and the level of carbon dioxide at sunrise.

____

4. In aquariums 2 and 3, why might the carbon dioxide content in the aquarium water at sunrise differ from the carbon dioxide content at sunset?

____

____

Safety Precautions

The following safety precautions should be observed during Parts I and II of this activity.

- Wear safety goggles, protective gloves, and a lab apron.
- Confine loose hair and clothing.
- Do not touch or taste any chemicals. Wash off any solutions that get on skin or clothing at the sink while calling to the teacher. If you get a solution in your eyes, immediately flush it out at the eyewash station while calling to the teacher. Be careful not to spill any solutions. Call the teacher in the event of an acid or base spill.
- Dispose of any broken glass in a separate container. Never put broken glass in a regular waste container.
- Clean up the lab and all equipment after use. Dispose of solutions according to proper disposal methods. Wash hands thoroughly before leaving the lab after all lab work is finished.

Practice

Carbon dioxide (CO_2) can combine with water molecules to form carbonic acid (H_2CO_3). Thus, the presence of carbon dioxide in a water solution can be checked by measuring the solution's pH. Bromothymol blue (BTB) is a pH indicator. In solutions having a pH of 7.6 or higher, BTB is blue. In solutions having a pH of 6.0 or lower, BTB is yellow, an indication that carbonic acid is present in the solution. If BTB turns green in a solution, the solution contains a small amount of carbonic acid.

5. **Caution:** *Avoid getting BTB on skin or clothes.* Fill a small flask with about 100 mL of tap water. Add 10 drops of BTB indicator. What do your observations tell you about the carbon dioxide content of tap water?

____

6. **Caution:** *Be careful not to get BTB in your eyes. Be careful not to inhale any of the solution.* Blow through a straw into the tap water containing BTB for one or more minutes. Continue to blow into the water until you observe no further change.

a. Describe what happens to the water solution when you blow into it.

____

b. What do your observations tell you about the carbon dioxide content of the solution?

____

7. Discard the tap water solution in a sink, and rinse the flask. You may keep the other materials used during Part I for use in the remainder of this Performance-Based Assessment Activity.

8. Dispose of your materials according to the instructions provided by your teacher, and rinse each container before continuing.

Unit 4
Performance-Based Assessment Activity
Testing Pond Water for Carbon Dioxide
Part II–Experiment A

Objectives

- Formulate a hypothesis based on a given problem.
- Design and conduct an experiment to test your hypothesis.
- Record the data from your experiment.
- Evaluate your hypothesis based on your data.

Problem

Based on the information in Part I, consider the following **problem.**

Does *Elodea* give off carbon dioxide in the light?

Experiment

1. **Hypothesis** In the space below, write a hypothesis for the **problem** that is stated above.

____

____

2. **Materials** Go to the equipment and materials table, and examine which materials are available. Think about how you would design a control experiment to test your hypothesis using these materials. Do not remove any materials from the table yet. Make a list of the materials that you will use in your experiment.

3. **Procedure**
   a. Describe how you will test your hypothesis. Make sure to include a control in your experimental design.

____

____

____

____

   b. List the observations or measurements you will make during the experiment.

____

____

   c. Be sure to observe the **Safety Precautions** listed in Part I.

d. To conduct your experiment, gather *only* the materials and equipment you need from the materials table. Pour into separate glassware the amount of each solution you need for your experiment, and transfer the solution or solutions to your work station. Do not take an entire quantity of any solution. Others may also need access to the solutions that you will be using in your experiment. When you have gathered the needed materials, set up and conduct your experiment as planned.

Data

4. Records Record your data in the space below. To obtain accurate results, your experiment would have run for 24 hours. To simulate the results of your experiment, add several drops of simulated time-adjustment solution A to your test and control setups, and stir the solutions.

____

____

____

____

5. Interpretation Write your interpretation of the data you recorded.

____

____

6. Conclusion State whether your data support or refute your hypothesis.

____

____

7. Analysis List any factors that could have affected the accuracy of the data you obtained during your experiment.

____

____

Cleanup

After you turn in your paper, dispose of your material according to the instructions provided by your teacher. Clean up your work area, and wash your hands before leaving the lab.

Unit 4
Performance-Based Assessment Activity
Testing Pond Water for Carbon Dioxide
Part II–Experiment B

Objectives

- Formulate a hypothesis based on a given problem.
- Design and conduct an experiment to test your hypothesis.
- Record the data from your experiment.
- Evaluate your hypothesis based on your data.

Problem

Based on the information in Part I, consider the following **problem.**

Does *Elodea* take in carbon dioxide in the light?

Experiment

1. **Hypothesis** In the space below, write a hypothesis for the **problem** that is stated above.

____

____

2. **Materials** Go to the equipment and materials table, and examine which materials are available. Think about how you would design a control experiment to test your hypothesis using these materials. Do not remove any materials from the table yet. Make a list of the materials that you will use in your experiment.

3. **Procedure**

   a. Describe how you will test your hypothesis. Make sure to include a control in your experimental design.

____

____

____

____

   b. List the observations or measurements you will make during the experiment.

____

____

   c. Be sure to observe the **Safety Precautions** in Part I.

d. To conduct your experiment, gather only the materials and equipment you need from the materials table. Pour into separate glassware the amount of each solution you need for your experiment, and transfer the solution or solutions to your work station. Do not take an entire quantity of any solution. Others may also need access to the solutions that you will be using in your experiment. When you have gathered the needed materials, set up and conduct your experiment as planned.

Data

4. **Records** Record your data in the space below. To obtain accurate results, your experiment would have run for 24 hours. To simulate the results of your experiment, add several drops of simulated time-adjustment solution B to your test setup, and stir the solution.

5. **Interpretation** Write your interpretation of the data you recorded.

6. **Conclusion** State whether your data support or refute your hypothesis.

7. **Analysis** List any factors that could have affected the accuracy of the data you obtained during your experiment.

Cleanup

After you turn in your paper, dispose of your material according to the instructions provided by your teacher. Clean up your work area, and wash your hands before leaving the lab.

Unit 4
Performance-Based Assessment Activity
Testing Pond Water for Carbon Dioxide
Teacher Notes and Scoring Rubric

Assessment Time Required
One 50-minute class period

Concepts and Vocabulary
Before beginning this assessment, students should be familiar with the following concepts and vocabulary:
- bromothymol blue
- carbon cycle
- cellular respiration
- control experiment
- ecosystem
- pH indicator
- photosynthesis

Introduction
The assessment activity for Unit 4 consists of two parts. Part I involves interpreting a data table and formulating hypotheses. Part II of the assessment activity consists of two separate experiments in which students are assigned the task of determining either whether *Elodea* gives off or whether it takes in carbon dioxide in light.

Strategies
All students should complete Part I. For Part II, assign each student one of the two experiments (Experiment A or B). If Part I is completed the day before Part II is to be done, return students' papers for use during Part II of the assessment.

It is recommended that this assessment be used as an individual assessment. However, this activity also works well for teams of two, three, or four students. You may require students to submit one paper per team or individual papers.

Before allowing students to begin the assessment activity, make sure that they are aware of appropriate safety precautions. Remind students that this activity is a test situation and that the standard rules of test-taking apply. Stress that talking among team members must be done very quietly and must be relevant to the assessment activity. When all students have completed an assessment, conduct a class discussion of the students' different approaches to the Performance-Based Assessment Activity.

Student Safety Precautions
The safety precautions for this assessment activity are listed in Part I. Before allowing students to begin the activity, make sure they have read these safety precautions. Advise students that they will be given points for properly following the safety precautions during this activity.

Materials
You will need the following materials to prepare the solutions needed for this assessment activity:
- bromothymol blue powder (0.4 g is enough to make 1 L) or purchase a 0.4% or 0.1% aqueous solution of bromothymol blue
- distilled water (1 L is enough to make 1 L of bromothymol blue indicator)
- 0.1 M NaOH (amount needed will depend on the pH of your local tap water)
- 0.1 M HCl (amount needed will depend on the pH of your local tap water)
- several dropper bottles
- 2 large (500-mL or 1-L) beakers (for mixing solutions)

For each student or team, you will need the following material in the quantities indicated:

- 1 100-mL flask
- 1 dropper bottle of bromothymol blue indicator (BTB)
- 1 drinking straw
- 1 3–4 in. length of fresh *Elodea*
- 1 10-mL or 25-mL graduated cylinder
- 2 40–125 mL beakers or similar size containers (size is not critical)
- 12 mL of pH adjusted tap water
- 1 125-mL flask
- 1 dropper bottle of simulated time-adjustment solution A or B (Note: Only one-half of your students will use each time-adjustment solution.)

Preparation

1. **Caution** When you prepare acid or base solutions, work in a ventilated hood and wear safety goggles, a face shield, impermeable gloves, and a lab apron. In case of an acid or base spill, dilute first with water. Then mop up the spill with wet cloths designated for spill cleanup.
2. Use pH paper or a pH meter to determine the pH of each solution. To adjust the pH of a solution, add small increments of 0.1 M HCl to tap water to lower its pH. Add small increments of 0.1 M NaOH to water to raise its pH. Before discarding leftover solutions in the sink, collect them in a single container, and neutralize to a pH of 7 with 0.1 M NaOH or 0.1 M HCl.
3. To prepare a 0.04% BTB solution, measure 0.4 g of bromothymol blue powder, dissolve it in a 1 L beaker with 200 mL of distilled water, and dilute the solution to 1 L with distilled water. Or you can purchase a 0.4% or 0.1% aqueous solution of bromothymol blue indicator from a biological supply house, and dilute it to 0.04%.
4. Check the pH of your local tap water with bromothymol blue indicator. If the water becomes yellow or green when BTB is added, slowly add drops of 0.1 M NaOH until the solution turns blue. Prepare 1 L quantities of clear tap water. Label the adjusted tap water, and place it on the supply table. Instruct students to use this preparation for tap water in their experiments.
5. Place a container labeled "Solution Disposal Container" at the front of the room. Instruct students to put leftover solutions in this container.
6. To obtain results during a 50-minute testing session, you will need to provide simulated time-adjustment solutions. Problem A will require the addition of a few drops of tap water, which should effect no change. Problem B will require a 0.1 M NaOH solution, which will simulate the removal of carbon dioxide from the water by making the solution more basic.

Laboratory Setup

Place the following materials on a centrally located supply table:

(Parts I and II)

- 1 dropper bottle of bromothymol blue indicator per student or team in the class, marked "BTB"
- tap water, pH adjusted, marked "Tap Water"
- drinking straws
- 100-mL flasks

(Part II only)

- 3–4 in. lengths of *Elodea*
- 10-mL or 25-mL graduated cylinders
- 40–125 mL beakers or similar size containers (size is not critical)
- 125-mL flasks
- dropper bottles of simulated time-adjustment solution A or B (Note: Only one-half of your students will use each time-adjustment solution.)

Cleanup and Disposal
Before discarding leftover solutions in the sink, collect them in a single container, and neutralize them to a pH of 7 with 0.1 M NaOH. For student use during the assessment activity, provide a container marked "Solution Disposal Container." Avoid having students handle acid spills themselves. Instruct students to wash and dry the glassware they use and return it to the materials table before they leave the lab.

Scoring Rubric
The following pages are a Scoring Rubric that provides possible answers to this Performance-Based Assessment Activity. Preferred student responses are shown in italics. Suggested point values for each step in the activity appear in parentheses next to the responses. The total points earned by a student for each step in the activity can be recorded in the blanks to the right of the steps. The Scoring Rubric can be modified to accommodate your specific circumstances. You may wish to use this rubric only as a guide in scoring the assessment activity. If possible, make a copy of the rubric for each student or team, and return a copy of the rubric showing the scores earned by the students when you return their graded papers.

Unit 4
Performance-Based Assessment Activity
Testing Pond Water for Carbon Dioxide (50 points)
Scoring Rubric

_______________ Total Score

Possible answers to this Performance-Based Assessment Activity are provided below. Preferred student responses are shown in italics. Suggested point values for each step in the activity appear in parentheses next to the responses. The total points earned for each step in the activity are recorded in the blanks to the right of the steps.

Part I (17 points total)

_______________ Part I Score

Background (11 points total)

1. *The samples taken from the aquariums containing plant A (aquariums 2 and 3) showed a substantially different average amount of carbon dioxide at sunrise than at sunset.* (2 points) _______________

2. *The fishes give off carbon dioxide.* (3 points) _______________
   or
   Other appropriately stated hypothesis. (1 point)

3. *Plant A gives off carbon dioxide.* (3 points) _______________
   or
   Other appropriately stated hypothesis. (1 point)

4. *Plants perform cellular respiration, which produces carbon dioxide. During the daytime, plants perform photosynthesis, which uses carbon dioxide.* (3 points) _______________

Practice (6 points total)

5. *No carbon dioxide is present in tap water.* (2 points) _______________

6. a. *The solution turns yellow.* (2 points) _______________

   b. *The tap water now contains carbon dioxide.* (2 points) _______________

Part II–Experiment A (33 points total)

Part II Score

Experiment (23 points total)

1. Hypothesis

Elodea *does (or does not) give off carbon dioxide in the light.* (3 points)

2. Materials

Students should include only materials that are found on the supply table and that are needed for their particular experiment. (2 points)

3. Procedure (18 points total)

a. *Students should draw two beakers with tap water that tests negative for carbon dioxide (turns blue with BTB).* (8 points)

In the test beaker, students should place a sprig of Elodea. (2 points)

In the control beaker, students should have no plant. (2 points)

b. *Test for carbon dioxide at beginning of experiment, using BTB.* (This step may be omitted because tap water was shown not to have carbon dioxide in Part I of this assessment activity.) (no points)

Observe the color of the water in the test and control beakers. (2 points)

c. Use of appropriate safety precautions during the experiment (2 points)

d. Appropriate use of materials (2 points)

Data (10 points total)

4. Records (4 points total)

Water in the test setup remained blue. (2 points)

Water in the control setup remained blue. (2 points)

5. Interpretation

Elodea *did not give off carbon dioxide.* (2 points)

or

Any other interpretation that is consistent with the data (1 point)

6. Conclusion

The data support the hypothesis that Elodea *does not give off carbon dioxide in the light.* (3 points)

or

The data contradict the hypothesis that Elodea *does give off carbon dioxide in the light.* (3 points)

or

Other appropriate conclusion that is supported by the data and interpretation (3 points)

7. Analysis

Acceptable factors include any possible constraining factor, such as the source and freshness of the *Elodea*, unclean glassware, or inappropriate use of the time-adjustment solutions. (1 point)

Part II–Experiment B (33 points total)

Part II Score

Experiment (23 points total)

1. **Hypothesis**
   Elodea *does (or does not)* take in carbon dioxide. (3 points)

2. **Materials**
   Students should include only materials that are found on the supply table and that are needed for their particular experiment. (2 points)

3. **Procedure (18 points total)**
   a. Students should draw two beakers with tap water that tests *positive for carbon dioxide (turns yellow with BTB).* (Students can blow carbon dioxide into tap water with a straw or use water left over from Part I.) (8 points)
   In the test beaker, students should place a sprig of Elodea. (2 points)
   In the control beaker, students should have no plant. (2 points)

   b. *Test for carbon dioxide at beginning of experiment, using BTB.* (This step may be omitted because tap water was shown not to have carbon dioxide in Part I of this assessment activity.) (no points)
   Observe the color of the water in the test and control beakers. (2 points)

   c. Use of appropriate safety precautions during the experiment (2 points)

   d. Appropriate use of materials (2 points)

Data (10 points total)

4. **Records (4 points total)**
   Water in the test setup turned blue. (2 points)

   Water in the control setup remained yellow. (2 points)

5. **Interpretation**
   Elodea *took in carbon dioxide.* (2 points)
   or
   Any other interpretation that is consistent with the data. (1 point)

6. **Conclusion**
   The data support the hypothesis that Elodea takes in carbon dioxide in the light. (3 points)
   or
   The data contradict the hypothesis that Elodea does not take in carbon dioxide in the light. (3 points)
   or
   Other appropriate conclusion that is supported by the data and interpretation. (3 points)

7. **Analysis**
   Acceptable factors include any possible constraining factor, such as the source and freshness of the *Elodea*, unclean glassware, or inappropriate use of the time-adjustment solutions. (1 point)

Unit 5
Performance-Based Assessment Activity
Fermentation by Yeast
Part I–Identifying a Problem

Follow the guidelines below during this assessment activity.

- Remember not to talk to other students taking the test. If you are working with a team, talk only with those in your group and talk very quietly, so that you do not disturb other teams.
- When you conduct your experiment, you will be able to use only the materials and equipment on the supply table. Some of the materials on the table will not be needed.
- Most of the other students involved in today's activity have different problems to solve. If you observe other students using materials different from those that you are using, this is because their problems are not the same as yours.
- Remember that a problem often has several different acceptable solutions.

Background

For centuries, humans have used yeast to make bread, wine, and other food products. Yeast are living cells that can generate energy in the absence of oxygen by a process known as fermentation. When yeast cells are incubated, the cells produce enzymes that convert sugar into energy and other products. Your task in this activity will be to gather information that could be useful in determining which of these three equations——Equation *A*, Equation *B*, or Equation *C*——best describes fermentation by yeast.

Equation *A*
$$\underset{\text{glucose}}{O_2 + C_6H_{12}O_6} \xrightarrow{\text{yeast enzymes}} CO_2 + H_2O + \text{energy}$$

Equation *B*
$$\underset{\text{glucose}}{C_6H_{12}O_6} \xrightarrow{\text{yeast enzymes}} CO_2 + \underset{\text{alcohol}}{CH_3CH_2O} + \text{energy}$$

Equation *C*
$$\underset{\text{glucose}}{C_6H_{12}O_6} \xrightarrow{\text{yeast enzymes}} H_2O + \underset{\text{alcohol}}{CH_3CH_2O} + \text{energy}$$

1. In the following table, list the products and reactants of each of the proposed equations.

Reactants and Products That May Be Involved in Yeast Fermentation

Equation	Products	Reactants
Equation *A*		
Equation *B*		
Equation *C*		

2. Based on the Background, state a **problem** that might be investigated in the laboratory.

____

____

Safety Precautions

The following safety precautions should be observed during Parts I and II of this activity.

- Wear safety goggles, protective gloves, and a lab apron.
- Confine loose hair and clothing.
- Do not touch or taste any chemicals. Wash off any solutions that get on skin or clothing at the sink while calling to the teacher. If you get a solution in your eyes, immediately flush it out at the eyewash station while calling to the teacher.
- Be careful not to spill any solutions. Call the teacher in the event of an acid or base spill.
- Dispose of any broken glass in a separate container. Never put broken glass in a regular waste container.
- Clean up the lab and all equipment after use. Dispose of solutions according to proper disposal methods. Wash hands thoroughly before leaving the lab after all lab work is finished.

Practice

Carbon dioxide (CO_2) can combine with water molecules to form carbonic acid (H_2CO_3). Thus, the presence of carbon dioxide in a water solution can be checked by measuring the solution's pH. Bromothymol blue (BTB) is a pH indicator. In solutions having a pH of 7.6 or higher, BTB is blue. In solutions having a pH of 6.0 or lower, BTB is yellow, an indication that carbonic acid is present in the solution. If BTB turns green in a solution, the solution contains a small amount of carbonic acid.

3. **Caution** *Avoid getting solutions on skin or clothing.* Fill a small flask with about 100 mL of tap water. Add 10 drops of BTB indicator. What do your observations tell you about the content of carbon dioxide in tap water?

____

4. **Caution** *Be sure to wear your safety goggles, and be careful not to get BTB in your eyes. Be careful not to inhale any of the BTB solution.* Blow gently through a straw into the tap water containing BTB for one or more minutes. Continue to blow into the water until you observe no further change.

a. Describe what happens to the water solution when you blow into it.

____

b. What do your observations tell you about the carbon dioxide content of the solution?

____

5. Remove about 0.5 in. of glucose test tape from the tape dispenser. Place one drop of the glucose solution on the test tape. Watch for any color change over 1–2 minutes. Compare any change on the test tape to the color scale on the side of the dispenser. Describe what happens when glucose is added to the glucose test tape. What does this change mean?

____

6. On a clean 0.5 in. glucose test tape, place one drop of tap water. Again, watch for any color change over 1–2 minutes. Compare any change on the test tape to the color scale on the side of the dispenser. Describe what happens when tap water is added to the glucose test tape. What does this indicate?

____

7. Dispose of your materials according to the instructions provided by your teacher, and rinse each container thoroughly before continuing.

Unit 5
Performance-Based Assessment Activity
Fermentation by Yeast
Part II–Experiment A

Objectives

- Design and conduct an experiment to test a given hypothesis.
- Organize the data from your experiment in a table.
- Evaluate your hypothesis based on your data.

Experiment

1. **Hypothesis** Consider the following hypothesis.

   In the presence of air, yeast enzymes use glucose to produce carbon dioxide.

   Think about how confirming or refuting this hypothesis might help you determine which equation in Part I best describes fermentation by yeast. Write an equation to represent the process described in the above hypothesis.

   ____

2. **Materials** Go to the equipment and materials table, and examine which materials are available. **Note:** It usually takes several hours to obtain measurable results in this kind of experiment. To speed up your investigation, freshly prepared solutions as well as solutions that have been incubated are available. Think about how you would design an experiment to test your hypothesis using these materials. Do not remove any materials from the table yet. Make a list of the materials that you will use in your experiment.

3. **Procedure**

   a. Draw a sketch of how you will set up an experiment that will test the hypothesis at the top of this page. Make sure to include a control in your experimental design. Label all parts of your experimental and control setups.

b. List the observations or measurements you will make during the experiment.

c. Be sure to observe the **Safety Precautions** listed in Part I.

d. To conduct your experiment, gather *only* the materials and equipment you need from the materials table. Pour into separate glassware the amount of each solution you need for your experiment, and transfer the solution or solutions to your work station. Do not take an entire quantity of any solution. Others may also need access to the same solutions. When you have gathered the needed materials, set up and conduct your experiment as planned.

Data

4. Records Make a table in the space below. Make sure your table has descriptive heads. Record your data in your table.

5. Interpretation Write your interpretation of the data you recorded.

6. Conclusion State whether your data support or refute your hypothesis.

7. Analysis Discuss the strengths and weaknesses of the hypothesis used in this assessment activity to investigate fermentation by yeast.

Cleanup

After you turn in your paper, dispose of your materials according to the instructions provided by your teacher. Clean up your work area, and wash your hands before leaving the lab.

Unit 5
Performance-Based Assessment Activity
Fermentation by Yeast
Part II–Experiment B

Objectives

- Design and conduct an experiment to test a given hypothesis about fermentation by yeast.
- Organize the data from your experiment in a table.
- Evaluate your hypothesis based on your data.

Experiment

1. **Hypothesis** Consider the following hypothesis.

   In the absence of air, yeast enzymes do not use glucose to produce carbon dioxide.

   Think about how confirming or refuting this hypothesis might help you determine which equation in Part I best describes fermentation by yeast. Write an equation to represent the process tested in your hypothesis.

   ____

2. **Materials** Go to the equipment and materials table, and examine which materials are available. **Note:** It usually takes several hours to obtain measurable results in this kind of experiment. To speed up your investigation, freshly prepared solutions as well as solutions that have been incubated are available. Think about how you would design an experiment to test your hypothesis using these materials. Do not remove any materials from the table yet. Make a list of the materials that you will use in your experiment.

3. **Procedure**
   a. Draw a sketch of how you will set up an experiment that will test the hypothesis at the top of this page. Make sure to include a control in your experimental design. Label all parts of your experimental and control setups.

b. List the observations or measurements you will make during the experiment.

____

____

____

____

c. Be sure to observe the **Safety Precautions** in Part I.

d. To conduct your experiment, gather *only* the materials and equipment you need from the materials table. Pour into separate glassware the amount of each solution you need for your experiment, and transfer the solution or solutions to your work station. Do not take an entire quantity of any solution. Others may also need access to the same solutions. When you have gathered the needed materials, set up and conduct your experiment as planned.

Data

4. **Records** Make a table in the space below. Make sure your table has descriptive heads. Record your data in your table.

5. **Interpretation** Write your interpretation of the data you recorded.

____

____

6. **Conclusion** State whether your data support or refute your hypothesis.

____

____

7. **Analysis** What substance that is always produced by yeast fermentation could have been tested for in this activity?

____

____

Cleanup

After you turn in your paper, dispose of your materials according to the instructions provided by your teacher. Clean up your work area, and wash your hands before leaving the lab.

Unit 5
Performance-Based Assessment Activity
Fermentation by Yeast
Teacher Notes and Scoring Rubric

Teacher Notes

Assessment Time Required
One 50-minute class period

Concepts and Vocabulary
Before beginning this assessment, students should be familiar with the following concepts and vocabulary:
- cellular respiration
- control experiment
- enzyme
- fermentation
- glucose
- pH indicator
- photosynthesis
- yeast

Introduction
The Performance-Based Assessment Activity for Unit 5 consists of two parts. Part I involves comparing chemical equations and identifying a scientific problem. Part II consists of two separate experiments in which students are assigned the task of determining whether specific reactants and products are involved in yeast fermentation.

Strategies
All students should complete Part I. For Part II, assign each student either Experiment A or B. If Part I is completed the day before Part II is done, return students' papers for use during Part II of the assessment.

It is recommended that this assessment be used as an individual assessment. However, this activity also works well for small teams of 2, 3, or 4 students. You may require students working in teams to submit one paper per team, or you may require them to submit individual papers.

Before allowing students to begin the assessment activity, make sure that they are aware of appropriate safety precautions. Remind students that this activity is a test situation and that the standard rules of test-taking apply. Stress that talking among team members must be done very quietly and must be relevant to the assessment activity. When all students have completed an assessment, conduct a class discussion of the students' different approaches to this Performance-Based Assessment Activity.

Student Safety Precautions
The safety precautions for this assessment activity are listed in Part I. Before allowing students to begin the activity, make sure they have read these safety precautions. Advise students that they will be given points for properly following safety precautions during this activity.

Materials
You will need the following materials to prepare the solutions needed for this assessment activity:
- bromothymol blue powder (0.4 g is enough to make 1 L) or purchase a 0.4% or 0.1% aqueous solution of bromothymol blue
- distilled water (1 L is enough to make 1 L of bromothymol blue indicator; approximately 2 L is enough to make 1 L each of 0.1% glucose and 2% glucose solutions)
- 0.1 M NaOH in a dropper bottle
- 0.1 M HCl in a dropper bottle
- glucose crystals (21 g is enough to make 1 L of 2% glucose solution and 1 L of 0.1% glucose solution)

For each student or team, you will need the following materials in the quantities indicated:

- 1 100-mL flask
- 2 pasteur pipettes
- 1 dropper bottle of bromothymol blue indicator (BTB)
- 1 drinking straw
- 1 10-mL or 25-mL graduated cylinder
- 2 40–125 mL beakers or similar size containers (size is not critical)
- 5 mL of tap water (pH adjusted to approximately 7.6)
- 1 125-mL flask
- 4 5-mL test tubes
- 4 rubber test-tube stoppers (Note that only one-half of your students will need stoppers.)
- test-tube rack
- 1 glass marking pencil
- 1 small funnel
- dispenser of glucose test tape or test strips with a color scale
- 5 mL of 0.1% glucose solution (pH adjusted to 4)
- 15 mL of 2% glucose solution (pH adjusted to 8)
- 2–3 mL of tap water (pH adjusted to 7.6)

Preparation

1. **Caution** When you prepare acid or base solutions, work in a ventilated hood and wear safety goggles, a face shield, impermeable gloves, and a lab apron. In case of an acid or base spill, dilute first with water. Then mop up the spill with wet cloths designated for spill cleanup.
2. Use pH paper or a pH meter to determine the pH of each solution. To adjust the pH of a solution, add small increments of 0.1 M HCl to tap water to lower its pH. Add small increments of 0.1 M NaOH to water to raise its pH. Before discarding leftover solutions in the sink, collect them in a single container, and neutralize to a pH of 7 with 0.1 M NaOH or 0.1 M HCl.
3. To prepare a 0.04% BTB solution, measure 0.4 g of bromothymol blue powder, dissolve it in a 1 L beaker with 200 mL of distilled water, and dilute the solution to 1 L with distilled water. Or you can purchase a 0.4% or 0.1% aqueous solution of bromothymol blue indicator from a biological supply house, and dilute it to 0.04%.
4. Check the pH of your local tap water with bromothymol blue indicator. If the water becomes yellow or green when BTB is added, slowly add drops of 0.1 M NaOH until the solution turns blue. Prepare 1 L quantities of clear tap water. Label the adjusted tap water, and place it on the supply table. Instruct students to use this preparation for tap water in their experiments.
5. To obtain results in a 50-minute period, you will need to provide time-adjusted solutions for students to test. Use a 0.1% glucose solution of pH 4 to simulate an incubated glucose solution with yeast. To prepare 0.1% glucose solution, measure about 1 g of glucose (also listed under the name dextrose in some biological supply catalogs) and dissolve it in 100–200 mL of distilled water. Then dilute the solution to 1 L. To adjust the pH to 4, add drops of 0.1 M HCl until the pH becomes 4. Provide at least two containers of this solution. Label one container "Glucose Solution With Yeast; Incubated in Open Air," and label the other container "Glucose Solution With Yeast; Incubated in Sealed Container."
6. To prepare 2% glucose solution, dissolve about 20 g of glucose in 100–200 mL of distilled water. Dilute the solution to 1 L. To adjust the pH to 8, add drops of 0.1 M NaOH until the pH becomes 8. Divide this solution among four separate containers. Label the containers "Fresh Glucose Solution With Yeast," Fresh Glucose Solution Without Yeast," "Glucose Solution Without Yeast; Incubated in Open Air," and "Glucose Solutions Without Yeast; Incubated in Sealed Container."
7. Place a container labeled "Solution Disposal Container" in a ventilated hood. Instruct students to put leftover acid or base solutions in this container.
8. You can vary the problems in this assessment by having students test for alcohol production by yeast. To test for the presence of ethyl alcohol in a fermenting glucose solution, add about four drops of a 10% sodium hydroxide solution to 5 mL of fermenting glucose solution. Then add Lugol's iodine solution (iodine potassium iodide) one drop at a time until the solution becomes slightly yellow. Allow the mixture to stand for about two minutes, before shaking it. A layer of iodoform will settle towards the bottom of the test tube in the presence of ethyl alcohol.

Laboratory Setup
Place the following materials on a centrally located supply table:

(Parts I and II)
- dropper bottles of bromothymol blue indicator per student or team in the class, marked "BTB"
- 3 beakers of tap water, pH adjusted, marked "Tap Water"
- drinking straws
- 100-mL flasks
- several dispensers of glucose test tape or test strips
- 1 L of 2% glucose solution labeled "Glucose Solution." (Divide the 1000 mL into at least 3 labeled containers to speed up student dispensing of the liquid.)
- 2 pasteur pipettes

(Part II only)
- test tubes
- test-tube stoppers
- test-tube rack
- 1 glass marking pencil
- 1 small funnel
- 0.1% glucose solution (pH adjusted to 4), identified as "Glucose Solution With Yeast; Incubated in Open Air"
- 0.1% glucose solution (pH adjusted to 4), identified as "Glucose Solution With Yeast; Incubated in Sealed Container"
- 2% glucose solution (pH adjusted to 8), identified as "Fresh Glucose Solution With Yeast"
- 2% glucose solution (pH adjusted to 8), identified as "Fresh Glucose Solution Without Yeast"
- 2% glucose solution (pH adjusted to 8), identified as "Glucose Solutions Without Yeast; Incubated in Open Air"
- 2% glucose solution (pH adjusted to 8), identified as "Glucose Solutions Without Yeast; Incubated in Sealed Container"
- several thermometers, to serve as decoys

Cleanup and Disposal
Before discarding leftover solutions in the sink, collect them in a single container, and neutralize them to a pH of 7 with 0.1 M NaOH. For student use during the assessment activity, provide a container marked "Solution Disposal Container." Avoid having students handle acid spills themselves. Instruct students to wash and dry the glassware they use and return it to the materials table before they leave the lab.

Scoring Rubric
The following pages are a Scoring Rubric that provides possible answers to this Performance-Based Assessment Activity. Preferred student responses are shown in italics. Suggested point values for each step in the activity appear in parentheses next to the responses. The total points earned by a student for each step in the activity can be recorded in the blanks to the right of the steps. The Scoring Rubric can be modified to accommodate your specific circumstances. You may wish to use this rubric only as a guide in scoring the assessment activity. If possible, make a copy of the rubric for each student or team, and return a copy of the rubric showing the scores earned by the students when you return their graded papers.

Unit 5
Performance-Based Assessment Activity
Testing the Effects of Acid Rain (50 points)
Scoring Rubric

_______________ Total Score

Possible answers to this Performance-Based Assessment Activity are provided below. Preferred student responses are shown in italics. Suggested point values for each step in the activity appear in parentheses next to the responses. The total points earned for each step in the activity are recorded in the blanks to the right of the steps.

Part I (12 points total)

_______________ Part I Score

Background (7 points total)

1. (6 points total; 1 point per table cell) _______________

Reactants and Products That May Be Involved in Yeast Fermentation

Equation	Products	Reactants
Equation *A*	*oxygen, glucose*	*carbon dioxide, water, energy*
Equation *B*	*glucose*	*carbon dioxide, alcohol, energy*
Equation *C*	*glucose*	*water, alcohol, energy*

2. *Which equation best describes fermentation by yeast?* (1 point) _______________
   or
   Any specific relevant problem that is stated as a question. (1 point) _______________

Practice (5 points total)

3. *No carbon dioxide is present in the tap water.* (1 point) _______________

4. a. *The solution turns yellow.* (1 point) _______________

   b. *The yellow color indicates that carbon dioxide is present in the solution.* (1 point) _______________

5. *The glucose test strip changes color, which indicates that glucose is present.* (1 point) _______________

6. *The glucose test strip does not change color, which indicates that no glucose is present in tap water.* (1 point) _______________

Part II–Experiment A (38 points total)

Part II Score

Experiment (19 points total)

1. **Hypothesis**
   Glucose ($C_6H_{12}O_6$) ⟶ Carbon dioxide (CO_2) (1 point)

2. **Materials**
   Students should include only materials that are found on the supply table and that are needed for their particular experiment. (1 point)

3. **Procedure (17 points total)**
   a. *Students should draw four test tubes in the open air.* (2 points)

      Two test tubes should contain fresh glucose solution, and may be labeled "beginning of experiment." (2 points)

      The other two test tubes should contain incubated glucose solution and may be labeled "end of experiment." (2 points)

      One test tube in each group should contain yeast. (2 points)

      One test tube in each group should not contain yeast. (2 points)
      or
      Any comparable arrangement (10 points)

   b. *At the beginning of the experiment, test both setups for glucose using test strips.* (1 point)

      At the beginning of the experiment, test both setups for carbon dioxide using BTB indicator. (1 point)

      At the end of the experiment, test both setups for glucose using test strips. (1 point)

      At the end of the experiment, test both setups for carbon dioxide using BTB indicator. (1 point)

   c. Use of appropriate safety precautions (2 points)

   d. Appropriate use of materials (1 point)

Data (19 points total)

4. **Records** (1 point per table cell; 15 points total)*

Tests	*Experimental setup results*	*Control setup results*
Test for glucose at start	*2%*	*2%*
Test for CO_2 at start	*blue BTB*	*blue BTB*
Test for glucose at end	*0.1 %*	*2%*
Test for CO_2 at end	*yellow BTB*	*blue BTB*

*Some students may organize their data differently. All answers that provide the above information, regardless of differences in format, are acceptable.

5. Interpretation
Glucose is broken down by the yeast to produce carbon dioxide. (2 points) ______
or
In the presence of air, the amount of glucose decreases and the amount of carbon dioxide increases in the glucose solution containing yeast. (2 points)
or
Any other interpretation that is consistent with the data. (2 points)

6. Conclusion
The hypothesis that, in the presence of air, yeast enzymes use glucose to produce carbon dioxide is supported by the data. (1 point) ______
or
Any relevant conclusion that is consistent with the data and the interpretation. (1 point)

7. Analysis
The hypothesis is weak because fermentation by yeast occurs only in the absence of oxygen. (1 point) ______

Part II–Experiment B (38 points total)

Part II Score

Experiment

1. Hypothesis

Glucose ($C_6H_{12}O_6$) ⟶ Carbon dioxide (CO_2) (1 point)

2. Materials

Students should include only materials that are found on the supply table and that are needed for their particular experiment. (1 point)

3. Procedure (17 points total)

a. *Students should draw four test tubes with stoppers.* (2 points)

Two test tubes should contain fresh glucose solution and may be labeled "beginning of experiment." (2 points)

The other two test tubes should contain incubated glucose solution, and may be labeled "end of experiment." (2 points)

One test tube in each group should contain yeast. (2 points)

One test tube in each group should not contain yeast. (2 points)

or

Any comparable arrangement (10 points)

b. *At the beginning of the experiment, test both setups for glucose using test strips.* (1 point)

At the beginning of the experiment, test both setups for carbon dioxide using BTB indicator. (1 point)

At the end of the experiment, test both setups for glucose using test strips. (1 point)

At the end of the experiment, test both setups for carbon dioxide using BTB indicator. (1 point)

c. Use of appropriate safety precautions (2 points)

d. Appropriate use of materials (1 point)

Data (19 points total)

4. Records (1 point per table cell; 15 points total)*

Tests	*Experimental setup results*	*Control setup results*
Test for glucose at start	*2%*	*2%*
Test for CO_2 at start	*blue BTB*	*blue BTB*
Test for glucose at end	*0.1 %*	*2%*
Test for CO_2 at end	*yellow BTB*	*blue BTB*

*Some students may organize their data differently. All answers that provide the above information, regardless of differences in format, are acceptable.

5. **Interpretation**
Glucose is converted to carbon dioxide by yeast in the absence of air.
(2 points) __________
or
In the absence of air, the amount of glucose decreases and the amount of carbon dioxide increases in the glucose solution containing yeast.
(2 points) __________
or
Any other interpretation that is consistent with the data. (2 points)

6. **Conclusion**
The hypothesis that yeast enzymes do not use glucose to produce carbon dioxide in the absence of air is contradicted by the data.
(1 point) __________
or
Any relevant conclusion that is consistent with the data and the interpretation. (1 point) __________

7. **Analysis**
The presence of alcohol could have been tested in this experiment.
(1 point) __________

Unit 6
Performance-Based Assessment Activity
Testing Enzyme Activity in Seeds
Part I–Identifying a Problem

Follow the guidelines below during this assessment activity.

- Remember not to talk to other students taking the test. If you are working with a team, talk only with those in your group and talk very quietly, so that you do not disturb other teams.
- When you conduct your experiment, you will be able to use only the materials and equipment on the supply table. Some of the materials on the table will not be needed.
- Other students involved in today's activity have different problems to solve. If you observe other students using materials different from those that you are using, this may be because their problems are not the same as yours.
- Remember that a problem often has several different acceptable solutions.

Objectives

After reading the Background, you will be expected to do the following:

- Identify a scientific problem that can be investigated in a laboratory.
- Perform laboratory tests for the presence of glucose and starch.

Background

The seeds of flowering plants store large reserves of food that will be used by plant embryos as they start to grow. Much of this food reserve is in the form of starch. Seeds also store varying amounts of sugar, fat, and protein. The food stored in seeds results from photosynthesis carried out by the leaves. Enzymes found in the cells of leaves control the production of sugars during photosynthesis. Sugars produced in the leaves of a plant travel through phloem cells to developing seeds. The seeds produce enzymes that convert glucose, which is one of the sugars that result from photosynthesis, to starch. The two reactions involved in sugar and starch production can be summarized as follows.

Reaction 1: $CO_2 + H_2O + \text{light energy} \xrightarrow{\text{enzymes}} C_6H_{12}O_6 + O_2$

Reaction 2: $C_6H_{12}O_6 \xrightarrow{\text{enzymes}} \text{starch}$

Certain compounds and metal ions found in pollutants that contaminate soil may be absorbed by plants growing in contaminated soil. These pollutants are known to interfere with an enzyme's ability to catalyze a reaction. Inside a plant, a substance that inhibits a particular enzyme interferes with the formation of the product usually produced in the reaction catalyzed by that enzyme.

1. Based on the preceding paragraphs, identify and state a **problem** that can be tested in a laboratory.

____

____

Safety Precautions

The following safety precautions should be observed during Parts I and II of this activity.

- Wear safety goggles, protective gloves, and a lab apron.
- Confine loose hair and clothing.
- Do not touch or taste any chemicals. Wash off any solutions that get on skin or clothing at the sink while calling to the teacher. If you get a solution in your eyes, immediately flush it out at the eyewash station while calling to the teacher.
- Be careful not to spill any solutions. Call the teacher in the event of a spill.
- Dispose of any broken glass in a separate container. Never put broken glass in a regular waste container.
- Clean up the lab and all equipment after use. Dispose of solutions according to proper disposal methods. After all lab work is finished, wash hands thoroughly before leaving the lab.

Practice

Simple tests for the presence of sugar and starch can be performed in a laboratory.

2. Go to the materials table, and obtain the following materials.
    - 1 soaked corn seed
    - 1 soaked bean seed
    - 1 mortar and pestle
    - 1 small test tube
    - 1 dropper bottle of Lugol's iodine solution
    - 1 dispenser of glucose test tape

3. Place the corn seed in the mortar. Add about 2 mL of tap water. Grind the seed and water with the pestle to break up the seed and release its stored molecules into the water. Following the directions on the glucose test tape dispenser, test the mixture for the presence of glucose.

    Does the corn seed contain glucose?

4. To test the same mixture for starch, add 1 drop of Lugol's iodine solution to the seed and water mixture. A blue-black color is a positive indication for the presence of starch. A yellow or yellow-orange color is a negative test for starch.

    Does the corn seed contain starch?

5. Dispose of the corn seed mixture according to the instructions provided by your teacher, rinse the mortar and pestle, and repeat steps 3 and 4 using the bean seed.

    a. Does the bean seed contain glucose? ______________________

    b. Does the bean seed contain starch? ______________________

6. In a clean, small test tube, test a new source of tap water for the presence of starch and glucose to be sure that any positive tests are due to the contents of the seed and not due to the tap water.

    a. Does tap water contain starch or glucose? ______________________

    b. What is the name for this type of test? ______________________

7. Dispose of your materials according to the instructions provided by your teacher, and rinse each container thoroughly before continuing.

Unit 6
Performance-Based Assessment Activity
Testing Enzyme Activity in Seeds
Part II–Experiment A

Objectives

After reading the Additional Background, you will be expected to do the following:

- Formulate a hypothesis for a scientific problem.
- Conduct an experiment to test the hypothesis.
- Record and interpret data to determine if the hypothesis is a suitable answer for the problem.

Additional Background

To learn whether a common enzyme-inhibiting substance interferes with either of the two reactions shown in Part I of this activity, a researcher grew several corn plants and bean plants in soil that was contaminated with that enzyme-inhibiting substance. The researcher also grew several corn plants and bean plants in uncontaminated soil. After the plants matured, the researcher harvested seeds from each of the plants and planned to test the seeds for any effect the enzyme-inhibiting substance might have had. In Part II of this activity, you will design and carry out an experiment that will help complete the researcher's work.

Problem

Consider the following problem based on the background information in Part I.

Does the enzyme-inhibiting substance that contaminated the soil in which the corn plants were grown affect the enzymes that control Reaction 1 or 2?

Experiment

1. **Hypothesis** In the space provided, state a hypothesis for this problem.

2. **Materials** Go to the equipment and materials table, and examine the available materials. Think about how you would design an experiment to test your hypothesis using these materials. Do not remove any materials from the table yet. In the space below, make a list of the materials that you will use in your experiment.

3. **Procedure** In the space provided, describe an experiment that would test your hypothesis.
   a. List the steps of your experiment.

b. Observations List the observations or measurements you will make during your experiment.

____

____

c. During your experiment, be sure to observe the **safety precautions** listed in Part I.

d. To conduct your experiment, gather *only* the materials and equipment you need from the materials table. Pour into separate glassware the amount of each solution you need for your experiment, and transfer the solution or solutions to your work station. Do not take an entire quantity of any solution. Others may need to use the same solutions. When you have gathered the needed materials, set up and conduct your experiment as planned.

Data

4. Records Record your data below in a neat table.

5. Interpretation Write your interpretation of the data you recorded.

____

____

6. Conclusion State whether your data support or refute your hypothesis.

____

____

Cleanup

After you turn in your paper, dispose of your materials according to the instructions provided by your teacher. Clean up your work area, and wash your hands before leaving the lab.

Unit 6
Performance-Based Assessment Activity
Testing Enzyme Activity in Seeds
Part II–Experiment B

Objectives

After reading the Additional Background, you will be expected to do the following:

- Formulate a hypothesis for a scientific problem.
- Conduct an experiment to test the hypothesis.
- Record and interpret data to determine if the hypothesis is a suitable answer for the problem.

Additional Background

To learn whether a common enzyme-inhibiting substance interferes with either of the two reactions shown in Part I of this activity, a researcher grew several corn plants and bean plants in soil that was contaminated with that enzyme-inhibiting substance. The researcher also grew several corn plants and bean plants in uncontaminated soil. After the plants matured, the researcher harvested seeds from each of the plants and planned to test the seeds for any effect the enzyme-inhibiting substance might have had. In Part II of this activity, you will design and carry out an experiment that will help complete the researcher's work.

Problem

Consider the following problem based on the background information in Part I.

Does the enzyme-inhibiting substance that contaminated the soil in which the bean plants were grown affect the enzymes that control Reaction 1 or 2?

Experiment

1. **Hypothesis** In the space provided, state a hypothesis for this problem.

____

____

2. **Materials** Go to the equipment and materials table, and examine the available materials. Think about how you would design an experiment to test your hypothesis using these materials. Do not remove any materials from the table yet. In the space below, make a list of the materials that you will use in your experiment.

3. **Procedure** In the space provided, describe an experiment that would test your hypothesis.

   a. List the steps of your experiment.

____

____

____

b. Observations List the observations or measurements you will make during your experiment.

c. During your experiment, be sure to observe the **safety precautions** listed in Part I.

d. To conduct your experiment, gather *only* the materials and equipment you need from the materials table. Pour into separate glassware the amount of each solution you need for your experiment, and transfer the solution or solutions to your work station. Do not take an entire quantity of any solution. Others may need to use the same solutions. When you have gathered the needed materials, set up and conduct your experiment as planned.

Data

4. Records Record your data below in a neat table.

5. Interpretation Write your interpretation of the data you recorded.

6. Conclusion State whether your data support or refute your hypothesis.

Cleanup

After you turn in your paper, dispose of your materials according to the instructions provided by your teacher. Clean up your work area, and wash your hands before leaving the lab.

Unit 6
Performance-Based Assessment Activity
Testing Enzyme Activity in Seeds
Teacher Notes and Scoring Rubric

Teacher Notes

Assessment Time Required (One 50-minute class period)

Concepts and Vocabulary

Before beginning this assessment, students should be familiar with the following concepts:

- catalysts
- enzymes
- photosynthesis
- scientific processes

Introduction

The Performance-Based Assessment Activity for Unit 6 consists of two parts. Part I involves reading two paragraphs of background information, determining a scientific problem that can be tested in a lab, and practicing the techniques needed for the assessment. Part II consists of two separate problems in which students are asked to test simulated extracts of corn seeds or bean seeds to determine whether a hypothetical enzyme-inhibiting substance might have interfered with one or both of the reactions shown in Part I of the activity.

Strategies

All students should complete Part I. You may want to have students complete Part I on the day before they complete Part II. For Part II, half of the students in the class should be assigned Experiment A, and the other half should be assigned Experiment B. If Part I is completed the day before Part II is to be done, students' papers for Part I should be returned to them for use during Part II of the assessment.

Because students should have completed several Performance-Based Assessment Activities before they get to Unit 6, make this activity an individual assessment if possible. If students work in teams, limit the group size to two students. You may require students working in teams to submit one paper per team, or you may require them to submit individual papers. If students work in teams, tell them that they may be assessed individually later in the year.

Remind students that this activity is a test and that standard test-taking rules apply. Allow talking among team members, but stress that talking must be done very quietly and must be relevant to the assessment activity. When all students have completed an assessment, conduct a class discussion of the students' approaches to the Performance-Based Assessment Activity.

Student Safety Precautions

The safety precautions for this assessment activity are listed in Part I. Before allowing students to begin either part of the activity, make sure they have read these safety precautions. Advise students that they will be given points for properly adhering to the safety precautions during this activity.

Materials

You will need the following materials to prepare the solutions needed for this assessment activity:

- potassium iodide, KI, crystals (100 g is enough to make 1 L of Lugol's iodine solution)
- iodine (I) crystals (50 g is enough to make 1 L of Lugol's iodine solution)
- soluble starch (0.03 g is enough to make 1 L of a 0.3% starch solution)
- granular glucose (10 g is enough to make 1 L of a 10% glucose solution)
- large (500 mL or 1 L) beakers (for mixing extracts)
- several dropper bottles (for Lugol's iodine solution)
- hot plate

For each student or team, you will need the following materials in the quantities indicated:

- 1 corn seed
- 1 bean seed
- 1 mortar and pestle (If a mortar and pestle is not available, students can simply mash the soft seeds with a spoon or other appropriate hard object.)
- 5–10 mL of tap water
- 2 small test tubes
- 1 test-tube rack
- 1 10-mL graduated cylinder
- 1 dispenser of glucose test tape (Purchase Urinary Glucose Test Strips at a pharmacy.)
- 1 dropper bottle of Lugol's iodine solution
- 2–3 mL of either corn or bean simulated extract labeled as "Extract of Corn (or Bean) Seeds From Plants Grown in Contaminated Soil"
- 2–3 mL of either corn or bean simulated extract labeled as "Extract of Corn (or Bean) Seeds From Plants Grown in Uncontaminated Soil"

Preparation

1. To soften corn and beans seeds for Part I, soak the seeds overnight in tap water. Seeds vary in their response to a glucose test depending on when they were harvested. Therefore, you should test your seeds with glucose test tape beforehand to determine if they contain glucose. If the seeds do not test positive for glucose, soak them overnight in a 10% glucose solution to assure a positive glucose test.
2. To prepare a 10% glucose solution, mix 10 g of glucose in 1 L of tap water.
3. Prepare a 0.3% starch solution by mixing 0.03 g of soluble starch with about 50 mL of tap water, and heat to boiling while stirring. Remove starch solution from heat, and dilute to 1 L with tap water. Stir well and pour into a stock bottle labeled as "0.03% Starch Solution."
4. To prepare the simulated extracts of corn seeds and bean seeds from plants grown in uncontaminated soil and the simulated extract of bean seeds grown in soil contaminated with an enzyme-inhibiting substance, dilute 100 mL of the 10% glucose solution to 1 L with the 0.03% starch solution. The resulting solution will test positive for both glucose and starch. You may need to make 3 to 5 L of this solution, depending on how many students you have. Divide the solution among 12-15 beakers or flasks. Label one-third as "Extract of Corn Seeds From Plants Grown in Uncontaminated Soil," label another one-third as "Extract of Bean Seeds From Plants Grown in Uncontaminated Soil," and label the remaining one-third as "Extract of Bean Seeds From Plants Grown in Soil Contaminated With an Enzyme-Inhibiting Substance."
5. To prepare the simulated extract of corn seeds grown in soil contaminated with an enzyme-inhibiting substance, dilute 100 mL of the 10% glucose solution to 1 L with tap water. To give the solution a slightly cloudy appearance similar to the starch solution, add a pinch of talc or chalk dust and mix thoroughly. The resulting solution will test positive for glucose and negative for starch. Divide the solution among 4-5 beakers or flasks labeled as "Extract of Corn Seeds From Plants Grown in Soil Contaminated With an Enzyme-Inhibiting Substance."
6. To prepare Lugol's iodine solution, dissolve 100 g potassium iodide, KI, in 1 L of tap water, and then add 50 g of iodine, I, crystals. Stir. Heat to dissolve if necessary. Pour into dropper bottles and label. Pour remainder of solution into a stock bottle and date. You can also purchase prepared Lugol's iodine solution from a biological supply house.
7. Place a container labeled "Solution Disposal Container" at the front of the room. Instruct students to put leftover solutions in this container.

Laboratory Setup

Place the following materials on a centrally located supply table:
(Part I only)

- 1 corn seed per student or team
- 1 bean seed per student or team
- 1 mortar and pestle per student or team

(Parts I and II)

- 2 small test tubes per student or team
- 1 test-tube rack per student or team
- 1 dispenser of glucose test tape per student or team
- 1 dropper bottle of Lugol's iodine solution per student or team
- 1 10-mL graduated cylinder per student or team

(Part II only)

- 4–5 containers of solution labeled as "Extract of Corn Seeds From Plants Grown in Soil Contaminated With an Enzyme-Inhibiting Substance"
- 4–5 containers of solution labeled as "Extract of Corn Seeds From Plants Grown in Uncontaminated Soil"
- 4–5 containers of solution labeled as "Extract of Bean Seeds From Plants Grown in Soil Contaminated With an Enzyme-Inhibiting Substance"
- 4–5 containers of solution labeled as "Extract of Bean Seeds From Plants Grown in Uncontaminated Soil"
- 1 petri dish per student or team (decoy)
- 1 dropper bottle of bromothymol blue indicator solution (decoy)

Cleanup and Disposal

For student use during the assessment activity, provide a container marked "Solution Disposal Container." Solutions used in this activity can safely be disposed of in the laboratory sink. Instruct students to dispose of macerated corn and bean seeds in the laboratory trash can. Avoid having students handle spills themselves. Instruct students to wash and dry the glassware they use and return it to the materials table before they leave the lab.

Scoring Rubric

The following pages are a Scoring Rubric that provides possible answers to this Performance-Based Assessment Activity. Preferred student responses are shown in italics. Suggested point values for each step in the activity appear in parentheses next to the responses. The total points earned by a student for each step in the activity can be recorded in the blanks to the right of the steps. The Scoring Rubric can be modified to accommodate your specific circumstances. You may wish to use this rubric only as a guide in scoring the assessment activity. If possible, make a copy of the rubric for each student or team, and return a copy of the rubric showing the scores earned by the students when you return their graded papers.

Unit 6
Performance-Based Assessment Activity
Testing Enzyme Activity in Seeds (50 points)
Scoring Rubric

______ Total Score

Possible answers to this Performance-Based Assessment Activity are provided below. Preferred student responses are shown in italics. Suggested point values for each step in the activity appear in parentheses next to the responses. The total points earned for each step in the activity are recorded in the blanks to the right of the steps.

Part I (9 points total)

______ Part I Score

Background (3 points total)

1. Problem
Does an enzyme-inhibiting substance affect the chemical reactions that produce sugar and starch in plants? (3 points)
or
Other appropriately stated problems that are relevant to the background information. (2 points)

Practice (6 points total)

3. Glucose Test on Corn Seed
yes (1 point) ______

4. Starch Test on Corn Seed
yes (1 point) ______

5. Glucose and Starch Tests on Bean Seed
(2 points total)
a. *yes* (1 point) ______

b. *yes* (1 point) ______

6. Glucose and Starch Tests on Tap Water
(2 points total)
a. *no* (1 point) ______

b. *control* (1 point) ______

Part II–Experiment A (41 points total)

Part II–A Score

Experiment (22 points total)

1. **Hypothesis** (2 points)
   The enzyme-inhibiting substance that contaminated the soil in which corn plants were grown affects the enzymes in Reaction 1.
   or
   The enzyme-inhibiting substance that contaminated the soil in which corn plants were grown affects the enzymes in Reaction 2.
   or
   The enzyme-inhibiting substance that contaminated the soil in which corn plants were grown affects the enzymes in both Reactions 1 and 2.
   or
   The enzyme-inhibiting substance that contaminated the soil in which corn plants were grown does not affect the enzymes in either Reaction 1 or 2.

2. **Materials**
   Students should include only materials that are found on the supply table and that are needed for their particular experiment. (2 points) ______

3. **Procedure** (18 points total)
   a. **Experimental Design** (12 points total)
      Experimental Setup: (6 points total)
      1. *Obtain 2-3 mL of the extract of corn seeds from plants that were grown in soil that contained the enzyme-inhibitng substance.* (2 points) ______
      2. *Use glucose test tape to test the extract for the presence of glucose.* (2 points) ______
      3. *Use Lugol's iodine solution to test the extract for the presence of starch.* (2 points) ______

      Control Setup: (6 points)
      1. *Obtain 2-3 mL of the extract of corn seeds from plants that were grown in soil that did not contain the enzyme-inhibiting substance.* (2 points) ______
      2. *Use glucose test tape to test the extract for the presence of glucose.* (2 points) ______
      3. *Use Lugol's iodine solution to test the extract for the presence of starch.* (2 points) ______

   b. **Observations** (4 points total)
      1. *In both the control and experimental setups, observe the color of the glucose test tape after it was placed in an extract, and compare the color with the color key on the glucose test tape dispenser to determine whether glucose is present.* (2 points) ______
      2. *In both the control and experimental setups, look for a color change from yellow-orange to blue or blue-black to determine whether starch is present.* (2 points) ______

c. **Safety Precautions** The student followed appropriate safety precautions. (1 point) ______

d. **Materials** The student made appropriate use of materials. (1 point) ______

Data (19 points total)

4. **Records** (15 points total)

Tests Made on Corn (1 point)	*Experimental* (2 points)	*Control* (2 points)
For Glucose (1 point)	*positive* (2 points)	*positive* (2 points)
For Starch (1 point)	*negative* (2 points)	*positive* (2 points)

5. **Interpretation** (2 points total)
In corn, the enzyme-inhibiting compound does not interfere with the enzymes for Reaction 1. (1 point) ______
and
In corn, the enzyme-inhibiting compound interferes with the enzymes for Reaction 2. (1 point) ______

6. **Conclusion** The student's conclusion must be consistent with the hypothesis stated in Step 1. Accept any one of the following: (2 points) ______

The hypothesis that the enzyme-inhibiting substance that contaminated the soil in which corn plants were grown affects the enzymes in Reaction 1 is refuted by the data.
or
The hypothesis that the enzyme-inhibiting substance that contaminated the soil in which corn plants were grown affects the enzymes in Reaction 2 is supported by the data.
or
The hypothesis that the enzyme-inhibiting substance that contaminated the soil in which corn plants were grown affects the enzymes in both Reaction 1 and Reaction 2 is refuted by the data.
or
The hypothesis that the enzyme-inhibiting substance that contaminated the soil in which corn plants were grown does not affect the enzymes in either Reaction 1 or Reaction 2 is refuted by the data.

Part II–Experiment B (41 points total)

Part II–B Score

Experiment (22 points total)

1. Hypothesis (2 points)

The enzyme-inhibiting substance that contaminated the soil in which bean plants were grown affects the enzymes in Reaction 1.

or

The enzyme-inhibiting substance that contaminated the soil in which bean plants were grown affects the enzymes in Reaction 2.

or

The enzyme-inhibiting substance that contaminated the soil in which bean plants were grown affects the enzymes in both Reactions 1 and 2.

or

The enzyme-inhibiting substance that contaminated the soil in which bean plants were grown does not affect the enzymes in either Reaction 1 or 2.

2. Materials

Students should include only materials that are found on the supply table and that are needed for their particular experiment. (2 points)

3. Procedure (18 points total)

a. Experimental Design (12 points total)

Experimental Setup: (6 points total)

1. *Obtain 2–3 mL of the extract of bean seeds from plants that were grown in soil that contained the enzyme-inhibiting substance.* (2 points)
2. *Use glucose test tape to test the extract for the presence of glucose.* (2 points)
3. *Use Lugol's iodine solution to test the extract for the presence of starch.* (2 points)

Control Setup: (6 points total)

1. *Obtain 2–3 mL of the extract of bean seeds from plants that were grown in soil that did not contain the enzyme-inhibiting substance.* (2 points)
2. *Use glucose test tape to test the extract for the presence of glucose.* (2 points)
3. *Use Lugol's iodine solution to test the extract for the presence of starch.* (2 points)

b. Observations (4 points total)

1. *In both the control and experimental setups, observe the color of the glucose test tape after it was placed in an extract, and compare the color with the color key on the glucose test tape dispenser to determine whether glucose is present.* (2 points)
2. *In both the control and experimental setups, look for a color change from yellow-orange to blue or blue-black to determine whether starch is present.* (2 points)

c. **Safety Precautions** The student followed appropriate safety precautions. (1 point) ________

d. **Materials** The student made appropriate use of materials. (1 point) ________

Data (19 points total)

4. **Records** (15 points total) ________

Tests Made on Bean (1 point)	*Experimental* (2 points)	*Control* (2 points)
For Glucose (1 point)	*positive* (2 points)	*positive* (2 points)
For Starch (1 point)	*positive* (2 points)	*positive* (2 points)

5. **Interpretation** (2 points total)
In beans, the enzyme-inhibiting compound does not interfere with the enzymes for either Reaction 1 or Reaction 2. (2 points) ________

6. **Conclusion** The student's conclusion must be consistent with the hypothesis stated in Step 1. Accept any one of the following: (2 points) ________

The hypothesis that the enzyme-inhibiting substance that contaminated the soil in which bean plants were grown affects the enzymes in Reaction 1 is refuted by the data.

or

The hypothesis that the enzyme-inhibiting substance that contaminated the soil in which bean plants were grown affects the enzymes in Reaction 2 is refuted by the data.

or

The hypothesis that the enzyme-inhibiting substance that contaminated the soil in which bean plants were grown affects the enzymes in both Reaction 1 and Reaction 2 is refuted by the data.

or

The hypothesis that the enzyme-inhibiting substance that contaminated the soil in which bean plants were grown does not affect the enzymes in either Reaction 1 or Reaction 2 is supported by the data.

Unit 7
Performance-Based Assessment Activity
Exploring Invertebrates
Part I–Experiment A

Follow the guidelines below during this assessment activity.

- Remember not to talk to other students taking the test. If you are working with a team, talk only with those in your group and talk very quietly, so that you do not disturb other teams.
- When you conduct your experiment, you will be able to use only the materials and equipment on the supply table. Some of the materials on the table will not be needed.
- Most of the other students involved in today's activity have different problems to solve. If you observe other students using materials different from those that you are using, this is because their problems are not the same as yours.
- Remember that a problem often has several different acceptable solutions.

Background

Daphnia are tiny, freshwater crustaceans from the order Cladorcera. They are also known as cladocerans or water fleas. *Daphnia* have a laterally compressed body that is enclosed in a transparent carapace. They can be found worldwide in lakes, ponds, and streams.

Safety Precautions

The following safety precautions should be observed during Parts I and II of this activity.

- Wear safety goggles, protective gloves, and a lab apron.
- Confine loose hair and clothing.
- Do not touch or taste any chemicals. Avoid inhaling chemical fumes. Wash off any solutions that get on skin or clothing at the sink while calling to the teacher. If you get a solution in your eyes, immediately flush it out at the eyewash station while calling to the teacher.
- Be careful not to spill any solutions. Call the teacher in the event of an acid or base spill.
- Dispose of any broken glass in a separate container. Never put broken glass in a regular waste container.
- Be sure electrical equipment is turned off before plugging it in. Turn it off again before unplugging it. Do not allow electrical cords to dangle from work stations. The area under and around electrical equipment should be dry, and your hands should be dry when using electrical equipment.
- Clean up the lab and all equipment after use. Dispose of solutions according to proper disposal methods. Turn off and unplug all electrical equipment before leaving the laboratory. After all lab work is finished, wash hands thoroughly before leaving the lab.

Practice

Caution *Do not use electrical equipment with a frayed or kinked cord.* Go to the supply table and obtain the following items. Do not try to transport all of these items in one trip.

1 compound microscope
1 desk or microscope lamp
2 microscope slides
2 slide coverslips
1 medicine dropper
1 piece of thread about 2—3 inches long
1 pair of scissors
1 culture of *Daphnia*

1. Set up the microscope and lamp. Using the medicine dropper, place 2—3 drops of *Daphnia* culture on a microscope slide. Cut the piece of thread into three of four pieces, each about 0.5 cm long. Place three of the thread pieces into the culture on the microscope slide before adding a coverslip. (The threads will prevent the cover slip from crushing the *Daphnia*.) View the *Daphnia* on low power. Identify the heart, compound eye, legs, and antennae.

Notice that the heart beats very fast. To estimate the heart rate, tap on a piece of scratch paper with a pencil in the rhythm of the heart beat. Tap in a continuous line back and forth for 15 seconds. Then count the number of dots, and multiply by four to get the approximate heart rate per minute. What is the approximate heart rate per minute?

---

2. Estimate the heart rate of two different *Daphnia* using the method described above. Determine the average heart rate per minute by adding your three estimates together and dividing the sum by 3. Show your work in the space below.

Unit 7
Performance-Based Assessment Activity
Exploring Invertebrates
Part II–Experiment A

Objectives

- Formulate a hypothesis based on a given problem.
- Design and conduct an experiment to test your hypothesis.
- Organize the data from your experiment into a table, and make a graph of it.
- Evaluate your hypothesis based on your data.

Problem

Based on the information in Part I, consider the following **problem.**

What effect do increasing concentrations of alcohol have on the heart rate of *Daphnia*?

Experiment

1. **Hypothesis** In the space below, write a hypothesis for the **problem** that is stated above.

____

____

2. **Materials** Go to the equipment and materials table, and examine the available materials. Think about how you would design an experiment to test your hypothesis using these materials. Do not remove any materials from the table yet. Make a list of the materials that you will use in your experiment.

3. **Procedure**

   a. Describe how you will test your hypothesis. Be sure to include at least one control in your experimental design.

____

____

____

____

____

____

b. List the observations or measurements you will make during the experiment.

____

____

c. Be sure to observe the **Safety Precautions** listed in Part I during your experiment.

d. To conduct your experiment, gather *only* the materials and equipment you need from the materials table. Pour into separate glassware the amount of each solution you need for your experiment, and transfer the solution or solutions to your work station. Do not take an entire quantity of any solution. Others may also need access to the same solutions. When you have gathered the needed materials, set up and conduct your experiment as planned. **Caution** *Alcohol solutions are flammable. Keep all alcohol solutions away from open flames or heat sources. Avoid inhaling alcohol fumes.*

Data

4. Records

a. Make a table in the space below. Make sure your table has descriptive heads. Record your data in your table.

b. Make a graph of your data. Label each axis, and title the completed graph.

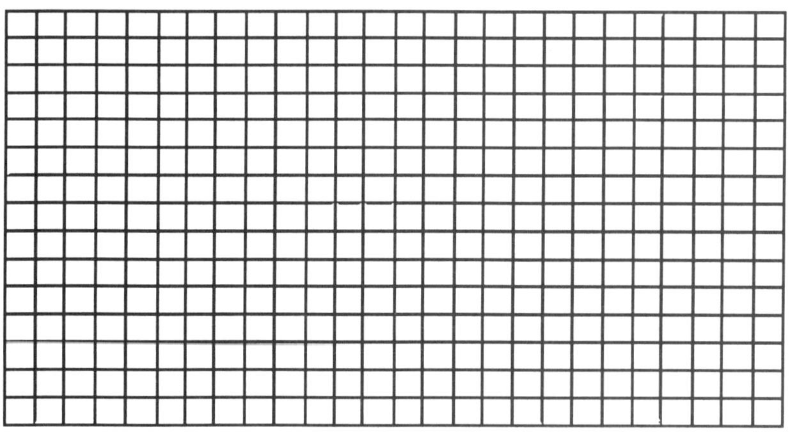

Unit 7 - Figure 1

5. Interpretation Write your interpretation of the data you recorded.

____

____

6. Conclusion State whether your data support or refute your hypothesis.

____

7. Analysis List any factors that could have affected the accuracy of the data you obtained during your experiment.

____

____

____

Cleanup

After you turn in your paper, dispose of your materials according to the instructions provided by your teacher. Clean up your work area, and wash your hands before leaving the lab.

Unit 7
Performance-Based Assessment Activity
Exploring Invertebrates
Part I–Experiment B

Follow the guidelines below during this assessment activity.

- Remember not to talk to other students taking the test. If you are working with a team, talk only with those in your group and talk very quietly, so that you do not disturb other teams.
- When you conduct your experiment, you will be able to use only the materials and equipment on the supply table. Some of the materials on the table will not be needed.
- Most of the other students involved in today's activity have different problems to solve. If you observe other students using materials different from those that you are using, this is because their problems are not the same as yours.
- Remember that a problem often has several different acceptable solutions.

Background

Planaria are small, nonparasitic flatworms of the phylum Platyhelminthes. They are generally aquatic and can be found on the undersides of logs or stones that are submerged in ponds and lakes. Planarians have two dark anterior eyespots that contain photosensitive cells.

1. Give one example of another organism that has an eyespot. What function does the eyespot serve in this organism?

____

____

2. What might be a possible function of a planarian's eyespots?

____

____

Safety Precautions

The following safety precautions should be observed during Parts I and II of this activity.

- Wear safety goggles, protective gloves, and a lab apron.
- Confine loose hair and clothing.
- Do not touch or taste any chemicals. Wash off any solutions that get on skin or clothing at the sink while calling to the teacher. If you get a solution in your eyes, immediately flush it out at the eyewash station while calling to the teacher.
- Be careful not to spill any solutions. Call the teacher in the event of an acid or base spill.
- Dispose of any broken glass in a separate container. Never put broken glass in a regular waste container.
- Be sure electrical equipment is turned off before plugging it in. Turn it off again before unplugging it. Do not allow electrical cords to dangle from work stations. The area under and around electrical equipment should be dry, and your hands should be dry when using electrical equipment.
- Clean up the lab and all equipment after use. Dispose of solutions according to proper disposal methods. Turn off and unplug all electrical equipment before leaving the laboratory. After all lab work is finished, wash hands thoroughly before leaving the lab.

Practice

Caution *Do not use electrical equipment with a frayed or kinked cord.* Go to the supply table and obtain the following items. Do not try to transport all of these items in one trip.

1 stereomicroscope or magnifying glass
1 culture of *Planaria* in a covered petri dish

3. Examine the culture of planarians in the petri dish. Note their distribution in the petri dish. Are all of the planarians clumped together or are they evenly dispersed?

____

____

4. View the *Planaria* culture using the stereomicroscope or magnifying glass. Pick out one planarian in the petri dish, and sketch it in the space below. Be sure to draw and label its eyespots.

Unit 7
Performance-Based Assessment Activity
Exploring Invertebrates
Part II–Experiment B

Objectives

- Formulate a hypothesis based on a given problem.
- Design and conduct an experiment to test your hypothesis.
- Organize the data from your experiment.
- Evaluate your hypothesis based on your data.

Problem

Based on the information in Part I, consider the following **problem.**

How does *Planaria* respond to light? Are planarians attracted by light, or do they avoid it?

Experiment

1. **Hypothesis** In the space below, write a hypothesis for the **problem** that is stated above:

____

____

2. **Materials** Go to the equipment and materials table, and examine the available materials. Think about how you would design an experiment to test your hypothesis using these materials. Do not remove any materials from the table yet. Make a list of the materials that you will use in your experiment.

3. **Procedure**
   a. Draw a sketch of how you will set up an experiment that will test the hypothesis at the top of this page. Be sure to include a control in your experimental design. Label all parts of your experimental and control setups.

b. List the observations or measurements you will make during the experiment.

c. Be sure to observe the **Safety Precautions** that are listed in Part I.

d. To conduct your experiment, gather *only* the materials and equipment you need from the materials table. Pour into separate glassware the amount of each solution you need for your experiment, and transfer the solution or solutions to your work station. Do not take the entire quantity of any solution. Others may also need access to the same solutions. When you have gathered the needed materials, set up and conduct your experiment as planned.

Data

4. Records Record your data below.

5. Interpretation Write your interpretation of the data you recorded.

6. Conclusion State whether your data support or refute your hypothesis.

7. Analysis

a. List any factors that could have affected the accuracy of the data you obtained during your experiment.

b. How might its reaction to light benefit *Planaria*?

c. Discuss other stimuli that could be used to test the responses of *Planaria*.

Cleanup

After you turn in your paper, dispose of your materials according to the instructions provided by your teacher. Clean up your work area, and wash your hands before leaving the lab.

Unit 7
Performance-Based Assessment Activity
Exploring Invertebrates
Teacher Notes and Scoring Rubric

Assessment Time Required
One 50-minute class period

Concepts and Vocabulary
Before beginning this assessment students should be familiar with the following concepts and vocabulary:
- compound microscope
- coverslip
- control experiment
- *Daphnia*
- response
- microscope slide
- *Planaria*
- wet mount
- stimulus

Introduction
The Performance-Based Assessment Activity for Unit 7 consists of two parts. Part I consists of two separate activities, which involve gaining experience in working with an invertebrate organism (either *Daphnia* or *Planaria*) that will be used in Part II. Part II of the assessment activity consists of two separate problems in which students are asked to test either the effect of alcohol on the heart rate of *Daphnia* or the effect of light on *Planaria.* Students should be familiar with proper microscope use and slide preparation before beginning this activity.

Strategies
Assign each student one of the two problems (Experiment A or B). All students should complete both Part I and Part II of the same problem. If Part I is completed the day before Part II is to be done, return students' papers for use during Part II of the assessment.

It is recommended that this assessment be used as an individual assessment. However, this activity also works well for teams of two or three students. You may require students working in teams to submit one paper per team, or you may require them to submit individual papers.

Before allowing students to begin the assessment activity, make sure that they are aware of appropriate safety precautions. Remind students that this activity is a test situation and that the standard rules of test-taking apply. Stress that talking among team members must be done very quietly and must be relevant to the assessment activity.

When all students have completed an assessment, conduct a class discussion of the students' different approaches to the Performance-Based Assessment Activity. You may want to discuss the advantage of using the same *Daphnia* as a control and a test in Experiment A. Stress that individual differences among organisms can affect the results of the experiment. However, an initial heart rate that is counted under normal conditions can be used as a standard of comparison (control) for heart rates that are counted when alcohol is added. You may also want to lead students to recognize the potential problems associated with using a light source that produces heat in Experiment B. Remind students that light and heat may affect the planarians' responses in this situation.

Student Safety Precautions
The safety precautions for this assessment activity are listed in Part I. Before allowing students to begin the activity, make sure they have read these safety precautions. Advise students that they will be given points for properly following the safety precautions during this activity.

Use only electrical equipment equipped with a three-wire cord and a three-prong plug. Each electrical socket in the laboratory must be of the three-hole type with a GFI (ground-fault interrupter) circuit. Be sure electrical equipment is turned off before plugging it in. Turn it off again before unplugging it. Do not allow electrical cords to dangle from work stations; dangling cords are a tripping and shock hazard. Do not use electrical equipment with frayed or kinked cords. The area under and around electrical equipment should be dry. Cords should not lie in puddles of spilled liquid. Students' hands should be dry when using electrical equipment. Turn off and unplug all electrical equipment before leaving the laboratory.

Materials

You will need the following materials to prepare the solutions needed for this assessment activity:

- ethyl alcohol, 95% by volume (105 mL is enough to prepare 100 mL of each of the following concentrations of ethyl alcohol solutions: 10%, 20%, 30%, and 40%)
- distilled water (approximately 300 mL to prepare 100 mL of all four ethyl alcohol solutions)
- 50-mL graduated cylinder
- 100-mL beakers (1 beaker for each 100 mL of solution needed)
- dropper bottles (one-half of your students will need 4 dropper bottles each)

For each student or team, you will need the following materials in the quantities indicated:

- 1 compound microscope (Note that only one-half of your students will require a compound microscope.)
- 1 stereomicroscope or magnifying glass (Note that only one-half of your students will require a stereomicroscope or magnifying glass.)
- 1 desk or microscope lamp (preferably one that does not generate much heat)
- 6–10 microscope slides
- 2–4 coverslips
- 1 medicine dropper
- 10 cm of thread
- 1 roll of transparent tape
- 1/2 piece of black construction paper
- 1 pair of scissors
- 1 covered petri dish of brown *Planaria* (at least 6–8 planarians), labeled "*Planaria*"
- 1 culture of *Daphnia* (at least 6–8 *Daphnia*), labeled "*Daphnia*" (Note: There are several species of *Daphnia*. The best species to use in this activity is *Daphnia magna,* which is the largest. Have extra *Daphnia* on hand to replace any *Daphnia* that may die.)
- 1 dropper bottle of 10% ethyl alcohol solution
- 1 dropper bottle of 20% ethyl alcohol solution
- 1 dropper bottle of 30% ethyl alcohol solution
- 1 dropper bottle of 40% ethyl alcohol solution

Preparation

1. **Caution** Alcohols are flammable. Make sure there are no open flames in the laboratory. When you prepare alcohol solutions, work in a ventilated hood and wear safety goggles and a lab apron. In case of a spill, mop up the spill with wet cloths designated for spill cleanup. To prepare 100 mL of 10% ethyl alcohol solution, mix 10.5 mL of 95% ethyl alcohol with approximately 89.5 mL of distilled water. Divide the 100 mL into dropper bottles. To prepare 100 mL of 20%, 30%, and 40% ethyl alcohol solutions, repeat the procedure mixing approximately 21.0 mL, 31.5 mL, and 42.0 mL of 95% ethyl alcohol with approximately 79 mL, 68.5 mL, and 58 mL of distilled water, respectively. Store the alcohol solutions in a cool place.
2. In Experiment B, students will require a light source to test the reaction of *Planaria* to light. If possible, provide small desk lamps that do not generate heat, or instruct students to use sunlight coming in a nearby window. If possible, avoid lamps that provide a lot of heat to reduce the chances of burns among students.
3. Provide a container, separate from the regular trash can, for glass disposal. If you wish, direct students to put used microscope slides and coverslips in this receptacle.

Laboratory Setup
Place the following materials on a centrally located supply table:

(Parts I and II)
- compound microscopes
- stereomicroscope or magnifying glass
- medicine droppers
- desk or microscope lamps
- several spools of thread
- microscope slides
- scissors
- coverslips
- cultures of *Daphnia*
- cultures of *Planaria* in a covered petri dish

(Part II only)
- dropper bottles of 10% ethyl alcohol solution
- dropper bottles of 20% ethyl alcohol solution
- dropper bottles of 30% ethyl alcohol solution
- dropper bottles of 40% ethyl alcohol solution
- several rolls of transparent tape
- black construction paper

Cleanup and Disposal
For student use during the assessment activity, provide a container marked "Glass Disposal." If you wish, direct students to put used microscope slides and coverslips in this receptacle. Instruct students to wash and dry the glassware they use and return it to the materials table before they leave the lab.

Scoring Rubric
The following pages are a Scoring Rubric that provides possible answers to this Performance-Based Assessment Activity. Preferred student responses are shown in italics. Suggested point values for each step in the activity appear in parentheses next to the responses. The total points earned by a student for each step in the activity can be recorded in the blanks to the right of the steps. The Scoring Rubric can be modified to accommodate your specific circumstances. You may wish to use this rubric only as a guide in scoring the assessment activity. If possible, make a copy of the rubric for each student or team, and return a copy of the rubric showing the scores earned by the students when you return their graded papers.

Activity and Scoring Rubric Art
These teacher notes are found in question 5 of the Unit 7 file. In the answer key for this question, you will find the art for Unit 7-Figure 1 and Unit 7-Figure 2, which are found in the Scoring Rubric for Part II–Experiment A in this assessment. Cut and paste this figure in the place indicated before photocopying the assessment for your students.

Unit 7
Performance-Based Assessment Activity
Exploring Invertebrates (50 points)
Scoring Rubric

______________ Total Score

Possible answers to this Performance-Based Assessment Activity are provided below. Preferred student responses are shown in italics. Suggested point values for each step in the activity appear in parentheses next to the responses. The total points earned for each step in the activity are recorded in the blanks to the right of the steps.

Part I–Experiment A (5 points total)

______________ Part I Score

Practice

1. *The heart rate of Daphnia ranges from about 200–350 beats per minute.* (The rate increases with temperature.) (2 points) ______________

2. *The average heart rate should be 200–350 beats per minute.* (3 points) ______________

Part II–Experiment A (45 points total)

______________ Part II Score

Experiment (22 points total)

1. **Hypothesis**
   Increasing concentrations of alcohol will increase the heart rate of Daphnia. (2 points) ______________
   or
   Increasing concentrations of alcohol will decrease the heart rate of Daphnia. (2 points) ______________
   or
   Increasing concentrations of alcohol will not affect the heart rate of Daphnia. (2 points) ______________

2. **Materials**
   Materials should include only those that are found on the supply table and that are needed for Experiment A. (1 point) ______________

3. **Procedure (19 points total)**
   a. *Prepare four wet mount slides of Daphnia. Count the heart rate of each Daphnia in the absence of alcohol. (There should be one control for each concentration of alcohol tested, with a total of four controls.) Add a drop of 10% ethyl alcohol to the edge of one of the slides, and count the heart rate of one Daphnia. Repeat this procedure by adding a drop of either 20%, 30%, or 40% ethyl alcohol solution, respectively, to each of the remaining slides.* (8 points) ______________
   or
   Prepare five wet mount slides of Daphnia. Count the heart rate of one Daphnia in the absence of alcohol. (This is the control. This step may be omitted if reference is made to the average heart rate of Daphnia calculated in Part I. If the slide prepared in Part I is used as a control, then only four wet mounts are needed.) Add a drop of 10% ethyl alcohol to the edge of one of the slides, and count the heart rate of one Daphnia. Repeat this procedure by adding a drop of either 20%, 30%, or 40% ethyl alcohol solution, respectively, to each of the remaining slides. (5 points) ______________

b. *Count the heart rate of Daphnia in culture in the absence of alcohol.* (1 point) ____________

Count the average heart rate of one Daphnia in each of the following concentrations of alcohol: 10%, 20%, 30%, and 40%. (1 point for each alcohol concentration) ____________

Compare the heart rate of each Daphnia before and after the addition of an alcohol solution. (3 points) ____________

c. Use of appropriate safety precautions (2 points) ____________

d. Appropriate use of materials (1 point) ____________

Data (23 points total)

4. Records (18 points total)

a. (10 points: 1 point for each piece of data, 2 points for table setup) ____________

Effect of Alcohol on the Heart Rate of *Daphnia**

Concentration of alcohol	Heart beats/minute of *Daphnia* in the presence of alcohol	Heart beats/minute of *Daphnia* in the absence of alcohol (controls)
10% alcohol	(1 point)	(1 point)
20% alcohol	(1 point)	(1 point)
30% alcohol	(1 point)	(1 point)
40% alcohol	(1 point)	(1 point)

* Beats per minute should decrease as the concentration of alcohol increases. Data will vary. Accept any reasonable data.

b. A sample graph is shown. Heart rate can be on either the vertical or the horizontal axis. Accept any graph that reflects the data recorded in the data table in **4a**. (8 points: 3 points for the curve, 2 points for axis labels, 2 points for appropriate numbering along the axes, 1 point for title) ____________

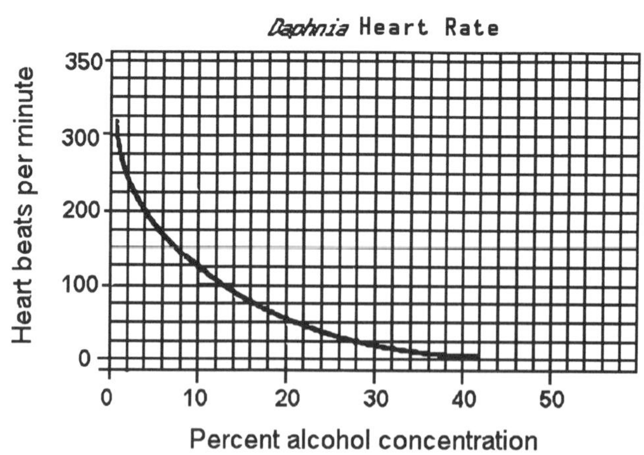

Unit 7 - Figure 2

5. Interpretation

Increasing concentrations of alcohol decrease the heart rate of Daphnia. (2 points) ________

or

Other interpretations that are consistent with the data. (2 points) ________

6. Conclusion

The data support the hypothesis that increasing concentrations of alcohol decrease the heart rate of Daphnia. (2 points) ________

or

The data contradict the hypothesis that increasing concentrations of alcohol increase the heart rate of Daphnia. (2 points) ________

or

The data contradict the hypothesis that increasing concentrations of alcohol will not affect the heart rate of Daphnia. (2 points) ________

7. Analysis

Acceptable factors include differences in heart rate among individual Daphnia *or contaminated alcohol solutions.* (1 point) ________

Unit 7
Performance-Based Assessment Activity
Exploring Invertebrates (50 points)
Scoring Rubric

_______________ Total Score

Possible answers to this Performance-Based Assessment Activity are provided below. Preferred student responses are shown in italics. Suggested point values for each step in the activity appear in parentheses next to the responses. The total points earned for each step in the activity are recorded in the blanks to the right of the steps.

Part I–Experiment B (5 points total)

_______________ Part I Score

Background

1. *Answers will vary. Euglena and some protozoans have eyespots. Eyespots function as light detectors.* (1 point) _______________

2. *Planarians use their eyespots to detect the intensity and direction of light. Accept any plausible answer.* (1 point) _______________

Practice

3. *Answers may vary. Planarians are usually evenly distributed in a covered petri dish.* (1 point) _______________

4. Accept any drawing that resembles a planarian. (1 point) _______________

   Eyespots should be labeled appropriately. (1 point) _______________

Part II–Experiment B (45 points total)

_______________ Part II Score

Experiment (25 points total)

1. **Hypothesis**
   Planarians are attracted to light. (3 points) _______________
   or
   Planarians avoid light. (3 points) _______________
   or
   Planarians do not respond to light. They are not attracted to light, nor do they avoid it. (3 points) _______________

2. **Materials**
   Materials should include only those that are found on the supply table and that are needed for Experiment B. (1 point) _______________

3. **Procedure** (21 points total)
   a. Drawings should include two petri dishes that contain cultures of approximately equal numbers of planarians. One culture should be labeled the test setup. The other culture should be labeled the control setup. (2 points) _______________

   Test setup: Cover half of one of the petri dishes containing planarians with black construction paper, and expose the entire dish to sunlight or lamp light for five minutes. (4 points) _______________

   Control setup: Expose one entire petri dish containing planarians to light or cover it with black construction paper for five minutes. (4 points) _______________

b. *Observe the movement of planarians in the test culture.* (4 points) ________

Observe the distribution of planarians in the control. (4 points) ________

c. Use of appropriate safety precautions (2 points) ________

d. Appropriate use of materials (1 point) ________

Data (20 points total)

4. Records

In the test culture setup, planarians slowly move away from the light into the shaded area. (4 points) ________

In the control setup, planarians move about randomly and are randomly distributed. (4 points) ________

5. Interpretation

Planarians move toward the shaded area. (2 points) ________

or

Planarians move away from the light source. (2 points) ________

6. Conclusion

The data contradict the hypothesis that planarians are attracted to light. (2 points) ________

or

The data support the hypothesis that planarians avoid light. (2 points) ________

or

The data contradict the hypothesis that planarians do not respond to light. (2 points) ________

7. Analysis

a. *Planarians may be avoiding any heat produced by the light source instead of the light.* (3 points) ________

or

Any other plausible factor (2 points) ________

b. *Planarians are more visible to predators in the light. By using their eyespots, planarians can avoid the light and protect themselves from predators.* (3 points) ________

or

Any other plausible benefit (3 points) ________

c. *Stimuli include food (such as liver), chemicals, temperature, pH, touch, etc.* (2 points) ________

Unit 8
Performance-Based Assessment Activity
The Challenge of Water Retention
Part I–Background

Follow the guidelines below during this assessment activity.

- Remember not to talk to other students taking the test. If you are working with a team, talk only with those in your group and talk very quietly, so that you do not disturb other teams.
- When you conduct your experiment, you will be able to use only the materials and equipment on the supply table. Some of the materials on the table will not be needed.
- Most of the other students involved in today's activity have different problems to solve. If you observe other students using materials different from those that you are using, this is because their problems are not the same as yours.
- Remember that a problem often has several different acceptable solutions.

Background

Living organisms require water balance to sustain life. For example, many vertebrates are made up of about two-thirds water. If their water content becomes too low, they can die of dehydration. If their water content becomes too high, their tissues may become swollen with the excess water, a condition called edema. Dehydration and edema can be serious threats to vertebrates. Thus, animals have adaptations that protect them from losing or gaining too much water. Animals living on land require protection against water evaporation, while animals living in marine environments may require protection against losing or gaining too much water through osmosis.

1. Describe two serious conditions that can result from loss of water balance.

____

____

2. Suggest some possible ways that vertebrates, such as fishes and reptiles, might maintain appropriate water balance.

____

Safety Precautions

The following safety precautions should be observed during Parts I and II of this activity.

- Wear safety goggles and a lab apron.
- Confine loose hair and clothing.
- Do not touch or taste any chemicals. Wash off any solutions that get on skin or clothing at the sink. If you get a solution in your eyes, immediately flush it out at the eyewash station while calling to the teacher.
- Be careful not to spill any solutions. Call the teacher in the event of an acid or base spill.
- Dispose of any broken glass in a separate container. Never put broken glass in a regular waste container.
- Clean up the lab and all equipment after use. Dispose of solutions according to proper disposal methods. After all lab work is finished, wash hands thoroughly before leaving the lab.

Practice

Dialysis tubing is selectively permeable. It is often used to simulate biological processes that involve a selectively permeable membrane, such as diffusion through a cell membrane.

Go to the supply table and obtain the following items:

- 1 15-cm strip of dialysis tubing
- 2 20-cm piece of string
- 1 test tube
- 1 test-tube rack
- 1 small funnel
- 60 mL of 20% salt solution
- 200 mL of distilled water
- 1 500-mL beaker

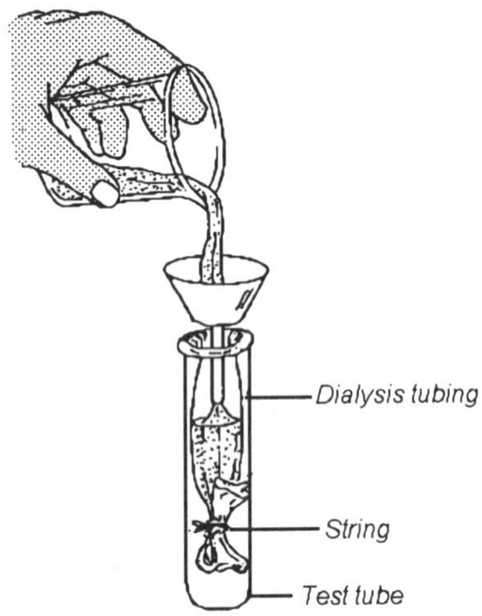

Making a Dialysis Tubing Cell

Unit 8 - Figure 1

3. Tie one end of the strip of dialysis tubing shut with string, and gently place the tubing in a test tube. Then stand the test tube in a test-tube rack. Using the funnel, fill the dialysis tubing about three-fourths full with 20% salt solution. Then securely tie the dialysis tubing shut with the other piece of string, forming a sac. Try not to allow any air bubbles in your dialysis tubing sac.

   Pour about 200 mL of distilled water into a beaker, and submerge your dialysis tubing sac in the distilled water. What kind of a solution is inside the dialysis tubing with respect to the solution outside the tubing?

   ____

4. Observe your dialysis tubing sac for at least 10 minutes. What happens to the sac?

   ____

   ____

5. Offer an explanation for your observations.

   ____

   ____

Unit 8
Performance-Based Assessment Activity
The Challenge of Retaining Water
Part II–Experiment A

Objectives

- Formulate a hypothesis about the effects of evaporation on water balance in amphibians.
- Design and conduct an experiment to test your hypothesis.
- Organize the data from your experiment in a table, and make a graph of it.
- Evaluate your hypothesis based on your data.

Additional Background

Amphibians spend time both in water and on land. Unlike reptiles, which have watertight scales covering their skin, frogs have thin, smooth skin that is permeable to water. Frogs may spend hours in the sun on dry land; they may also spend a lot of time in water. Thus, frogs must guard against excessive water loss by means of evaporation when on land and osmosis when in water. A frog's tissue fluid is made up of about 99% water and 1% dissolved salt. Most frogs secrete a slippery mucus that coats their skin. This covering is what makes some amphibians feel slimy.

Problem

Consider the following **problem** based on the background information in Parts I and II.

Is the mucous covering secreted by some amphibians effective in preventing the loss of water through evaporation?

1. **Hypothesis** In the space below, write a hypothesis for the problem stated above.

____

____

2. **Materials** Go to the equipment and materials table, and examine the available materials. Think about how you would design an experiment to test your hypothesis using these materials. Do not remove any materials from the table yet.

   a. **Revised Hypothesis** Since there are no living animals at the materials table, rewrite your hypothesis in a form that takes into account the available materials.

   ____

   ____

   b. Make a list of the materials that you will use in your experiment.

3. Procedure

a. Describe how you will test your hypothesis. Be sure to include a control in your experimental design.

____

____

____

____

____

b. List the observations or measurements you will make during the experiment.

____

____

c. Be sure to observe the **Safety Precautions** listed in Part I during your experiment.

d. To conduct your experiment, gather *only* the materials and equipment you need from the materials table. Pour into separate glassware the amount of each solution you need for your experiment, and transfer the solution or solutions to your work station. Do not take an entire quantity of any solution. Others may also need access to the same solutions. When you have gathered the needed materials, set up and conduct your experiment as planned. **Note** It usually takes at least 20 minutes to observe measurable results in experiments of this type.

Data

4. Records

a. Make a table in the space below. Make sure your table has descriptive heads. Record your data in your table.

b. Make a graph of the data you collected from your test and control setups. Label each axis, and title the completed graph.

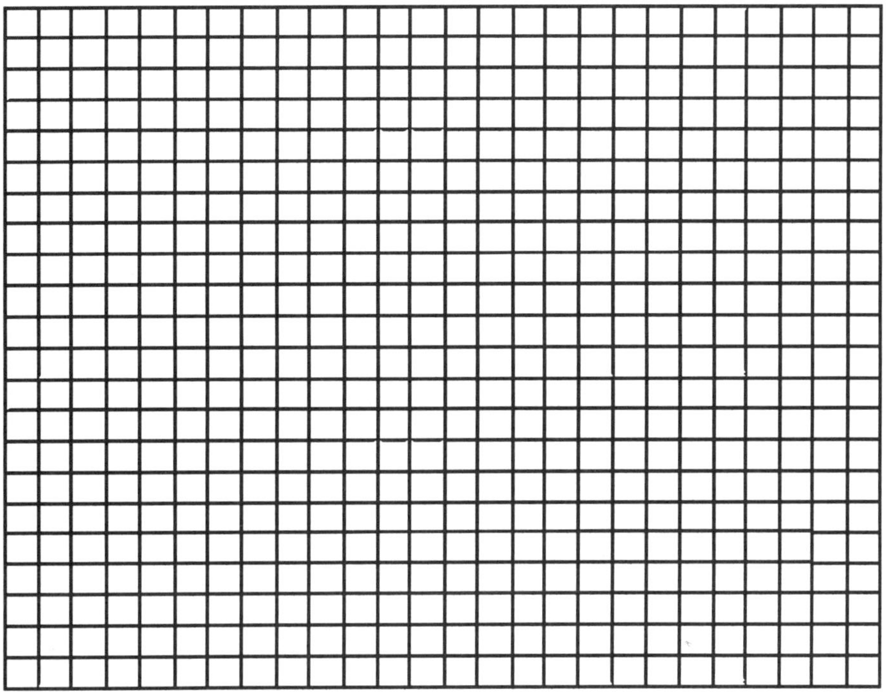

Unit 8 - Figure 2

5. Interpretation Write your interpretation of the data you recorded.

____

____

6. Conclusion State whether your data support or refute your hypothesis.

____

____

7. Analysis List any factors that could have affected the accuracy of the data you obtained during your experiment.

____

____

____

Cleanup

After you turn in your paper, dispose of your materials according to the instructions provided by your teacher. Clean up your work area, and wash your hands before leaving the lab.

Unit 8
Performance-Based Assessment Activity
The Challenge of Retaining Water
Part II–Experiment B

Objectives

- Formulate a hypothesis about the effects of osmosis on water balance in marine fishes.
- Design and conduct an experiment to test your hypothesis.
- Organize the data from your experiment in a table, and make a graph of it.
- Evaluate your hypothesis based on your data.

Additional Background

Although fishes are surrounded by water, they must maintain the appropriate water balance within their tissues. For example, many fishes living in fresh water, which is almost 100% water, have tissues composed of 99% water and 1% dissolved salts. The difference in salinity between the two water environments results in the diffusion of water into freshwater fishes, primarily through the gills. Fishes living in sea water, which is about 96% water and 4% salt, have a body tissue composition of about 99% water and 1% dissolved salts.

Problem

Consider the following **problem** based on the background information in Parts I and II.

Does water diffuse into or out of fishes living in sea water?

1. **Hypothesis** In the space below, write a hypothesis for the problem stated above.

____

____

2. **Materials** Go to the equipment and materials table, and examine the available materials. Think about how you would design an experiment to test your hypothesis using these materials. Do not remove any materials from the table yet.

   a. **Revised Hypothesis** Since there are no living animals at the materials table, rewrite your hypothesis in a form that takes into account the available materials.

   ____

   ____

   b. Make a list of the materials that you will use in your experiment.

3. **Procedure**

a. Describe how you will test your hypothesis. Be sure to include a control in your experimental design.

____

____

____

____

____

b. List the observations or measurements you will make during the experiment.

____

____

c. Be sure to observe the **Safety Precautions** listed in Part I during your experiment.

d. To conduct your experiment, gather *only* the materials and equipment you need from the materials table. Pour into separate glassware the amount of each solution you need for your experiment, and transfer the solution or solutions to your work station. Do not take an entire quantity of any solution. Others may also need access to the same solutions. When you have gathered the needed materials, set up and conduct your experiment as planned. **Note:** It usually takes at least 20 minutes to observe measurable results in experiments of this type.

Data

4. **Records**

a. Make a table in the space below. Make sure your table has descriptive heads. Record your data in your table.

b. Make a graph of the data you collected from your test and control setups. Label each axis, and title the completed graph.

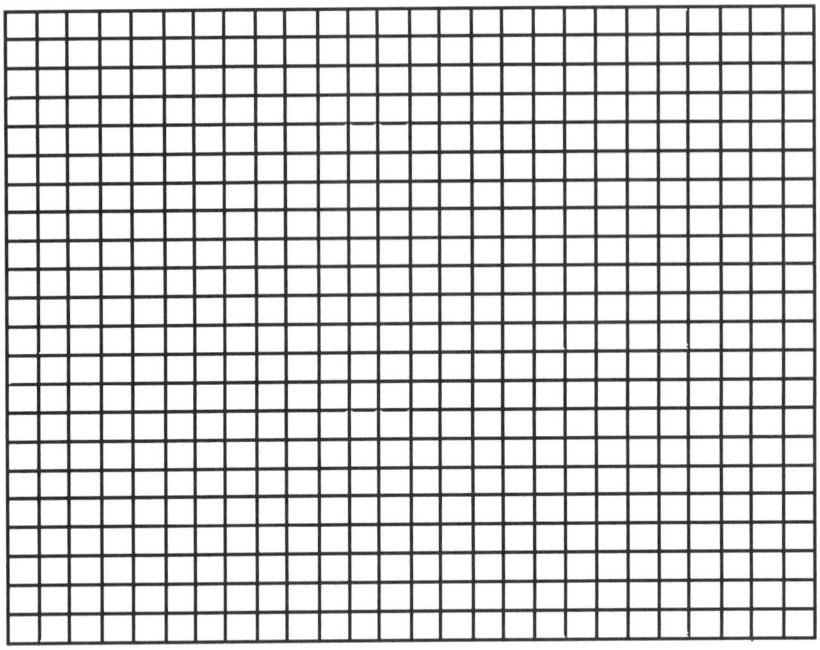

Unit 8 - Figure 4

5. **Interpretation** Write your interpretation of the data you recorded.

____

____

6. **Conclusion** State whether your data support or refute your hypothesis.

____

7. **Analysis** List any factors that could have affected the accuracy of the data you obtained during your experiment.

____

____

____

Cleanup

After you turn in your paper, dispose of your materials according to the instructions provided by your teacher. Clean up your work area, and wash your hands before leaving the lab.

Unit 8
Performance-Based Assessment Activity
The Challenge of Retaining Water
Teacher Notes and Scoring Rubric

Assessment Time Required

One 50-minute class period

Concepts and Vocabulary

Before beginning this assessment, students should be familiar with the following concepts:

- dehydration
- diffusion
- dialysis tubing
- edema
- evaporation
- osmosis
- selectively permeable membrane

Introduction

The assessment activity for Unit 8 consists of two parts. Part I is an introduction to the materials and concepts that students will be working with in Part II. Part II of the assessment activity consists of two separate experiments in which students are assigned the task of simulating environmental challenge to maintaining water balance in either an amphibian or a fish.

Strategies

All students should complete Part I. For Part II, assign each student one of the two experiments (Experiment A or B). Part II requires making sacs out of dialysis tubing and observing them over a 20-minute period of time. To allow students time to complete their experiments, you may wish to allot an entire class period to Part II. If Part I is completed the day before Part II is to be done, return students' papers for use during Part II of the assessment.

It is recommended that this assessment be used as an individual assessment. However, this activity also works well for teams of two or three students. Teams may be able to complete the assessment activity more quickly than students working independently. You may require students to submit one paper per team, or you may require them to submit individual papers.

Before allowing students to begin the assessment activity, make sure that they are aware of the appropriate safety precautions. Remind students that this activity is a test situation and that the standard rules of test-taking apply. Stress that talking among team members must be done very quietly and must be relevant to the assessment activity. When all students have completed an assessment, conduct a class discussion of the students' different approaches to the Performance-Based Assessment Activity.

Student Safety Precautions

The safety precautions for this assessment activity are listed in Part I. Before allowing students to begin the activity, make sure they have read these safety precautions. Advise students that they will be given points for properly adhering to the safety precautions during this activity.

Materials

You will need the following materials to prepare the solutions needed for this assessment activity:

- table salt (About 150 g is enough to prepare 1 L of 1% salt solution, 1 L of 4% salt solution, and 500 mL of 20% salt solution.)
- distilled water (2.5 L is enough to prepare 1 L of 1% salt solution, 1 L of 4% salt solution, and 500 mL of 20% salt solution.)

For each student or team, you will need the following materials in the quantities indicated:

- 1 test tube (2 or 3 cm in diameter or large enough to fit a dialysis tubing sac)
- 1 test-tube rack
- 3 15-cm strips of dialysis tubing (35–43 mm flat width)
- 6 20-cm pieces of string or thread
- 1 small funnel
- 100 mL of distilled water
- 2 400–500-mL beakers (Note that only one-half of your students will need to submerge their dialysis tubing sacs in Part II.)

- 2 100-mL flasks or beakers
- 1–2 mL of petroleum jelly
- masking or labeling tape
- 25–50 mL of 20% salt solution
- 200 mL of distilled water
- 100–150 mL 1% salt solution
- 25–50 mL 4% salt solution (Note that only one-half of your students will use this solution.)
- 1 balance, accurate to the nearest 0.1 g (If you do not have a balance, provide students with small graduated cylinders.)

Preparation

1. To prepare 1 L of 1% salt solution, dissolve 10 g of regular table salt in about 500 mL of distilled water. Dilute the solution to 1 L. Label this solution "99% Water/1% Salt Solution."
2. To prepare 1 L of 4% salt solution, dissolve 40 g of regular table salt in about 500 mL of distilled water. Dilute the solution to 1 L. Label this solution "96% Water/4% Salt Solution."
3. To prepare 500 mL of 20% salt solution, dissolve 100 g of regular table salt in about 300 mL of distilled water. Dilute the solution to 500 mL. Label this solution "20% Salt Solution."
4. Put regular petroleum jelly in small containers. Label each container "Amphibian Mucous Secretion."

Laboratory Setup

Place the following materials on a centrally located supply table:

(Parts I and II)
- test tubes
- test-tube racks
- strips of dialysis tubing
- string or thread
- small funnels
- distilled water
- 20% salt solution
- 400–500 mL beakers
- 100 mL flasks or beakers

(Part II only)
- balances or graduated cylinders
- 1% salt solution
- 4% salt solution
- petroleum jelly, labeled "Amphibian Mucous Secretion"
- masking or labeling tape

Cleanup and Disposal

Instruct students to wash and dry the glassware they use and return it to the materials table before they leave the lab.

Scoring Rubric

The following pages are a Scoring Rubric that provides possible answers to this Performance-Based Assessment Activity. Preferred student responses are shown in italics. Suggested point values for each step in the activity appear in parentheses next to the responses. The total points earned by a student for each step in the activity can be recorded in the blanks to the right of the steps. The Scoring Rubric can be modified to accommodate your specific circumstances. You may wish to use this rubric only as a guide in scoring the assessment activity. If possible, make a copy of the rubric for each student or team, and return a copy of the rubric showing the scores earned by the students when you return their graded papers.

Activity and Scoring Rubric Art

These teacher notes are found in question 4 of the Unit 8 file. In the answer key for this question, you will find all of the art for Unit 8. Cut and paste these figures in the places indicated before photocopying the assessment for your students.

Unit 8
Performance-Based Assessment Activity
The Challenge of Water Retention (50 points)
Scoring Rubric

_______________ Total Score

Possible answers to this Performance-Based Assessment Activity are provided below. Preferred student responses are shown in italics. Suggested point values for each step in the activity appear in parentheses next to the responses. The total points earned for each step in the activity are recorded in the blanks to the right of the steps.

Part I (9 points total)

_______________ Part I Score

Background

1. *Dehydration and edema are serious conditions that can result from water imbalance.* (2 points) _______________

2. *Answers will vary. Students may suggest that fishes have kidneys or that reptiles have watertight skin.* (2 points) _______________

Practice

3. *The solution inside the tubing is hypertonic with respect to the solution outside the tubing.* (2 points) _______________

4. *The sac becomes larger.* (1 point) _______________

5. *Water moves from an area of lower salt concentration (the water in the beaker) to an area of greater salt concentration (the salt solution inside the sac).* (2 points) _______________

Part II-Experiment A (41 points total)

Part II Score

Experiment (22 points total)

1. **Hypothesis**
The mucous covering secreted by some amphibians is effective at preventing water loss through evaporation. (1 point)
or
The mucous covering secreted by some amphibians is not effective at preventing water loss through evaporation. (1 point)

2. **Materials** (4 points total)
   a. **Revised Hypothesis**
   Less water evaporates from a sac that is made of dialysis tubing and that is coated with amphibian mucous secretion than from a similar sac that is not coated with mucous secretion. (2 points)
   or
   Amphibian mucous secretion reduces the evaporation of water from a dialysis tubing sac. (2 points)
   or
   Amphibian mucous secretion does not reduce the evaporation of water from a dialysis tubing sac. (2 points)
   or
   Any other plausible and relevant hypothesis (2 points)
   b. Materials should include only those that are found on the supply table and that are needed for Experiment A. (2 points)

3. **Procedure** (17 points total)
   a. *Using dialysis tubing, prepare two sacs of similar size. Fill both of the sacs with "99% Water/1% Salt Solution."* (2 points)

   Coat the outside of the test sac with "Amphibian Mucous Secretion." (2 points)

   Do not coat the outside of the control sac with "Amphibian Mucous Secretion." (2 points)

   Expose both setups to air for about 20 minutes. (2 points)
   b. *Weigh the test sac at the beginning of the experiment. (Or measure the volume of the liquid that is poured into the sac.)* (1 point)

   Weigh the control sac at the beginning of the experiment. (Or measure the volume of the liquid that is poured into the sac.) (1 point)

   Weigh the test sac at the end of the experiment. (Or measure the volume of the liquid that is left in the sac.) (1 point)

   Weigh the control sac at the end of the experiment. (Or measure the volume of the liquid that is left in the sac.) (1 point)

   Observe any changes in the control and test sacs for about 20 minutes. (1 point)

   c. Use of appropriate safety precautions (2 points)
   d. Appropriate use of materials (2 points)

Data (19 points total)

4. Records (12 points total)

a. (6 points: 1 point for each data box; 2 points for appropriate heads) ________________

Effects of Amphibian Mucous Secretion on Water Evaporation

Time of Measurement	*Weight of Sac Coated With Amphibian Mucous Secretion*	*Weight of Sac Without Amphibian Mucous Secretion (Control)*
Beginning of experiment	*19.4 g**	*19.6 g**
End of experiment	*19.1 g**	*17.4 g**

*Weight of the test sac and volume of the liquid inside the test sac should decrease. Data will vary. Accept any reasonable data.

b. A sample graph is shown. Accept any graph that reflects the data recorded in the data table in **4a**. (6 total points: 1 point for each curve, 3 points for appropriate labels and numbers along the axes, and 1 point for the title) ________________

Effects of Amphibian Mucous Solution on Water Evaporation

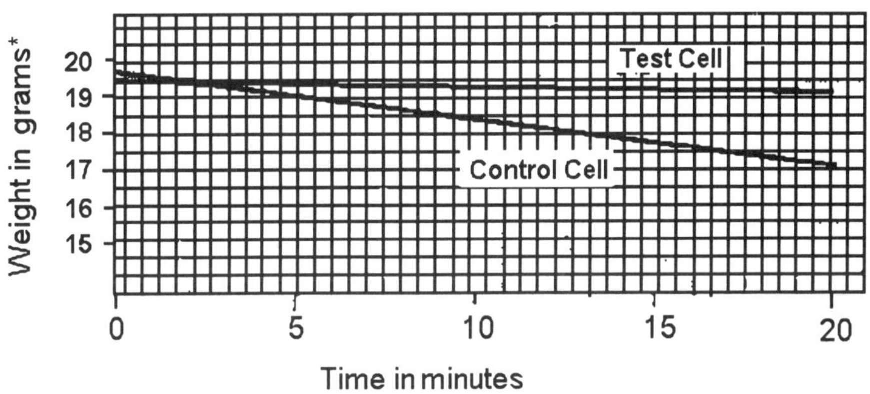

* Volume in milliliters is also acceptable

Unit 8 - Figure 3

5. **Interpretation**
Over a 20-minute period, less water evaporated from the dialysis tubing sac that was coated with amphibian mucous secretion than from the sac without the mucous secretion. (3 points) ______
or
Other interpretations that are consistent with the data (3 points) ______

6. **Conclusion**
The data support the hypothesis that less water evaporates from a sac that is made of dialysis tubing and that is coated with amphibian mucous secretion than from a similar sac that is not coated with mucous secretion. (2 points) ______
or
The data support the hypothesis that amphibian mucous secretion reduces the evaporation of water from a dialysis tubing sac. (2 points) ______
or
The data contradict the hypothesis that amphibian mucous secretion does not reduce the evaporation of water from a dialysis tubing sac. (2 points) ______
or
Any relevant conclusion that is consistent with the data and interpretation (2 points) ______

7. **Analysis**
Acceptable factors include differences in the construction of the dialysis tubing sacs, such as air bubbles or tears. (2 points) ______

Part II–Experiment B (41 points total)

Part II Score

Experiment (23 points total)

1. **Hypothesis**

In a saltwater environment, water will diffuse into fishes. (1 point)

or

In a salt water environment, water will diffuse out of fishes. (1 point)

or

Water diffuses neither into nor out of fishes living in saltwater. (1 point)

2. **Materials** (4 points total)

a. **Revised Hypothesis**

When placed in a solution that is similar to sea water, water will diffuse into a sac that is made of dialysis tubing and that contains 1% salt solution. (2 points)

or

When placed in a solution that is similar to sea water, water will diffuse out of a sac that is made of dialysis tubing and that contains 1% salt solution. (2 points)

or

When placed in a solution that is similar to sea water, water will diffuse neither into nor out of a sac that is made of dialysis tubing and that contains 1% salt solution. (2 points)

or

Any other plausible and relevant hypothesis (2 points)

b. Materials should include only those that are found on the supply table and that are needed for Experiment B. (2 points)

3. **Procedure** (18 points total)

a. *Using dialysis tubing, prepare two sacs of similar size. Both sacs should contain 1% salt solution.* (2 point)

Test setup: Pour "99% Water/1% Salt Solution" into the test sac. Submerge the sac in about 100–200 mL of "96% Water/4% Salt Solution" for about 20 minutes. (3 points)

Control setup: Pour "99% Water/1% Salt Solution" into the test sac. Submerge the sac in about 100–200 mL of "99% Water/1% Salt Solution" for about 20 minutes. (3 points)

or

Control setup: Pour "96% Water/4% Salt Solution" into the test sac. Submerge the sac in about 100–200 mL of "96% Water/4% Salt Solution" for about 20 minutes. (1 point)

b. *Weigh the test sac at the beginning of the experiment. (Or measure the volume of the liquid that is poured into the sac.)* (1 point)

Weigh the control sac at the beginning of the experiment. (Or measure the volume of the liquid that is poured into the sac.) (1 point)

Weigh the test sac at the end of the experiment. (Or measure the volume of the liquid that is left in the sac.) (1 point)

Weigh the control sac at the end of the experiment. (Or measure the volume of the liquid that is left in the sac.) (1 point)

Observe any changes in the control or test sacs for about 20 minutes. (1 point)

c. Use of appropriate safety precautions (2 points) ______

d. Appropriate use of materials (2 points) ______

Data (18 points total)

4. Records (12 points total)

a. (6 points: 1 point for each data box; 2 points for appropriate heads) ______

Effects of 4% Salt Solution on a Dialysis Tubing Sac Containing 1% Salt

Time of Measurement	*Weight of Sac Submerged in a 4% Salt Solution*	*Weight of Sac Submerged in a 1% Salt Solution (Control)*
Beginning of experiment	*20.4 g**	*20.8 g**
End of experiment	*16.2 g**	*20.9 g**

*Weight of the test sac and volume of the liquid inside the test sac should decrease. Data will vary. Accept any reasonable data.

b. A sample graph is shown. Accept any graph that reflects the data recorded in the data table in **4a**. (6 total points: 1 point for each curve, 3 points for appropriate labels and numbers along the axes, and 1 point for the title) ______

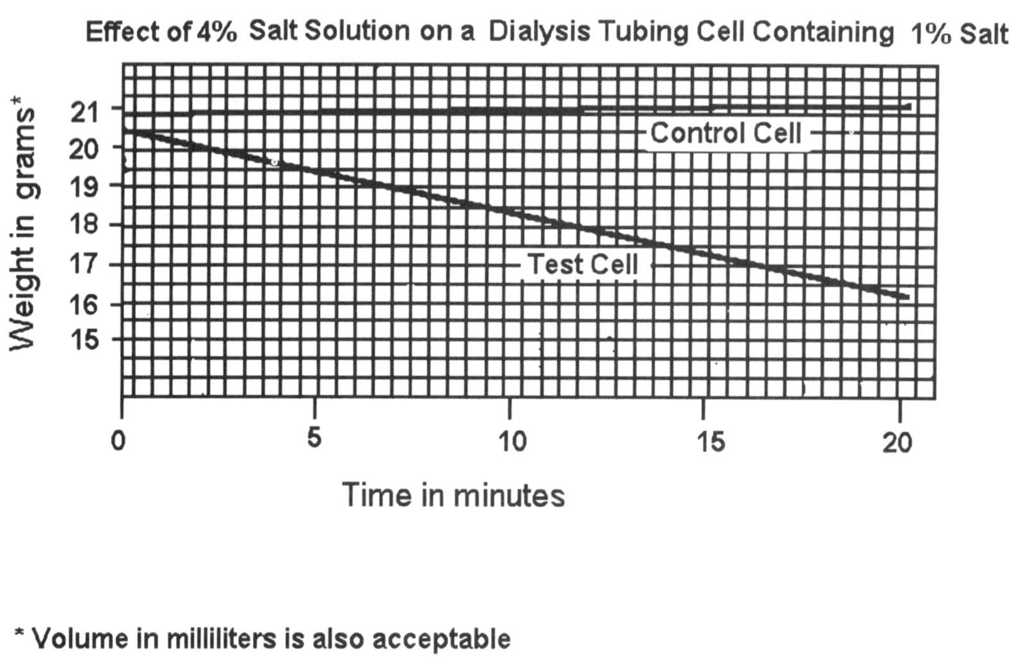

Unit 8 - Figure 5

5. Interpretation

Over a 20-minute period, more water diffused out of the dialysis tubing sac that was placed in 4% salt solution than diffused out of the control sac.
(2 points) ______
or
Other interpretations that are consistent with the data (2 points) ______

6. Conclusion

The data contradict the hypothesis that water will diffuse into a sac that is made of dialysis tubing and that contains 1% salt solution when it is placed in a 4% salt solution. (2 points) __________

or

The data support the hypothesis that water will diffuse out of a sac that is made of dialysis tubing and that contains 1% salt solution when it is placed in a 4% salt solution. (2 points) __________

or

The data contradict the hypothesis that water will diffuse neither into nor out of a sac that is made of dialysis tubing and that contains 1% salt solution when it is placed in a 4% salt solution. (2 points) __________

or

Any relevant conclusion that is consistent with the data and interpretation (2 points) __________

7. Analysis

Acceptable factors include differences in the construction of the dialysis tubing sacs, such as air bubbles or tears. (2 points) __________

Unit 9
Performance-Based Assessment Activity
Testing Human Digestive Enzymes
Part I–Background

Follow the guidelines below during this assessment activity.

- Remember not to talk to others also taking the test. If you are working with a team, talk only with those in your group and talk very quietly so that you do not disturb other teams.
- When you conduct your experiment, you will be able to use only the materials and equipment on the supply table. Some of the materials on the table will not be needed.
- Most of the other students involved in today's activity have different problems to solve. If you observe other students using materials different from those that you are using, this may be because their problems are not the same as yours.
- Remember that a problem often has several different acceptable solutions.

Background

A variety of different enzymes are involved in human digestion. One set of enzymes catalyzes the breakdown of protein. Other enzymes catalyze the breakdown of carbohydrates or fat molecules. Digestive enzymes are like other enzymes in that they catalyze specific chemical reactions. For example, the digestive enzyme amylase, which is found in saliva, catalyzes the breakdown of starch to glucose, but it cannot catalyze the breakdown of fat molecules.

Because of their shapes, enzymes bind to specific substrates, or reactants. In the chemical reaction in which amylase catalyzes the breakdown of starch into glucose, starch is the substrate. At an optimum temperature and pH, the shape of amylase fits the shape of starch, starch binds to amylase, and the chemical reaction that converts starch to glucose is accelerated.

In this assessment you will be investigating some properties of the human digestive enzyme pepsin or enzymes produced by the pancreas. One or both of these enzyme mixtures catalyze the breakdown of fat molecules into three fatty acid molecules and one glycerol molecule. As with other enzymes, the effectiveness of pepsin and pancreatic enzymes is determined by the enzyme's shape.

1. Write an equation to represent the chemical reaction that is possibly catalyzed by pepsin.

____

2. List two factors that can change the shape of an enzyme and thus affect its activity.

____

Safety Precautions

The following safety precautions should be observed during Parts I and II of this activity.

- Wear safety goggles, protective gloves, and a lab apron. Use tongs or appropriate insulated holders when heating objects.
- Confine loose hair and clothing.
- Do not touch or taste any chemicals. Avoid inhaling chemical fumes. Wash off any solutions that get on skin or clothing at the sink while calling to the teacher. If you get a solution in your eyes, immediately flush it out at the eyewash station while calling to the teacher.
- Be careful not to spill any solutions. Call the teacher in the event of an acid or base spill.
- Dispose of any broken glass in a separate container. Never put broken glass in a regular waste container.

- Be sure electrical equipment is turned off before plugging it in. Turn it off again before unplugging it. Do not allow electrical cords to dangle from work stations. The area under and around electrical equipment should be dry, and your hands should be dry when using electrical equipment.
- When heating chemicals in a test tube, never point the test tube toward anyone. Never pick up an object with your hand unless you are certain that it is not hot. Heated objects often do not look hot.
- Never leave a heated hot plate unattended, even for a minute.
- Clean up the lab and all equipment after use. Dispose of solutions according to proper disposal methods. Turn off and unplug all electrical equipment before leaving the laboratory. After all lab work has been completed, wash hands thoroughly before leaving the lab.

Practice

Phenol red is a pH indicator. In solutions having a pH of 6.8 or lower, phenol red is yellow or tan. In solutions having a pH of 8.4 or higher, phenol red is red or pink. In solutions having a pH between 6.8 and 8.3, phenol red is peach.

3. Cream, milk, and canned milk are sources of liquid fat. Pour 3–5 mL of canned milk into a test tube. **Caution:** *Avoid getting phenol red on skin or clothes.* Then add 2 drops of phenol red indicator to the test tube. What is the resulting color?

____

4. Does the canned milk sample contain acid?

____

5. **Caution:** *Be careful not to get acid on your skin or clothes.* Add the solution labeled "Simulated Fatty Acids" one drop at a time to the mixture of canned milk and phenol red. Stir the mixture between drops, and observe the color with the addition of each drop. Continue to add drops of the acid until no additional color changes occur. Describe your observations below.

____

____

6. What does the color change indicate?

____

7. Discard the milk mixture in the sink, and clean the test tube with soap and water.

Unit 9
Performance-Based Assessment Activity
Testing Human Digestive Enzymes
Part II–Experiment A

Objectives

- Formulate a hypothesis about the activity of pepsin and pancreatic enzymes.
- Design and conduct an experiment to test your hypothesis.
- Organize the data from your experiment in a table.
- Evaluate your hypothesis based on your data.

Problem

Consider the following **problem** based on the background information in Part I.

> **Do pepsin and pancreatic enzymes both convert fat to fatty acids at human body temperature (37°C)?**

Experiment

1. **Hypothesis** In the space below, write a hypothesis for the **problem** that is stated above.

   ____

   ____

2. **Materials** Go to the equipment and materials table, and examine the available materials. Think about how you would design a control experiment to test your hypothesis using these materials. Do not remove any materials from the table yet. Make a list of the materials that you will use in your experiment. **Note** The reaction time of enzyme reactions in which there is a 1:1 enzyme-to-substrate ratio is commonly 7–15 minutes at 37°C.

3. Procedure

a. Draw a sketch of how you will set up an experiment that will test your hypothesis, and describe the procedure you will follow. Make sure to include a control in your experimental design. Label all parts of your experimental and control setups.

b. List the observations or measurements you will make during the experiment.

____

____

____

c. Be sure to observe the **Safety Precautions** listed in Part I.

d. To conduct your experiment, gather *only* the materials and equipment you need from the materials table. For transportation to your work station, transfer the amount of solution(s) you need in separate glassware. Do not take an entire quantity of any solution. Others may also need access to the solutions that you will be using in your experiment. When you have gathered the needed materials, set up and conduct your experiment as planned.

Data

4. **Records** Make a table in the space below. Make sure your table has descriptive heads. Record your data in your table.

5. **Interpretation** Write your interpretation of the data you recorded.

6. **Conclusion** State whether your data support or refute your hypothesis.

7. **Analysis**
   a. List any factors that could have affected the accuracy of the data you obtained during your experiment.

   b. Based on the results of your experiment, suggest a new problem that could be pursued as an extension of this investigation.

Cleanup

After you turn in your paper, dispose of your material according to the instructions provided by your teacher. Clean up your work area, and wash your hands before leaving the lab.

Unit 9
Performance-Based Assessment Activity
Testing Human Digestive Enzymes
Part II–Experiment B

Objectives

- Formulate a hypothesis about the effects of increasing temperature on pancreatic enzyme activity.
- Design and conduct an experiment to test your hypothesis.
- Organize the data from your experiment in a table.
- Evaluate your hypothesis based on your data.

Problem

Consider the following **problem** based on the background information in Part I.

Does the rate of pancreatic enzyme catalysis increase with temperature?

Experiment

1. **Hypothesis** In the space below, write a hypothesis for the **problem** that is stated above.

____

____

2. **Materials** Go to the equipment and materials table, and examine the available materials. Think about how you would design a control experiment to test your hypothesis using these materials. Do not remove any materials from the table yet. Make a list of the materials that you will use in your experiment. **Note:** It is recommended that you test 1:1 enzyme-to-substrate ratios at the following temperatures: 18–22°C, 35–40°C, and 80–90°C.

3. Procedure

a. Draw a sketch of how you will set up an experiment that will test your hypothesis, and describe the procedure you will follow. Make sure to include appropriate controls in your experimental design. Label all parts of your experimental and control setups.

b. List the observations or measurements you will make during the experiment.

____

____

____

____

c. Be sure to observe the **Safety Precautions** listed in Part I.

d. To conduct your experiment, gather *only* the materials and equipment you need from the materials table. Pour into separate glassware the amount of each solution you need for your experiment, and transfer the solution or solutions to your work station. Do not take an entire quantity of any solution. Others may also need access to the solutions that you will be using in your experiment. When you have gathered the needed materials, set up and conduct your experiment as planned.

Data

4. **Records** Make a table in the space below. Make sure your table has descriptive heads. Record your data in your table.

5. **Interpretation** Write your interpretation of the data you recorded.

6. **Conclusion** State whether your data support or refute your hypothesis.

7. **Analysis** Consider the results of your experiment. Discuss why digestive enzymes might be more effective at certain temperatures than at others.

Cleanup

After you turn in your paper, dispose of your material according to the instructions provided by your teacher. Clean up your work area, and wash your hands before leaving the lab.

Unit 9
Performance-Based Assessment Activity
Testing Human Digestive Enzymes
Teacher Notes and Scoring Rubric

Assessment Time Required
One or two 50-minute class periods

Concepts and Vocabulary
Before beginning this assessment, students should be familiar with the following concepts and vocabulary:
- catalyst
- digestion
- enzyme
- enzyme activity
- fatty acid
- glycerol
- pancreatic enzyme
- pepsin
- pH indicator
- substrate

Introduction
The assessment activity for Unit 9 consists of two parts. Part I is an introduction to the enzymes that students will be working with in Part II. Part II of the assessment activity consists of two separate experiments in which students are assigned the task of testing digestive enzyme activity at certain temperatures.

Strategies
All students should complete Part I. For Part II, assign each student one of the two experiments (Experiment A or B). Part II requires heating solutions to specific temperatures and observing them over a 15-minute period of time. To allow students time to complete their experiments, you may wish to allot an entire class period to Part II. If Part I is completed the day before Part II is to be done, return students' papers for use during Part II of the assessment.

It is recommended that this assessment be used as an individual assessment. However, this activity also works well for teams of two, three, or four students. You may require students to submit one paper per team, or you may require them to submit individual papers.

Before allowing students to begin the assessment activity, make sure that they are aware of the appropriate safety precautions. Remind students that this activity is a test situation and that the standard rules of test-taking apply. Stress that talking among team members must be done very quietly and must be relevant to the assessment activity. When all students have completed an assessment, conduct a class discussion of the students' different approaches to the Performance-Based Assessment Activity.

Student Safety Precautions
The safety precautions for this assessment activity are listed in Part I. Before allowing students to begin the activity, make sure they have read these safety precautions. Advise students that they will be given points for properly following the safety precautions during this activity.

Be sure that hot plates and water baths have an "on-off" switch and indicator light. Students should not be allowed to leave operating hot plates unattended.

Use only electrical equipment equipped with a three-wire cord and a three-prong plug. Each electrical socket in the laboratory must be of the three-hole type with a GFI (ground-fault interrupter) circuit. Be sure electrical equipment is turned off before plugging it in. Turn it off again before unplugging it. Do not allow electrical cords to dangle from work stations; dangling cords are a tripping and shock hazard. Do not use electrical equipment with frayed or kinked cords. The area under and around electrical equipment should be dry. Cords should not lie in puddles of spilled liquid. Students' hands should be dry when using electrical equipment. Turn off and unplug all electrical equipment before leaving the laboratory.

Materials
You will need the following materials to prepare the solutions needed for this assessment activity:
- phenol red powder (0.2 g is enough to make 500 mL of indicator.)
- canned milk (200 mL will make 1 L of canned milk solution.)
- 0.1 M NaOH (You may require 10 mL or more to adjust the pH of the canned milk solution.)
- pancreatin (2.4 g will make 300 mL of pancreatic enzyme solution.)

- pepsin (1 g will make 300 mL of pepsin enzyme solution; note that pepsin is optional because distilled or tap water with a pH of 8 or more will produce the same results in Experiments A and B. See **Preparation Note 3**.)
- HCl, diluted (The concentration is not critical. Students can use 1%, 0.1 M, or any dilute HCl solution that you may have on hand.)
- distilled water (You will need 800 mL of distilled water for each liter of canned milk solution, about 500 mL for each 500 mL of phenol red indicator, and about 300 mL for each 300 mL of pancreatic enzyme solution that you plan to prepare. Additional amounts of distilled water may also be needed for the dilute HCl solution and the pepsin enzyme solution.)
- 1 L and 500 mL beakers
- 5–10 mL graduated cylinder
- balance
- pH paper or meter

For each student or team, you will need the following materials in the quantities indicated:
- 3–12 10-mL test tubes (Note: Experiment A requires only three test tubes; Experiment B calls for as many as 12 test tubes.)
- 2 50-mL beakers (for transporting water and other solutions)
- 2 500-mL or larger beakers (to serve as water baths)
- 3–6 glass stirring rods (or straws)
- 2 thermometers
- 1 10-mL or 25-mL graduated cylinder
- 1 roll of masking or labeling tape
- 1 or 2 test-tube holders
- 1 test-tube rack
- 1 dropper bottle of 0.04% phenol red indicator
- 1 dropper bottle of dilute HCl
- 5–15 mL of distilled water (Be sure the pH is high enough to produce the intended results in students' controls.)
- 20–35 mL of canned milk solution
- 5–20 mL of pancreatic enzyme solution
- 5 mL of pepsin enzyme solution (Note that only one-half of your students will use this solution in Part II.)
- 1 pair of insulated gloves or metal tongs (If students are working in teams, each student should have their own pair of gloves or tongs when possible.)
- 1 hot plate or hot-water bath (To reduce the number of hot plates and hot-water baths required, you may wish to direct students who are performing Experiment A to use tap water that is 37°C. However, test your tap water first to ensure that its pH is not acidic. If your tap water is acidic, students' controls will not produce the intended results.)

Preparation Notes

1. To prepare 500 mL of 0.04% phenol red indicator, mix 0.2 g of phenol red powder with 500 mL of distilled water. Stir well. Prepare one dropper bottle of phenol red indicator for each testing station. Label each dropper bottle "Phenol Red Indicator."
2. **Caution:** *When you prepare solutions containing acids or bases, wear safety goggles, a face shield, impermeable gloves, and a lab apron, and work in a ventilated hood. In case of an acid spill, dilute first with water. Then mop up the spill with wet cloths designated for spill cleanup.* The "Simulated Fatty Acids" solution is a dilute (approximately 1%) solution of HCl. The concentration is not critical. Do not use concentrated HCl.
3. To make the pepsin enzyme solution, dissolve 1 g of pepsin powder in 300 mL of distilled water. Place mixture in at least three flasks labeled "Pepsin Enzymes." **Note**: Pepsin converts proteins to proteoses and peptones; thus, it will not break down the fat in milk to fatty acids. Because students' tests for pepsin enzyme effectiveness at catalyzing the breakdown of fat into fatty acids will be negative, you can provide a distilled or tap water solution labeled "Pepsin Enzymes," instead of a true pepsin solution. The pH may have to be adjusted to 8 or greater to prevent positive results.
4. To prepare 300 mL of pancreatic enzyme solution, dissolve 2.4 g of pancreatin (a mixture of amylopsin, trypsin, and steapsin enzymes) in 300 mL of distilled water. Provide at least three flasks of the solution labeled "Pancreatic Enzymes."

5. To make 1 L of canned milk solution, mix 200 mL of canned milk and 800 mL of distilled water. Be sure to test the pH of the milk solution. Place about 5 mL of the milk solution, 5 mL of the pancreatic enzyme solution, and 1 mL of the pepsin enzyme solution in a test tube. Then add 2–3 drops of phenol red indicator. If phenol red becomes yellow or tan, the canned milk solution is acidic. **Caution:** *When you prepare solutions containing acids or bases, wear safety goggles, a face shield, impermeable gloves, and a lab apron, and work in a ventilated hood. In case of an acid spill, dilute first with water. Then mop up the spill with wet cloths designated for spill cleanup.* Adjust the pH of the milk solution by adding 0.1 M NaOH solution 1 mL at a time. Test the solution with phenol red indicator after each addition of NaOH. The final milk solution should produce a vivid pink color when tested with phenol red. Provide at least three flasks of milk solution labeled "Canned Milk." Keep the solution refrigerated when not in use.
6. Decoy materials, or unnecessary materials, can also be placed on the central supply table. This causes students to make informed choices about the materials they use in their experiments. The following materials are suggested decoys for this activity: solutions labeled "Starch Solution," "Protein Solution," "Amylase Enzyme Solution," and dropper bottles of Lugol's iodine solution.

Laboratory Setup
Place the following materials on a centrally located supply table.

(Parts I and II)
- 10-mL test tubes
- test-tube racks
- dropper bottles of phenol red indicator
- dropper bottles of "Simulated Fatty Acids"
- canned milk solution
- glass stirring rods
- pepsin enzyme solution
- pancreatic enzyme solution

(Part II only)
- 50-mL beakers
- 500-mL or larger beakers
- thermometers
- 10-mL or 25-mL graduated cylinders
- several rolls of masking or labeling tape
- test-tube holders
- distilled water
- insulated gloves or metal tongs
- hot plates or hot-water baths

Cleanup and Disposal
Before discarding leftover acidic solutions in the sink, collect them in a single container, and neutralize them to a pH of 7 with 0.1 M NaOH. For student use during the assessment activity, provide a container marked "Solution Disposal Container." Avoid having students handle acid spills themselves. Instruct students to wash and dry the glassware they use and return it to the materials table before they leave the lab.

Scoring Rubric
The following pages are a Scoring Rubric that provides possible answers to this Performance-Based Assessment Activity. Preferred student responses are shown in italics. Suggested point values for each step in the activity appear in parentheses next to the responses. The total points earned by a student for each step in the activity can be recorded in the blanks to the right of the steps. The Scoring Rubric can be modified to accommodate your specific circumstances. You may wish to use this rubric only as a guide in scoring the assessment activity. If possible, make a copy of the rubric for each student or team, and return a copy of the rubric showing the scores earned by the students when you return their graded papers.

Unit 9
Performance-Based Assessment Activity
Testing Human Digestive Enzymes (50 points)
Scoring Rubric

_______________ Total Score

Possible answers to this Performance-Based Assessment Activity are provided below. Preferred student responses are shown in italics. Suggested point values for each step in the activity appear in parentheses next to the responses. The total points earned for each step in the activity are recorded in the blanks to the right of the steps.

Part I (8 points total)

_______________ Part I Score

Background

1. $\textit{fat} \xrightarrow{\textit{pepsin}} \textit{3 fatty acids + 1 glycerol}$ (2 points) _______________

2. *Temperature and pH can affect the shape of an enzyme.* (2 points) _______________

Practice

3. *The phenol red turns pink.* (1 point) _______________

4. *The canned milk sample does not contain acid.* (1 point) _______________

5. *As drops of acid are added, the solution turns from pink to lighter pink and then to peach, light tan, or cream.* (1 point) _______________

6. *The color change indicates the presence of acid (or fatty acids).* (1 point) _______________

Part II–Experiment A (42 points total)

Part II Score

Experiment (24 points total)

1. **Hypothesis**
   Both enzymes convert fat to fatty acids at 37° C. (2 points)
   or
   Only pepsin converts fat to fatty acids at 37° C. (2 points)
   or
   Only pancreatic enzymes convert fat to fatty acids at 37° C. (2 points)
   or
   Neither pepsin nor pancreatic enzymes convert fat to fatty acids at 37° C. (2 points)

2. **Materials**
   Materials should include only those that are found on the supply table and that are needed for Experiment A. (1 point)

3. **Procedure** (21 points total)
   a. *First, all ingredients are separately heated to 37° C in a hot-water bath. Then they are mixed together. Otherwise, the reaction may take place before the solutions reach 37° C.* (2 points)

      Test setup: Two test setups are needed, one for each enzyme solution. Each test tube should contain equal parts of enzyme solution and canned milk solution, as well as a few drops of phenol red indicator. One test setup should contain pepsin enzyme. The other test setup should contain pancreatic enzymes. (7 points)

      Control setup: The control setup should contain equal parts of milk and water, as well as a few drops of phenol red indicator. (4 points)

      All three test tubes contain only enough drops of phenol red to turn the solution pink. (2 points)

   b. *Measure the temperature of the solutions before combining them. They should be at 37° C.* (1 point)

      Observe the color of phenol red in each test tube over a period of at least 7–15 minutes. (2 points)

   c. Appropriate use of safety precautions (2 points)

   d. Appropriate use of materials (1 point)

Data (18 points total)

4. Records (10 points total)

*Reaction Time of Pepsin and Pancreatic Enzymes at 37° C**

	Test setup		*Control setup*
Time in minutes	***Color of phenol red in presence of pepsin***	***Color of phenol red in presence of pancreatic enzymes***	***Color of phenol red in the absence of enzymes***
0	*pink*	*pink*	*pink*
1	*pink*	*pink*	*pink*
2	*pink*	*pink*	*pink*
3	*pink*	*pink*	*pink*
4	*pink*	*light pink*	*pink*
5	*pink*	*light pink*	*pink*
6	*pink*	*peach*	*pink*
7	*pink*	*peach*	*pink*
8	*pink*	*peach*	*pink*
9	*pink*	*light peach*	*pink*
10	*pink*	*light peach*	*pink*
11	*pink*	*light peach*	*pink*
12	*pink*	*light peach*	*pink*
13	*pink*	*light peach*	*pink*
14	*pink*	*yellow*	*pink*
15	*pink*	*yellow*	*pink*

*Phenol red should change color only in the test tube containing pancreatic enzymes. Data may vary slightly.

Breakdown of scoring for the table above is as follows:
Appropriate column heads (5 points) __________

Recording color changes (2 points) __________

Data indicate that phenol red remains pink in the presence of pepsin. (1 point) __________

Data indicate that phenol red becomes peach or yellow in the presence of pancreatic enzymes. (1 point) __________

Data indicate that phenol red remains pink in the control setup. (1 point) __________

5. Interpretation

Only pancreatic enzymes convert the fat in milk to fatty acids. (3 points) __________

or

Any other relevant statement that is consistent with the data (2 points) __________

6. Conclusion

The data contradict the hypothesis that both enzymes convert fat to fatty acids at 37° C. (3 points) ______

or

The data contradict the hypothesis that only pepsin converts fat to fatty acids at 37° C. (3 points) ______

or

The data support the hypothesis that only pancreatic enzymes convert fat to fatty acids at 37° C. (3 points) ______

or

The data contradict the hypothesis that neither pepsin nor pancreatic enzymes convert fat to fatty acids at 37° C. (3 points) ______

or

Any other conclusion that is consistent with the data and interpretation (2 points) ______

7. Analysis

a. *Factors include changes in pH or temperature that could denature the enzymes.* (1 point) ______

b. *What function does pepsin play in digestion?* (1 point) ______

or

Which chemical reactions in digestion does pepsin catalyze? (1 point) ______

or

Which molecules serve as substrates for the enzyme pepsin? (1 point) ______

Part II–Experiment B (42 points total)

Part II Score ____________

Experiment (22 points total)

1. **Hypothesis**

   The rate of pancreatic enzyme catalysis increases with temperature. (2 points) ____________

   or

   The rate of pancreatic enzyme catalysis does not increase with temperature. (2 points) ____________

2. **Materials**

   Materials should include only those that are found on the supply table and that are needed for Experiment B. (1 point) ____________

3. **Procedure**

   a. Test setup: *Prepare one test setup for each of the following temperature ranges: 18–20° C, 35–40° C, and 80–90° C. Each test setup consists of one test tube that contains pancreatic enzyme solution and one test tube that contains an equal amount of canned milk and a few drops of phenol red (so that it turns pink).* (5 points) ____________

   Control setup: *Prepare one control setup for each of the following temperature ranges: 18–20° C, 35–40° C, and 80–90° C. Each control setup consists of one test tube that contains water and one test tube that contains an equal amount of canned milk and a few drops of phenol red (so that it turns pink). (5 points)* ____________

   Procedure: *Place the test and control setups in a hot-water bath. When the temperature reaches 18–20° C, combine the pancreatic enzyme and canned milk solutions of one of the test setups. Also combine the water and the canned milk of one of the control setups. Using the remaining setups, repeat the procedure at the other two temperature ranges. (2 points)* ____________

   b. *Measure the actual temperature of the test and control solutions when they are combined.* (1 point) ____________

   Observe the color of phenol red in each test and control setup. (1 point) ____________

   Record the amount of time it takes for any color changes to occur in each test and control setup. (2 points) ____________

   c. Appropriate use of safety precautions (2 points) ____________

   d. Appropriate use of materials (1 point) ____________

Data (20 points total)

4. Records (14 points total) ____________

*Reaction Time of Pancreatic Enzymes With Increasing Temperature**

Time in minutes	*Color of phenol red at 18–20° C*		*Color of phenol red at 35–40° C*		*Color of phenol red at 80–90° C*	
	Test	*Control*	*Test*	*Control*	*Test*	*Control*
0	*pink*	*pink*	*pink*	*pink*	*pink*	*pink*
1	*pink*	*pink*	*pink*	*pink*	*pink*	*pink*
2	*pink*	*pink*	*pink*	*pink*	*pink*	*pink*
3	*pink*	*pink*	*light pink*	*pink*	*pink*	*pink*
4	*pink*	*pink*	*light pink*	*pink*	*pink*	*pink*
5	*pink*	*pink*	*peach*	*pink*	*pink*	*pink*
6	*pink*	*pink*	*peach*	*pink*	*pink*	*pink*
7	*pink*	*pink*	*light peach*	*pink*	*pink*	*pink*
8	*light pink*	*pink*	*light peach*	*pink*	*pink*	*pink*
9	*light pink*	*pink*	*light peach*	*pink*	*pink*	*pink*
10	*light pink*	*pink*	*light tan*	*pink*	*pink*	*pink*
11	*light pink*	*pink*	*light tan*	*pink*	*pink*	*pink*
12	*light pink*	*pink*	*light tan*	*pink*	*pink*	*pink*
13	*light pink*	*pink*	*light tan*	*pink*	*pink*	*pink*
14	*light pink*	*pink*	*light tan*	*pink*	*pink*	*pink*
15	*light pink*	*pink*	*light tan*	*pink*	*pink*	*pink*

*Data may vary slightly

Breakdown of scoring for the table above is as follows:
Appropriate column heads (5 points) ____________

At least four color observations recorded for each test and control setup (4 points) ____________

Data indicate that phenol red changes color slightly at 18–20° C. (1 point) ____________

Data indicate that phenol red becomes peach or yellow at 35–45° C. (1 point) ____________

Data indicate that phenol red shows no or slight color change at 80–90° C. (1 point) ____________

Data indicate that phenol red shows no or slight color change in all controls. (2 points) ____________

5. Interpretation

Pancreatic enzymes catalyzed the breakdown of fat in canned milk into fatty acids in 10 minutes or less at 35–40° C. (1 point) ____________

At 80–90° C, pancreatic enzymes were unable to catalyze the breakdown of fat in canned milk into fatty acids over a 15-minute period. (1 point) ____________

At 18–20° C, only pancreatic enzymes were able to begin catalyzing the breakdown of fat in canned milk into fatty acids. (1 point) ____________

6. Conclusion

The data contradict the hypothesis that the rate of pancreatic enzyme catalysis increases with temperature. (2 points) ______

or

The data support the hypothesis that the rate of pancreatic enzyme catalysis does not increase with temperature. (2 points) ______

or

Any other statement that is consistent with the data and interpretation (1 point) ______

7. Analysis

According to the data, pancreatic enzymes are most effective at catalyzing the breakdown of fat at temperatures near normal body temperature. Thus, pancreatic enzymes are most effective at temperatures at which digestion occurs. (1 point) ______